Photocatalysts for Energy and Environmental Sustainability

Online at: https://doi.org/10.1088/978-0-7503-5697-8

Photocatalysts for Energy and Environmental Sustainability

Edited by
Vijay B Pawade
Department of Applied Physics, Laxminarayan Innovation Technological University, Nagpur 440033, India

Bharat A Bhanvase
Department of Chemical Engineering, Laxminarayan Innovation Technological University, Nagpur 440033, India

IOP Publishing, Bristol, UK

ISBN   978-0-7503-5697-8 (ebook)
ISBN   978-0-7503-5693-0 (print)
ISBN   978-0-7503-5700-5 (myPrint)
ISBN   978-0-7503-5694-7 (mobi)

DOI   10.1088/978-0-7503-5697-8

Version: 20240301

IOP ebooks

British Library Cataloguing-in-Publication Data: A catalogue record for this book is available from the British Library.

Published by IOP Publishing, wholly owned by The Institute of Physics, London

IOP Publishing, No.2 The Distillery, Glassfields, Avon Street, Bristol, BS2 0GR, UK

US Office: IOP Publishing, Inc., 190 North Independence Mall West, Suite 601, Philadelphia, PA 19106, USA

Contents

5 Photocatalysts for hydrogen evolution   **5-1**

Somnath C Dhawale and Bhaskar R Sathe

6 Photocatalysis for organic degradation using perovskite materials   **6-1**

Phuong Hoang Nguyen, Thi Minh Cao and Viet Van Pham

7 Emerging trends and future prospects in photocatalysis-based environmental remediation and hydrogen production   **7-1**

İbrahim Hakkı Karakaş and Zeynep Karcıoğlu Karakaş

Preface

In recent years, research into energy and environmental sustainability has received more importance in interdisciplinary areas of science and engineering due to the vast increase in industrial globalization. So, many research organizations and academic institutes are promoting these frontier areas as a way of developing highly efficient, environmentally friendly technology to achieve sustainability goals. Thus, photocatalysis is an emerging, simple, and low-cost technique that has the potential to resolve issues related to hydrogen generation and the photocatalytic degradation of pollutants under sunlight illumination. This textbook summarizes the fundamental mechanisms, properties, and applications of different types of photocatalysts. It contains seven chapters that cover the current progress and future scope of new and advanced photocatalytic materials, written by well-known authors in these fields. Therefore, this textbook is designed to be of benefit in undergraduate as well as postgraduate courses in science and technology. As per the global scope of environmental research, this book can provide an ideal platform for the reader to understand the concepts presented in a more systematic way, increasing their interest in the content of the book. So, we thank all the contributing authors for their efforts to enhance the depth of this book and their expertise in making this textbook attractive among the other books. We also thank IOP Publishing for introducing this textbook on new and advanced photocatalytic materials and their sustainable approach for the betterment of mankind. We sincerely hope this book can ultimately make a significant contribution to research and development activities in the field of photocatalysis.

Editor biographies

Vijay B Pawade

Dr Vijay B Pawade is an assistant professor (Sr.Gr) in the Department of Applied Physics at the Laxminarayan Innovation Technological University, Nagpur, India. His research focuses on rare-earth-doped oxide materials and their applications in light-emitting diodes (LEDs), solar cell devices and photocatalytic processes. He has published 45 research papers in respected international peer-reviewed journals and acts as a reviewer for journals published by Elsevier, Springer, Wiley, Taylor and Francis, the Royal Society of Chemistry, and the American Chemical Society. He has contributed 12 book chapters on different themes such as nanomaterial synthesis and characterization, the applications of nanomaterials in energy conversion and storage devices, quantum dots (QDs), the spectroscopy of lanthanides, etc. He is the author of books titled Phosphor for Energy Saving and Conversion Technology (*CRC Press—Taylor and Francis*), Optical Properties of Phosphate and Pyrophosphate Compounds, and Lanthanide-Doped Aluminate Phosphors (*Woodhead Publishing—Elsevier*). He has edited five books on Nanomaterials for Green Energy (*Elsevier*), Spectroscopy of Lanthanide-Doped Oxide Materials (*Woodhead Publishing—Elsevier*), Multifunctional Nanostructured Metal Oxides for Energy Harvesting and Storage Devices (*CRC Press—Taylor and Francis*), Handbook of Nanomaterials for Wastewater Treatment (*Elsevier*), Nanoscale Compound Semiconductors and their Optoelectronics Applications (*Woodhead Publishing—Elsevier*), Phosphor Handbook: Process, Properties and Applications (*Woodhead Publishing—Elsevier*).

Bharat A Bhanvase

Dr Bharat A Bhanvase is currently working as professor and head of the Chemical Engineering Department at the Laxminarayan Innovation Technological University, Nagpur, India. His research interests are focused on wastewater treatment, cavitation-based synthesis of nanomaterials and nanocomposites, solid waste processing, process intensification, microfluidics, nanofluids, etc. He obtained his PhD in Chemical Engineering from the University of Pune. He has published 103 articles in international journals, and four in national journals; he has presented 17 papers in international conferences and 12 in national conferences. He has written 55 book chapters in internationally renowned books, nine edited books, and one authored book. He has obtained four Indian patents and applied for six Indian patents. He received IIChE Awards in 2021 (the Chemical Weekly Award, the IIChE NRC Award, and the Kuloor Memorial Award) for the best paper published in the 'Indian Chemical Engineer' in its 2020 issues. He is a fellow of the Maharashtra Academy of Sciences (MASc) and a fellow of IIChE. He was the recipient of the Best Scientist Award from Rashtrasant Tukadoji Maharaj Nagpur University in 2017. He also received a Young Scientists (Award) start-up research grant from the Science and Engineering Research Board, New Delhi (India) in 2015. He has guided 28 M.Tech. students and one PhD student; two M.Tech. and three PhD students are currently working with him.

List of contributors

Timur Sh Atabaev
Department of Chemistry, Nazarbayev University, Astana 010000, Kazakhstan

B A Bhanvase
Department of Chemical Engineering, Laxminarayan Innovation Technological University, Nagpur 440033, India

Shubham Bonde
Department of Chemical Technology, Laxminarayan Innovation Technological University, Nagpur 440033, India

Thi Minh Cao
HUTECH University, 475A Dien Bien Phu Street, Binh Thanh District, Ho Chi Minh City, Viet Nam

Somnath C Dhawale
Department of Chemistry, Dr Babasaheb Ambedkar Marathwada University, Chatrapati Sambhajinagar 431004, MH, India

Darya Goponenko
Department of Chemistry, Nazarbayev University, Astana 010000, Kazakhstan

G A Suganya Josephine
Department of Chemistry, Center for Nanotechnology Research, Aarupadai Veedu Institute of Technology—Vinayaka Mission Research Foundation, Rajiv Gandhi Salai, Paiyanoor, Kanchipuram 603104, India

Gauri Kallawar
Department of Chemical Technology, Dr Babasaheb Ambedkar Marathwada University, Aurangabad 431004, MS, India

İbrahim Hakkı Karakaş
Department of Food Engineering, Bayburt University, Bayburt, Turkey

Zeynep Karcıoğlu Karakaş
Department of Environmental Engineering Atatürk University, Erzurum, Turkey

S Rubesh Ashok Kumar
Department of Chemistry, Center for Nanotechnology Research, Aarupadai Veedu Institute of Technology—Vinayaka Mission Research Foundation, Rajiv Gandhi Salai, Paiyanoor, Kanchipuram 603104, India

D Vasvini Mary
Department of Chemistry, Center for Nanotechnology Research, Aarupadai Veedu Institute of Technology—Vinayaka Mission Research Foundation, Rajiv Gandhi Salai, Paiyanoor, Kanchipuram 603104, India

Phuong Hoang Nguyen
HUTECH University, 475A Dien Bien Phu Street, Binh Thanh District, Ho Chi Minh City, Viet Nam

V B Pawade
Department of Applied Physics, Laxminarayan Innovation Technological University, Nagpur 440033, India

Viet Van Pham
HUTECH University, 475A Dien Bien Phu Street, Binh Thanh District, Ho Chi Minh City, Viet Nam

Bhaskar R Sathe
Department of Chemistry, Dr Babasaheb Ambedkar Marathwada University, Chatrapati Sambhajinagar 431004, MH, India

Kamila Zhumanova
Department of Chemistry, Nazarbayev University, Astana 010000, Kazakhstan

IOP Publishing

Photocatalysts for Energy and Environmental Sustainability

Vijay B Pawade and Bharat A Bhanvase

Chapter 1

An introduction to photocatalysts and their applications

Vijay B Pawade and Bharat A Bhanvase

This chapter introduces the fundamentals of photocatalysts and their role in the development of sustainable technologies. It explains the basic principles, mechanisms, and workings of photocatalysts for wastewater treatment and hydrogen generation. It also explores the different types of photocatalysts, including their characteristics and features at both the nanoscale and the microscale. The methods of synthesis and the importance of green synthesis compared to other conventional routes are discussed in detail. In addition, this chapter discusses some other important parameters reported in the research literature, such as the reusability and stability of photocatalysts, factors affecting the photocatalytic performance of photocatalysts, and the need for new and advanced strategies to improve the photocatalytic efficiency of photocatalysts for the production of energy and the development of environmentally friendly technology.

1.1 Introduction

In the last few decades, a major worldwide focus has been placed on the development of sustainable technology to protect the environment and maintain the harmony of nature on our mother planet. It is our social responsibility to protect and balance the environment through the development of new and advanced sustainable technologies. Nowadays, there is tremendous growth and strong competition everywhere due to the supply of, and demand for products in the global market due to the vast increase in populations in underdeveloped countries. Thus, to meet the need for low-cost products and the need to recycle cheaper raw materials, many new small- and large-scale industries have been set up to fulfill the global demands for materials and related products. During the recycling of technologically outdated products and devices, many kinds of toxic gases and heavy elements are

doi:10.1088/978-0-7503-5697-8ch1

produced and dispersed into the air, water, and soil; these not only affect the quality of the air, water, and soil but also increase their toxicity levels. Thus, among these three sources of pollution, the prevention of air and water pollution are the top priorities, as they impact all living things on our planet. They also have a major impact. In terms of the social and economic development of nations, further increases in pollutant levels in fresh air and water increase the potential risks to people's health and cause many health issues related to respiratory disorders, dermatitis, asthma, mutagenicity, cancer, etc [1]. Photocatalysis is a more promising and sustainable way to resolve to resolve such global environmental issues related to air and water pollution than other techniques. Because it can effectively convert and utilize solar energy, it has received more attention in recent years for its prospective use in photocatalytic wastewater treatment and advanced water splitting processes for the production of H_2, which is considered to be a clean and environmentally friendly source of energy [2, 3]. Basically, a photocatalyst is a material involved in specific chemical reactions that take place under exposure to light radiation, in which it converts the solar energy into other useful forms. For the photocatalytic process, sunlight is the most prominent inexhaustible and clean source of driving energy that leads to slow reaction conditions, high energy of the active species, and a deep oxidation effect during the photocatalytic reaction. This reaction can be caused by the absorption of sunlight in different regions of the spectrum, such as UV, visible light, and infrared radiation; the specific region involved usually depends on the photocatalyst material [4–7]. Thus, the photocatalytic process has the ability to resolve the problems related to the environment and energy without the utilization of excessive fossil fuels. Photocatalysts are capable and operate well under natural sunlight, but more effort is needed for the development of highly efficient visible-light-driven photocatalyst materials [8]. Khan *et al* [9] reported that research and development into the photocatalytic process exhibits a broad scope for widespread application in the near future. However, a few parameters of photocatalytic materials, such as their efficiency, thermal stability, purity, environmental compatibility, and low efficiency in photocatalytic reactors, are major hurdles that restrict their application at scale. In recent years, inorganic bandgap semiconductors such as ZnO, CdS, ZnS, ZnSe, CdSe, ZnTe, etc. have been widely studied for the photocatalytic process due to their unique chemical properties and high stability [10–13]. Among these, zinc oxide (ZnO) is the most favorable, environmentally friendly, and economically viable for large-scale wastewater treatment. In the following section, we discuss the details, principles, and working mechanisms of the photocatalytic materials and their types that are proposed for the advanced water treatment and hydrogen generation processes.

1.2 The principles and mechanism of photocatalysis

The photocatalytic process is based on the absorption of light by the photocatalyst, for which metal-oxide semiconductors are preferable because of their suitability for the formation of electron–hole pair creation in the conduction band (CB) and the valence band (VB) [14]. Thus, during the absorption of light, electrons in the VB are

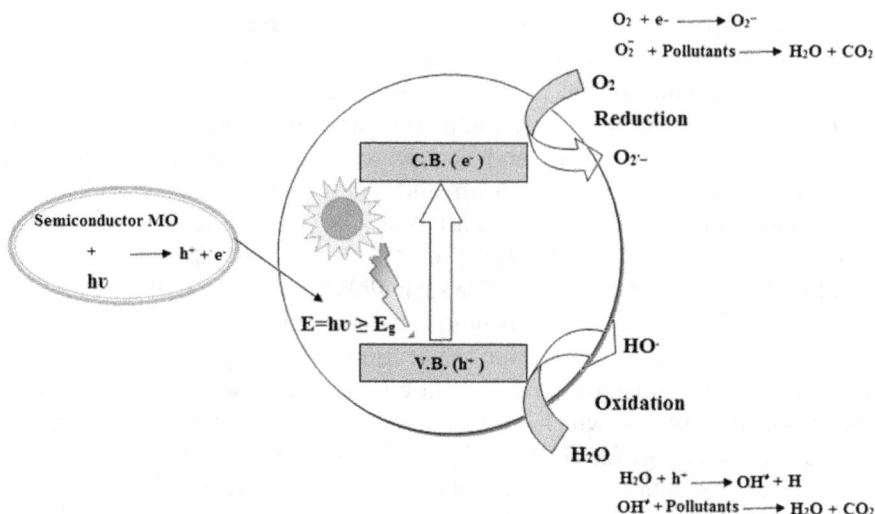

Figure 1.1. The photocatalytic mechanism under solar illumination.

excited into the CB where they form electron (e^-) and hole (h^+) pairs. There are two photochemical reactions that involve the photoinduced electrons and holes, which are continuously generated. A schematic of the photocatalytic process and its mechanism of pollutant degradation under solar illumination are shown in figure 1.1. In general, photocatalytic materials play a major role in initiating the oxidation and reduction reactions in the presence of solar energy. The following steps are followed during the photocatalytic process:

Step **I. The** generation of hole/electron pairs

Step **II.** The separation of charge carriers and their diffusion towards the electrode surface

Step **III.** Photooxidation and -reduction reactions take place at the surface of the photocatalyst.

Here, photoinduced holes and electrons react with O_2 and H_2O on the photocatalyst's surface, which leads to the formation of O^{-2} and $OH^{\bullet}$ radicals. These radicals have strong redox potentials, and hence, when they react with pollutants, photodegradation takes place. A possible photocatalysis reaction is shown in figure 1.1.

When active species are adsorbed by the photocatalyst's surface, the electron transfer process becomes more prominent [15]. Further, during the water cleaning process, oxygen acts as a common electron acceptor. When photogenerated electrons react with oxygen, they reduce to O^-_2 and can be transformed into various oxygen-activated species, such as $HO^{\bullet}$, H_2O_2, $HO_2^{\bullet}$ and HO_2 anions [16, 17], which involve the oxidation of the electron donor [18], while the generated holes can oxidize the electron donor. At the same time, reactive oxidizing species and free carriers react with absorbed surface impurities, and the degradation of pollutants takes place. The efficiency of the photocatalysis process depends on the ability of the photocatalyst to produce a large number of holes and electrons, which results in the production of reactive free radicals. A shift in the light absorption range of the photocatalyst into the visible spectral range helps to generate a large number of electron–hole pairs, thereby improving the degradation response of the photo-catalyst [19]. Hence, the absorption range plays a key role for highly active photocatalysts.

1.2.1 Types of photocatalyst

As compared to conventional water treatment processes, advanced oxidation processes are assigned great importance due to their stronger oxidation capabilities, faster reaction times, and production of smaller amounts of secondary pollutants. These processes are generally categorized into homogeneous and heterogeneous processes, depending on the type of reaction medium. Further, they can be classified into energy- and non-energy-related categories [20]. The abovementioned heteroge-neous photocatalytic oxidation technique is widely accepted for the degradation of pollutants in wastewater [21]. This technique has some advantages, such as flexibility, simplicity, low cost, the use of an environmentally friendly catalyst, and high photocatalytic efficiency. In the recent years, many new types of photo-catalysts have been proposed and used for the removal of organic pollutants from wastewater [22, 23], Some of these are discussed below.

1.2.1.1 Homojunction semiconductor photocatalysts

Homojunction semiconductor photocatalysts are synthesized by incorporating semiconductor interfaces that have compatible bandgap energies and chemical compositions and specific dimensions [24]. They also possess particular physical, electrical, and optical properties and exhibit superior photocatalytic activities for the photodegradation of waterborne organic pollutants [25]. Nanoscale photocatalytic devices have been fabricated by using homojunction photocatalysts and are applicable in various disciplines [26, 27]. Further, homojunction photocatalysts help to improve photocatalytic efficiency in the production of hydrogen via the water splitting process [28]. But the use of semiconductors for photocatalytic applications has some limitations.

1.2.1.2 Hetrojunction semiconductor photocatalysts

Heterogeneous semiconductor photocatalysis is another type of advanced oxidation process that has received great attention due to its prospective use in resolving energy and environmental issues by, for example, generating hydrogen through the water splitting process and degrading organic pollutants through redox reactions [29]. Semiconductor heterojunctions are constructed by combining two semicon-ductors, and they have been demonstrated to be one of the most efficient ways to spatially separate photoexcited electron–hole (e^-/h^+) pairs [30, 31]. When a hetero-junction photocatalyst is illuminated by a light source, photoexcited charge carriers are forced to move between the two semiconductors, building up an electric field and hence inhibiting the recombination of the charge carriers. The formation of a built-in electric field at the semiconductor heterojunction interface and the transfer rate of a photoexcited charge carrier depend on the semiconductivity of the materials, the work function, and the ratio of the CB to VB potentials of the semiconductors.

1.2.2 Single-atom photocatalysts

Single-atom photocatalysts (SACs) are considered to be low-cost, high-efficiency photocatalysts and have been assigned more importance in the field of catalysis [32]. Qiao *et al* [33] were the first to report the concept of 'single-atom catalysis' [34]. SACs have been synthesized by loading a single metal atom onto a suitable support; further electrons are exchanged with the support to form single-atom active sites, enhancing the photocatalytic performance of the material [35]. Due to continuous research into, and development of the preparation of SACs, many preparation techniques have come into existence. Thus, the flexible pairing of metal centers and charge carriers facilitates the preparation of environmentally friendly and sustainable single-atom photocatalysts with high catalytic efficiency [36, 37]. SACs are also used to produce both homogeneous and heterogeneous catalysts [38, 39]. The absorption range and charge separation efficiency of SACs are high [40]. As a result of these characteristics, SACs are emerging materials in the photocatalytic field for the evolution of photocatalytic H_2 and the removal of toxic contaminants from wastewater [41, 42].

1.2.3 Quantum-dot-based photocatalysts

Quantum dot (QD)-based composite catalysts are considered to be promising candidates for resolving issues related to energy and environmental sustainability. QDs are zero-dimensional spherical nanoparticles, and their physical dimension is smaller than the exciton's Bohr radius [43, 44]. Colloidal semiconductor nanocrystal QDs 2–10 nm in size may contain 10–50 atoms within their volume [45]. Recently, Kandi *et al* [46] discussed the scope and advantages of quantum dots in the photocatalytic hydrogen production process. There are some characteristics of QDs that make them suitable for enhanced H_2 production compared to other types of nanostructured materials with superior properties; these characteristics play a significant role in enhancing photocatalytic activity. Some of the important properties of QDs are given below:

  (i) The capability to absorb light in the visible spectral range
  (ii) A better multiple-exciton generation rate under solar illumination due to the quantum confinement effect.
  (iii) Better charge transport and separation characteristics.
  (iv) Size-dependent tuneable optical properties.
  (v) Their visible-light absorption edge can be enhanced by doping them with wide-bandgap semiconductors.

At present, many research groups are working on the development of highly efficient QDs based on hybrid systems that have the above characteristics and properties and are applicable for effective photocatalysis.

1.2.4 Perovskite-based photocatalysts

Perovskite-structured materials represented by the chemical formula ABX_3 belong to a ternary family of crystalline structures in which the A-site contains metal cations, rare earth ions, or alkaline earth metal ions with larger ionic radii and the B-site contains transition-metal ions with smaller ionic radii, while X indicates the oxygen atoms available in the host structure. Perovskite-structured materials exhibit many interesting properties; further, perovskite nanomaterials show excellent photocatalytic efficiency due to their characteristics such as superior chemical and thermal stability, nontoxicity, cost-effectiveness, tuneable properties, adjustable bandgap, large charge carrier lifetime, etc [47]. The shape- and size-dependent properties of a perovskite nanostructure depend on the method of synthesis and its structural characteristics. Today, perovskite nanoparticles have the potential to be used in a variety of applications, such as chemical sensing, catalysis, water splitting, and the photodegradation of organic pollutants. But single-component perovskite materials have a broader bandgap, and hence recombination of the charge carrier takes place much faster, which restricts their performance in visible-light-driven photocatalysis. For effective photocatalysis under solar illumination, strong absorption near 520 nm is needed. Further challenges still remain, such as resolving the problems of the separation and recycling of perovskite materials in treated water. To overcome these challenges, further research activity and strategies, such as modification of their surface, doping with metal ions, coupling with metal nanoparticles, the synthesis of nanocomposites, etc. [48], are required to improve the photocatalytic performance of these materials.

1.2.5 S-scheme photocatalysts

S-scheme heterojunction photocatalysts exhibit characteristics such as a superior light absorption ability, a high charge carrier separation efficiency, a strong redox potential, and a diverse range of both inorganic and organic semiconductors. Considering the related advantages and disadvantages of conventional heterojunction photocatalysts, the step-scheme (S-scheme) is a novel semiconductor catalyst that fulfills the current need for efficient photocatalysts [49]. Here, the heterojunction is formed by contact between two different semiconductors, which helps to increase the absorption band edge of the semiconductors, further improving the separation and migration rate of the photogenerated carriers [50]. As a result of the close contact between the different semiconductors, the electrons from the reduced semiconductor spontaneously migrate toward the oxidized semiconductor and build an electric field that is directed toward the oxidized semiconductor. At the same time, the holes in the VB of the reduced semiconductor and the electrons in the CB of the oxidized semiconductor combine, which results in the accumulation of a greater number of negatively charged carriers (i.e. electrons) in the CB of the reduced semiconductor and a greater number of positively charged carriers (i.e. holes) in the VB of the oxidized semiconductor. Thus, the S-scheme heterojunction photocatalyst shows strong redox capacity [51] and has a wide range of potential applications. S-scheme heterojunction photocatalysts can be categorized as inorganic–inorganic [52], inorganic–organic [53], or organic-organic composites [54]. The inorganic–inorganic types of S-scheme heterojunction photocatalysts are of great interest for the photocatalysis process.

1.2.6 rGO-based composite photocatalysts

Graphene is a zero-bandgap material, which restricts its applications, particularly in the electronics field. Therefore, doping graphene with heteroatoms can form a number of localized energy levels in its bandgap, and hence it can exhibit tuneable properties that make it responsive in visible light. However, there are some restrictions on the use of GO, such as its toxicity and corrosiveness, the explosive nature of the reducing agents, etc [55]. However, reduced graphene oxide (rGO) also exhibits excellent properties such as tunable electrical properties, transparency, and the ability to integrate with various photoactive surfaces to enhance their efficiencies. Further, it has good electrical conductivity and a large surface area and is a better substitute for pure graphene which can be synthesized at low production costs using a simple reduction process [56]. Thus, it makes rGO a desirable candidate for solar-driven photocatalytic applications. Coupling rGO with oxide materials promotes electron separation, boosting their photodriven activity in the visible spectral region and promoting the degradation of some harmful dyes. Thus, 0D, 1D, and 2D nanostructured semiconductors coupled with rGO nanocomposites play an important role in improving photocatalytic activity, hydrogen generation, nitroaromatic reduction, etc. In the case of semiconductor composites, coupling the semiconductor with rGO helps to separate the photogenerated charge carriers at the catalyst/RGO interface. Here, the nature of the semiconductor/rGO interface and defects in the rGO play an important role in enhancing the photocatalytic activity. Recently, Witjaksono *et al* synthesized and reported visible-light-driven N-doped rGO with reduced bandgap energy, i.e. from 3.4 to 2.2 eV [57]. This material reduced the electron–hole pair recombination rate by exhibiting the characteristics and features of a visible-light-driven photocatalyst [58]. Similarly, ferrite-based rGO nanostructures are magnetic materials and have good absorption in the visible spectral range. Further, they are strongly responsive to applied magnetic fields and can be readily recovered using conventional magnetic bars [59]. Today, rGO is one of the benchmark materials for improving the performance of some advanced materials used in the development of sustainable technology [59]. Recyclability and recovery are also important aspects of the development of new photocatalyst materials. Researchers have made many efforts to resolve current issues and future challenges in these fields.

1.2.7 Semiconductor photocatalysts

Semiconductor photocatalysts contain different materials such as metal oxides, nitrides, or sulfides (e.g. TiO_2 and MoS_2) [60, 61] as well as metal-free semi-conductors such as C_3N_4. Other materials, such as copper, gold, and silver metal nanoparticles, exhibit strong localized surface plasmon resonance (LSPR) properties under visible-light irradiation. These nanoparticles were assigned great importance at the beginning of 21st century, but they have higher costs, which restricts their wider industrial scope. However, semiconductor photocatalysts are comparatively lower in cost and have been a topic of research for more than 50 years, but they suffer from the issue of a low absorption band in visible light and hence lower degradation efficiency. Based on some characteristics, features, and innovative approaches, they are considered promising materials for use in industrial applications at enhanced photocatalytic efficiencies. Semiconducting nanomaterial-based photocatalysts such as ZnO, CdS, ZnS, ZnSe, CdSe, ZnTe, etc. have been studied many times in the last few decades, and they are widely accepted due to their unique properties and good stability, which also promote strong redox reactions [10, 11]. ZnO is the most favorable, environmentally friendly, and economical catalyst for the large-scale treatment of wastewater due to its direct bandgap energy, which is of the order of 3.37 eV. It also seems to have an excellent degradation response under UV light illumination. The available wavelength spectrum of solar radiation contains only 4% of UV light but 43% of the visible-light component. To shift the response of ZnO under visible light, there is a need to alter the optical properties of ZnO by adding a narrow energy gap semiconductor, which improves its absorption capacity in visible light and also reduces the e^-/h^+ recombination rate [62]. ZnSe/ZnO is a well-known example of a composite catalyst, in which ZnO is combined with ZnSe, which acts as a narrower-bandgap semiconductor (2.67 eV). The bandgap of ZnSe is well aligned with that of ZnO, hence, it improves the photocatalytic degradation efficiency of ZnO [63]. However, much more investigation is still needed to explore the future prospects of advanced semiconductor materials to fully meet the needs of energy and environmental sustainability.

1.3 Synthesis methods

1.3.1 Chemical methods

In recent years, many conventional methods have been used to prepare nano-structures of different dimensions [64]. Thus, the top-down and bottom-up approaches are popular methods for the preparation of oxides and other forms of materials. The bottom-up approach is well suited for the fabrication of defect-free nanostructured materials of specific shapes and sizes. Some of the techniques, such as combustion, the hydrothermal method, the solvothermal method, the sonochemical method, the sol–gel method, etc. fall into the category of bottom-up approaches, as shown in figure 1.2. These are well-known methods used in materials synthesis. Among these techniques, great importance it attached to the green synthesis route, which involves the use of nonacid mediums as well as nontoxic materials for the

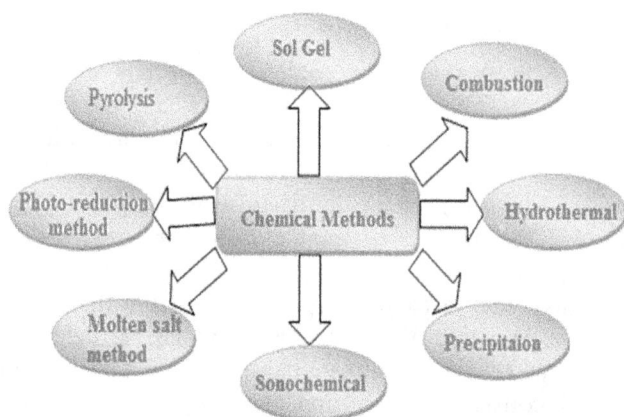

Figure 1.2. Chemical methods used for the synthesis of materials.

preparation of materials [64, 65]. Thus, the chemical approach is well suited for the preparation of inorganic nanostructured photocatalytic materials because of its ability to produce materials in the desired shape and size, and the methods used in the bottom-up approach are also cheaper than those of the top-down approach. But there is always a need to take care in the selection of nontoxic raw materials and the formation of the final pure-phase product while avoiding the presence of the acid medium, impurities, etc. High-temperature solid-state diffusion is also a well-known technique that corresponds to the top-down approach used to prepare perovskite oxide nanostructures, but this technique leads to the formation of an impure phase that contains inhomogeneous perovskite oxide nanomaterials due to repeatedly grinding, crushing, and preheating them before calcination. This leaves some defects on the surface of the nanostructured materials, and the presence of such defects may affect the properties of the nanostructured materials. In contrast, using the sol–gel method, the coprecipitation method, or the combustion method, it is possible to obtain pure perovskite oxide nanomaterials with a high surface area and an ideal nanostructure size. These are the simplest and most efficient techniques with which perovskite oxide materials are synthesized. Hydrothermal methods are also well-known synthesis techniques that allow the nanostructure's shape and size to be controlled. However, these techniques require precursors that readily mix well in aqueous solution at high temperature and constant pressure [66]. Therefore, the hydrothermal method is an important branch of inorganic synthesis that depends on the solubility of material in hot water under high pressure [67]. During the hydrothermal reaction, parameters such as the type of solvent, temperature, and time of reaction play an important role; in such reactions, nucleation and grain mechanisms form nanosized crystallites.

1.3.2 Green synthesis

Today, green synthesis is an environment-friendly technique for the preparation of nontoxic metal-oxide nanoparticles and has received great interest in the field of nanotechnology. During the synthesis of nanomaterials by chemical methods, some toxic gases are liberated during chemical reactions, and the harmful chemical species present are also adsorbed on the surfaces of nanoparticles. Thus, considering this drawback, green synthesis is an emerging technique for the production of NPs. Further, this method is clean, safe, and a cost-effective way to deploy environmentally friendly processes [68]. It is possible to synthesize nanoparticles of different shapes and sizes using this technique. The use of different fuels or reducing agents also plays an important role in obtaining the pure phase and the desired nanostructure morphology [69].

1.4 The reusability and stability of photocatalysts

Efficiency and reusability are two basic parameters that play an important role in the practical use of photocatalysts. They can be recycled at least five times, and the photocatalytic materials must remain stable during this process. The recovery of the materials can be carried out through the centrifugation technique after each cycle. The separated materials are then rinsed more than two times with deionized water. Later, the recovered materials are reused in the photocatalytic dye degradation process. Metal-oxide nanostructures such as acid protease functionalized silver nanoparticles (APTs–AgNPs) exhibit excellent catalytic performance that removes up to 95% of methylene blue (MB) dye from wastewater, and the reuse of the materials does not significantly alter their efficiency after each run. Hence, they can be reused more often with a minimal reduction in their efficiencies. But there is a need for more research to resolve the issues of the maintenance of effective photocatalysis and NP reusability associated with structural effects such as the porosity and surface area of metal oxides [70]. $NaYF_4$:(Gd, 1% Si)/TiO_2 is another well-known metal-oxide-coupled phosphor composite photocatalyst; it can be reused and is stable for up to three cycles, but the degradation efficiency of this material decreases from 95% to 60% and 40% in the 2nd and 3rd cycles, respectively. However, $NaYF_4$: (Gd, Si)/TiO_2 composites can be considered to be low-cost and efficient photocatalysts for the removal of pollutants from wastewater [71].

1.5 Factors affecting photocatalytic activity

1.5.1 The bandgap

In recent decades, wide-bandgap semiconductor photocatalysts have been assigned more importance due to their use in environmentally friendly wastewater treatment processes. Among the different types of oxides, ZnO and TiO_2 are the most promising materials; they have been studied many times because of their characteristics, such as high chemical stability, low cost, and nontoxic nature [72–74]. However, due to their large bandgap, these types of large-bandgap semiconductors can absorb only light in the UV region and are unable to absorb visible light. They also have a fast recombination rate of photogenerated electron–hole pairs that restricts the effective degradation of pollutants [75]. These issues related to semiconductor photocatalysts could be resolved if their light absorption range in the UV region could be easily extended to the visible region by tuning their bandgap; this might be made possible by adding some impurities to their structure or by adopting some new methods of synthesis [76]. In general, semiconductor photocatalysts bandgaps that fall between 1.23 and 3 eV include the oxidation and reduction potential ranges of H_2O. The redox potentials of the photocatalytic water splitting process correspond to $EH^+/H_2 = -4.44$ eV and $EO_2/H_2O = -5.67$ eV, respectively [77]; the difference between these is 1.23 eV, which is therefore the minimum bandgap that must be present for a material to be considered an effective photocatalyst. Fujishima and Honda reported water splitting by the photocatalytic approach; they studied both large- and narrow-bandgap semiconductor materials theoretically and experimentally [78, 79]. Semiconductor materials with a large bandgap (greater than 3 eV) have band edge positions suitable for the overall water splitting and hydrogen evolution processes, but their light absorption range is not compatible with the visible spectrum. Thus, there is a need for more efforts to search out new materials or design innovative materials with a broad absorption range that covers the full spectral component of visible light.

1.5.2 Particle size

The shape, size, and morphology of a material can affect its photocatalytic properties. Consider the bandgaps of bulk materials: the atomic orbitals overlap, producing bands with a small energy gap as compared to those of nanomaterials. However, in nanoscale metals, the orbitals are discontinuous; they rather form discrete energy levels in the band structure, which can be tuned by changing the nanoparticle diameter, as shown in figure 1.3. Hence, as the particle size decreases, the electrons become more confined in the particle, and confined electrons have more energy. Thus, the atomic orbitals of nanoparticles become discrete or quantized. The presence of the quantum confinement effect in nanomaterials leads to tunable electrical and optical properties.

Therefore, the bandgaps of bulk and nanostructured materials are different, and changes in particle size affect the bandgap energy, the light absorption capacity, and

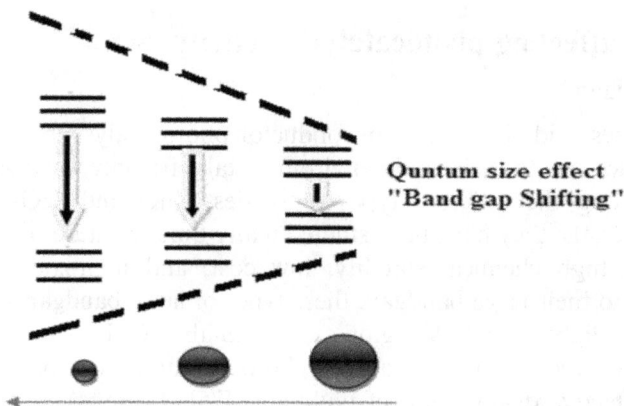

Figure 1.3. The effect of particle size on the bandgap.

the average free range of photogenerated charge carriers in the material. When the particle size of the material is smaller than the thickness of its space charge layer, which is negligible, then photogenerated charge carriers migrate from the bulk phase to the material's surface via the diffusion process and take part in surface redox reactions. The charge transfer rate of carriers and their separation efficiency can be improved by reducing the charge migration distance [80]. Therefore, small particles usually have a high specific area and show better adsorption properties that help to initiate the interaction between the catalyst and the reactants. Small particles thereby provide abundant active sites and a large light absorption area that hosts redox reactions and hence improves the photocatalytic performance [81].

1.5.3 Doping

Doping elements into a different host material is the most common strategy for improving the performance of metal-oxide semiconductors [82–85]. It can affect many parameters, such as the morphology, particle size, bandgap, binding energy, lattice defects, and other associated properties [86, 87]. For example, if the average particle size of pure SnO_2 particles is 52.3 nm, then the addition of Ag results in a reduction in the particle size of SnO_2; the new particle size may turn out to be 45.5 nm [88]. Further, Yakout *et al* [88] reported that the use of Co doping in an Ag/SnO_2 system was not effective in reducing the particle size. But the addition of higher amounts of dopants may affect the particle size, which may also result in a more uniform grain size distribution. Entradas *et al* also reported [89] that the particle size distribution was narrower when the amount of Co codoping was increased. This created agglomeration clusters of particles in pure SnO_2 due to the presence of the dopants [90].

1.6 Strategies for boosting photocatalytic efficiency

In the last few years, many researchers have made concerted efforts to design and develop new energy-efficient photocatalysts which are better at harvesting the maximum component of the solar spectrum, generating large numbers of charge carriers, and providing catalytic sites to support effective photocatalytic process [91]. To resolve the issues related to metal-oxide photocatalysts, various strategies have been proposed in the research literature to boost the photocatalytic efficiency of catalysts; some of these are discussed below.

1.6.1 Spatial separation of excitons

As discussed above, the photocatalytic efficiency of catalysts greatly depends on the separation rate of electron/hole pairs. In order to generate photoinduced charge carriers, a photocatalyst must be adequately excited by the incident photon flux energy. These positive and negative charge carriers can take femtoseconds (fs) to picoseconds (ps); they also subsequently take time to transit the surface of the photocatalyst (nanoseconds (ns) to microseconds (μs)) to reach the corresponding bands to initiate the redox reactions [92]. During the transit of charge carriers in the catalyst, there is a high probability of e−/h+ recombination that can release heat.

The recombination of e−/h+ pairs occurs within picoseconds to nanoseconds, which is much faster than their transfer rate to the surface of the catalyst (where they participate in oxidation and reduction reactions) [93]. Thus, to reduce the recombination rate, there is a need to develop new strategies to boost the separation of charge carriers and enhance the efficiency of photocatalysts.

1.6.1.1 The loading of cocatalysts

The loading of a cocatalyst improves the separation and transfer of excitons in the photocatalytic process, which helps to promote and stabilize the activity of the photocatalyst. Under light illumination, the photogenerated negative charge carriers in the CB of the catalyst are transferred toward the cocatalyst, which prevents the recombination of charge carriers in the VB [94]. Therefore, contact between the catalyst and the cocatalyst is essential for the transportation of charge carriers. When the charges reach the catalyst interface, the cocatalyst helps to improve the separation of charge carriers. When metallic cocatalysts are deposited on the surface of photocatalysts, they form a Schottky heterojunction that restricts the backward flow of electrons to the CB and thereby induces an electric field that enhances the separation of charge carriers, thereby promoting effective photocatalytic activity [95]. Semiconductor cocatalysts with narrow bandgaps also form heterostructures that improve charge separation. Similarly, cocatalysts consisting of transition-metal dichalcogenides have also exhibited better separation of electrons and holes because of their unique metallic and semiconductor structures. Among the different types of cocatalyst metals, cocatalysts such as Ag and Pt are excellent candidates for enhancing the photocatalytic efficiency of photocatalysts [96]. Recently, Sun et al [97] reported the photocatalytic response of CoP-loaded QDs as cocatalysts on CdS nanorods, confirming enhanced H_2 production under visible-light illumination.

1.6.1.2 Metal-oxide–coupled composites and their photocatalytic response

García et al [98] reported the photocatalytic performance of Bi codoped strontium aluminates blended with nanocrystalline TiO2. Here, small grains of TiO2 were made available on the surface of the strontium aluminate grains. With an increase in the surface area of the composite grains, an enhancement in photocatalytic activity was achieved. Figure 1.4 shows the photocatalytic degradation of MB achieved using the blended composite catalyst under UV light illumination. It can be seen that complete degradation of MB occurred after 210 min of UV exposure. Thus, the composite TiO_2–Bi codoped sample degraded 91.0% of the MB dye after 210 min of exposure time. Figure 1.4 shows that lower concentrations of Bi codoping in TiO_2–strontium aluminate composites exhibit a rapid photocatalytic degradation response as compared to pure TiO_2 powder.

1.6.1.3 Persistent phosphors and their photocatalytic response

Wang et al [99] reported the photocatalytic response of persistent phosphor used for the degradation of RhB. Figure 1.5 shows the persistent luminescence spectra of Zn^{2+} and Cr^{3+} codoped Ga_2O_3 phosphor. The photoluminescent excitation and emission spectra of the phosphor were observed in the UV–visible spectral range. When the

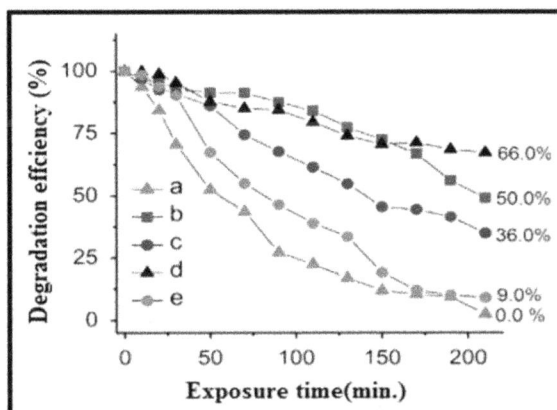

Figure 1.4. Photocatalytic degradation of MB as a function of exposure time under UV light illumination. The photodegradation responses of TiO_2–$Bi_{(2.0 \text{ and } 0.0 \text{ mol\%})}$ codoped strontium aluminate composites are shown in curves (a) and (b), and the response of pure TiO_2 is shown in curve (c); the responses of the TiO_2–Bi $_{(15.0 \text{ and } 1.0 \text{ mol\%})}$ codoped strontium aluminate composites are shown in curves (d) and (e), respectively. Reprinted from [98], Copyright (2015), with permission from Elsevier.

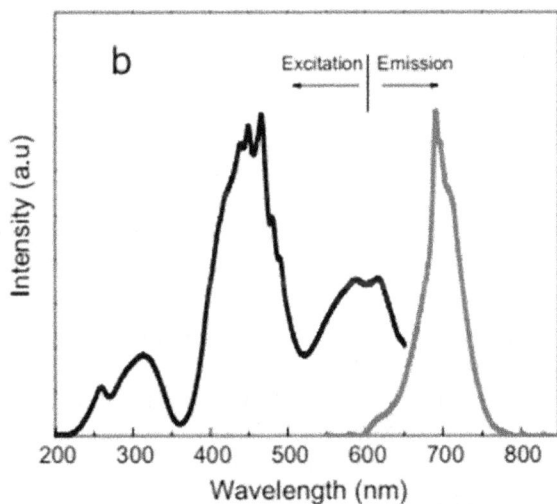

Figure 1.5. The persistent luminescence spectra of Ga_2O_3:$Cr^{3+}_{(0.01)}$ and $Zn^{2+}_{(0.005)}$ phosphors. Reprinted from [99], Copyright (2014), with permission from Elsevier.

phosphor is illuminated under ultraviolet light, a large number of charge carriers are created, and their energy is transferred to Cr^{3+} luminescence centers. These charge carriers are trapped by the oxygen vacancies, and after the irradiation source is cut off, the trapped electrons and holes are released and transferred to the luminescence centers by Cr^{3+} ions, which then emit characteristic persistent luminescence. Figure 1.6 shows the degradation of RhB as a function of the irradiation time.

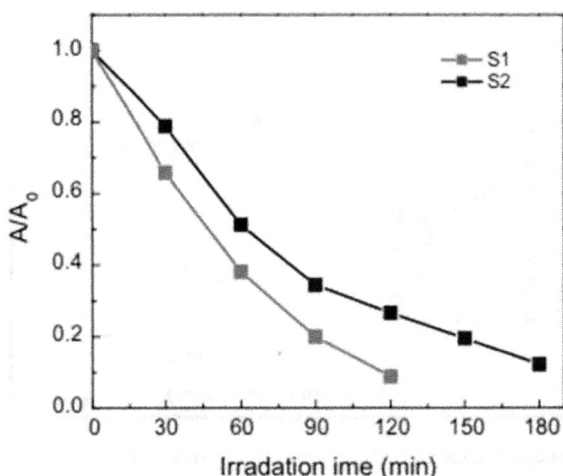

Figure 1.6. The absorption degradation of RhB as a function of irradiation time under ultraviolet light irradiation. Reprinted from [99], Copyright (2014), with permission from Elsevier.

It was found that the absorption of RhB photocatalyzed by Ga_2O_3:$Cr^{3+}_{(0.01)}$ reached 10% under UV light irradiation in 180 min, while the absorption of RhB photocatalyzed by the doping of $Zn^{2+}_{(0.005)}$ into Ga_2O_3:$Cr^{3+}_{(0.01)}$ took only 120 min. Thus, it can be seen that doping with Zn^{2+} can improve the photocatalytic properties of Ga_2O_3:$Cr^{3+}_{(0.01)}$ phosphor [100, 101].

1.6.1.4 The photocatalytic response of nanosized mixed metal oxides
It is well known that wide-bandgap nanosemiconductors such as TiO_2, ZnO, SnO_2, CeO_2, and NiO have excellent abilities to remove toxic dyes and organic pollutants from wastewater via photocatalytic degradation [102]. Nanosized titania (TiO_2) has a superior ability to remove textile dyes via degradation because of its effective generation of charge carriers under UV light illumination. However, due to its large bandgap, it does not work effectively under a visible-light source. Therefore, mixed metal-oxide systems have recently gained more importance due to their good photocatalytic degradation performance under visible or UV light sources. Various mixed metal-oxide systems that incorporate TiO_2, such as TiO_2–CeO_2, TiO_2–SnO_2, TiO_2–CuO, TiO_2–CdO, etc. have been shown to have excellent photocatalytic properties. These types of coupled metal-oxide systems exhibit better visible-light-driven photocatalytic activity and have higher dye degradation efficiency when used to remove toxic dyes and organic contaminants [103]. Recently, Rajendran *et al* [104] reported the photocatalytic response of a TiO_2/NiO composite catalyst used for the degradation of methyl orange. According to their experimental evidence, they observed 98% of methyl orange degradation within 60 min of irradiation. Here, the best performance (98%) of the composite catalyst was observed at pH = 7 (neutral), as shown in figure 1.7. In this composite system, the p–n junction takes a form in which Ni^{3+} states promote a large number of

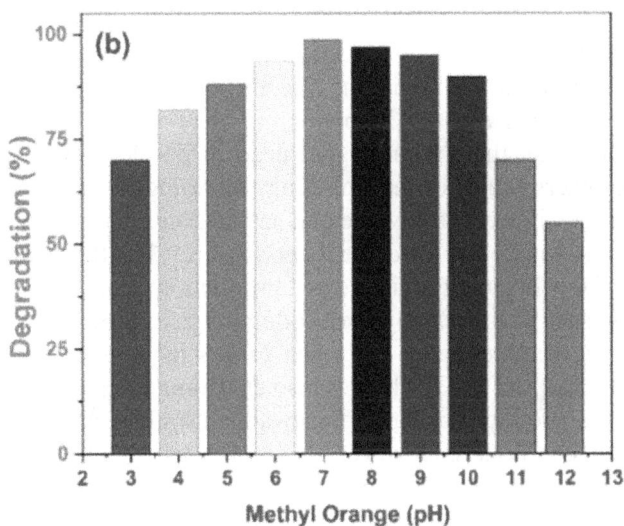

Figure 1.7. The degradation performance of the composite catalyst at different pH values. Reprinted from [104], Copyright (2020), with permission from Elsevier.

electrons, which reduces the recombination rate and helps to enhance the photo-catalytic degradation process under visible light.

1.7 Applications

Due to globalization in the industrial sector, issues related to energy and the environment are becoming more serious. Nowadays, water and air pollution are hot topics due to their adverse effects on human health. To control and resolve these issues, there is a need to adopt sustainable technology for the betterment of mankind. Photocatalysis is an evergreen and economically viable way of solving the problems associated with wastewater treatment and air pollution. Hence, this research topic has become more attractive in the fields of science and engineering. Considering the global need for sustainable energy and environmental technology to replace traditional polluting technologies, photocatalysis is one of the better approaches with which to explore the innovative idea of using clean and natural solar light energy [105]. In the 1970s, Honda and Fujishima published their important discoveries related to water splitting and hydrogen generation using TiO_2 semiconductors for photocatalysis [60]. The scope of photocatalysts in energy and environmental sustainability is discussed below.

1.7.1 Energy sustainability

In the last few decades, many countries have used fossil fuels such as coal, oil, and agricultural waste products to generate electricity in power plants to fulfill the demand for energy in all sectors; such fossil fuel uses have made a major contribution to air, water, and soil pollution all over the world. The utilization of oil and petroleum products in automobiles and heavy transportation vehicles is another major cause of air pollution. The use of this type of traditional power generation technology disperses many kinds of harmful pollutants into the atmosphere, so the level of toxic contaminants affecting the air quality index is increasing rapidly; these pollutants are also a major cause of global warming. To reduce environmental damage, there is an urgent need to develop and replace the traditional sources of energy and related polluting technologies with sustainable energy sources. Among the different renewable and sustainable energy sources, hydrogen is the most promising source for the current century, as it has the advantages of being an eco-friendly form of energy with lower production costs; however, it still has some challenges, such as increasing the production rate, storing the fuel, etc [106]. The photocatalytic production of hydrogen results in pure hydrogen that can be converted into energy and H_2O, which is also environmentally friendly [107]. In the past, hydrogen was produced using nonrenewable resources such as natural gas and petroleum-based technologies. However, these processes suffer from some disadvantages and liberate some other pollutants that are not much cheaper to deal with from a commercial perspective [108]. Hence, considering the previous drawbacks as well as the economic and environmental benefits of the new and advanced photocatalytic hydrogen generation process using solar light energy as the source of clean and low-cost sustainable energy technologies [109].

1.7.2 Environmental sustainability

In recent years, water pollution from industrial waste has been a very serious global issue. Contamination due to heavy metals and organic dye molecules in natural water resources accumulates for a long time, causing negative effects for organisms and human health.

To conserve natural water resources and control environmental pollution, there is an urgent need to develop environmentally friendly water purification technology and other sustainable energy technologies to avoid environmental issues in the air, water, and soil. From the literature database [100–110], it can be seen that approximately 10 000 types of dyes are used in industry for different purposes; therefore, the amounts of textile dyes and other dyes found in industrial wastewater are excessive. When these spread into fresh water resources, many types of toxic contamination can affect the quality of the water. The consumption and use of such contaminated water may cause health issues such as skin rashes, sinus infections, and cancer by entering the human ecosystem through water and animals [110]. Traditional methods including biological treatment, reverse osmosis, coagulation, adsorption, and ultrafiltration are ineffective water treatment processes [111]. Compared to these methods, the energy-efficient photocatalysis process has great potential to remove and degrade organic pollutants naturally using solar energy; it represents the lowest-cost and most favorable method that can utilize low-cost and nontoxic catalysts for the development of technology for water purification and environmental protection [112, 113]. Therefore, many research organizations are currently working in fundamental and applied research areas in the field of environmental sustainability to protect our planet and maintain the harmony of nature.

1.8 Conclusions and future prospects

Photocatalysis is a promising low-cost, energy-efficient, and environmentally friendly technique for resolving environmental and energy problems. There is a great demand for metal-oxide photocatalysts with a visible-light-driven photocatalytic response for the production of hydrogen and wastewater treatment. Further, research into large-scale production and treatment processes remains incomplete. So, in the future, more work is needed to develop this environmentally friendly technology that utilizes natural sunlight for the photocatalytic reaction. In this century, more research is focused on the development of sustainable technology, and photocatalysis has attracted great interest due to its attractive and broad scope for use in the near future due to the current global problems related to energy and the environment. There are still a few challenges in the development of photocatalytic materials, such as achieving high efficiency, thermal stability, purity, and environmental friendliness and overcoming low efficiency in photocatalytic reactors, which is also a major hurdle for their use on a large scale.

References

[1] Briffa J, Sinagra E and Blundell R 2020 Heavy metal pollution in the environment and their toxicological effects on humans *Heliyon* **6** E04691

[2] Hisatomi T and Domen K 2019 Reaction systems for solar hydrogen production via water splitting with particulate semiconductor photocatalysts *Nat. Catal.* **2** 387–99

[3] Kosco J *et al* 2020 Enhanced photocatalytic hydrogen evolution from organic semiconductor heterojunction nanoparticles *Nat. Mater.* **19** 559–65

[4] Meng X, Wang S, Zhang C, Dong C, Li R, Li B, Wang Q and Ding Y 2022 Boosting Hydrogen Evolution Performance of a CdS-Based Photocatalyst: In Situ Transition from Type I to Type II Heterojunction during Photocatalysis *ACS Catal.* **12** 10115–26

[5] Zhao X, Li J, Kong X, Li C, Lin B, Dong F, Yang G, Shao G and Xue C 2022 Carbon Dots Mediated In Situ Confined Growth of Bi Clusters on g-C_3N_4 Nanomeshes for Boosting Plasma-Assisted Photoreduction of CO_2 *Small* **18** 2204154

[6] Zhou Q, Guo Y, Ye Z, Fu Y, Guo Y and Zhu Y 2022 Carbon nitride photocatalyst with internal electric field induced photogenerated carriers spatial enrichment for enhanced photocatalytic water splitting *Mater. Today* **58** 100–9

[7] Collado L, Naranjo T, Gomez-Mendoza M, López-Calixto C, Oropeza F, Liras M, Marugán J and Peña O'Shea V 2021 Conjugated Porous Polymers Based on BODIPY and BOPHY Dyes in Hybrid Heterojunctions for Artificial Photosynthesis *Adv. Funct. Mater.* **31** 2105384

[8] Chen J, Tang T, Feng W, Liu X, Yin Z, Zhang X, Chen J and Cao S 2022 Largescale synthesis of p–n heterojunction Bi_2O_3/TiO_2 nanostructures as photocatalysts for removal of antibiotics under visible light *ACS Appl. Nano Mater.* **5** 1296–307

[9] Khan M M 2023 Chapter 8—Photocatalysis: laboratory to market *Theoretical Concepts of Photocatalysis* ed M Mansoob Khan (Amsterdam: Elsevier) 187–212

[10] Tong H, Ouyang S, Bi Y, Umezawa N, Oshikiri M and Ye J 2012 Nano-photocatalytic materials: possibilities and challenges *Adv. Mater.* **24** 229–51

[11] Ihsan A, Irshad A, Warsi M F, Din M I and Zulfiqar S 2022 NiFe$_2$O$_4$/ZnO nanoparticles and its composite with flat 2D rGO sheets for efficient degradation of colored and colorless effluents photocatalytically *Opt. Mater.* **134** 113213

[12] Irshad A, Farooq F, Warsi M F, Shaheen N, Elnaggar A Y, Hussein E E, ElBahy Z M and Shahid M 2022 Ag-doped FeCo$_2$O$_4$ nanoparticles and their composite with flat 2D reduced graphene oxide sheets for photocatalytic degradation of colored and colorless compounds *Flat Chem.* **31** 100325

[13] Irshad A, Warsi M F, Agboola P O, Dastgeer G and Shahid M 2022 Sol–gel assisted Ag doped NiAl$_2$O$_4$ nanomaterials and their nanocomposites with g-C$_3$N$_4$ nanosheets for the removal of organic effluents *J. Alloys Compd.* **902** 163805

[14] Santhi K, Manikandan P, Rani C and Karuppuchamy S 2015 Synthesis of nanocrystalline titanium dioxide for photodegradation treatment of remazol Brown dye *Appl. Nanosci.* **5** 373–8

[15] Matthews R W 1988 An adsorption water purifier with *in situ* photocatalytic regeneration *J. Catal.* **113** 549–55

[16] Henderson M A, Epling W S, Perkins C L, Peden C H F and Diebold U J 1999 Interaction of molecular oxygen with the vacuum-annealed TiO$_2$ (110) surface: molecular and dissociative channels *Phys. Chem.* B **103** 5328–37

[17] Karthikeyan C, Thamima M and Karuppuchamy S 2019 Structural and photocatalytic property of CaTiO$_3$ nanosphere *Mater. Sci. Forum* **979** 169–74

[18] Vinodgopal K, Stafford U, Gray K A and Kamat P V 1994 The role of oxygen and reaction intermediates in the degradation of 4-chlorophenol on immobilized TiO$_2$ particulate films *J. Phys. Chem.* **98** 6797–803

[19] Zhang X Y, Ling S Y, Ji H Y, Xu L, Huang Y, Hua M Q, Xia J X and Li H M 2017 Metal ion-containing ionic liquid assisted synthesis and enhanced photoelectrochemical performance of g-C3N$_4$/ZnO composites *Mater. Technol.* **3** 1–8

[20] Lin N, Gong Y, Wang R, Wang Y and Zhang X 2022 Critical review of perovskite-based materials in advanced oxidation system for wastewater treatment: design, applications and mechanisms *J. Hazard. Mater.* **424** 127637

[21] Verma V, Al-Dossari M, Singh J, Rawat M, Kordy M G M and Shaban M 2022 A review on green synthesis of TiO$_2$ NPs: photocatalysis and antimicrobial applications *Polym* **14** 1444

[22] Mittal M, Dana J, Lübkemann F, Ghosh H N, Bigall N C and Sapra S 2022 Insight into morphology dependent charge carrier dynamics in ZnSe–CdS nanoheterostructures *Phys. Chem. Chem. Phys.* **24** 8519–28

[23] Alfryyan N, Kordy M G M, Abdel-Gabbar M, Soliman H A and Shaban M 2022 Characterization of the biosynthesized intracellular and extracellular plasmonic silver nanoparticles using *Bacillus cereus* and their catalytic reduction of methylene blue *Sci. Rep.* **121** 14

[24] Pawar R C and Lee C S 2015 Basics of photocatalysis *Heterogeneous Nanocomposite-Photocatalysis for Water Purification* ed R C Pawar and C S Lee (Boston, MA: William Andrew Publishing) 1–23 ch 1

[25] Feng X, Hu G and Hu J 2011 Solution-phase synthesis of metal and/or semiconductor homojunction/heterojunction nanomaterials *Nanoscale* **3** 2099–117

[26] Li R, Weng Y, Zhou X, Wang X, Mi Y, Chong R, Han H and Li C 2015 Achieving overall water splitting using titanium dioxide-based photocatalysts of different phases *Energy Environ. Sci.* **8** 2377–82

[27] Zheng C, Huang L, Guo Q, Chen W, Li W and Wang H 2018 Facile one-step fabrication of upconversion fluorescence carbon quantum dots anchored on graphene with enhanced nonlinear optical responses *RSC Adv.* **8** 10267–76

[28] Martha S, Mansingh S, Parida K M and Thirumurugan A 2017 Exfoliated metal free homojunction photocatalyst prepared by a biomediated route for enhanced hydrogen evolution and Rhodamine B degradation *Mater. Chem. Front.* **1** 1641–53

[29] Navarrete-Magana M, Estrella-Gonzalez A, May-Ix L, Cipagauta-Diaz S and Gomeza R 2021 Improved photocatalytic oxidation of arsenic (III) with WO_3/TiO_2 nanomaterials synthesized by the sol-gel method *J. Environ. Manage.* **282** 111602

[30] Majhi D, Das K, Bariki R, Padhan S, Mishra A, Dhiman R, Dash P, Nayakc B and Mishra B G 2020 A facile reflux method for in situ fabrication of a non-cytotoxic $Bi_2S_3/\beta\text{-}Bi_2O_3/ZnIn_2S_4$ ternary photocatalyst: a novel dual Z-scheme system with enhanced multifunctional photocatalytic activity *J. Mater. Chem.* A **8** 21729–43

[31] Cheng T T, Gao H J, Sun X F, Xian T, Wang S F, Yi Z, Liu G R, Wang X X and Yang H 2021 An excellent Z-scheme $Ag_2MoO_4/Bi_4Ti_3O_{12}$ heterojunction photocatalyst: Construction strategy and application in environmental purification *Adv. Powder Technol.* **32** 951–62

[32] Chen F, Ma T, Zhang T, Zhang Y and Huang H 2021 Atomic-level charge separation strategies in semiconductor-based photocatalysts *Adv. Mater.* **33** 2005256

[33] Qiao B, Wang A, Yang X, Allard L F, Jiang Z, Cui Y, Liu J, Li J and Zhang T 2011 Single-atom catalysis of CO oxidation using Pt1/FeOx *Nat. Chem.* **3** 634–41

[34] Li X, Yang X, Huang Y, Zhang T and Liu B 2019 Supported noble-metal single atoms for heterogeneous catalysis *Adv. Mater.* **31** 1902031

[35] Shan J, Li M, Allard L F, Lee S and Flytzani-Stephanopoulos M 2017 Mild oxidation of methane to methanol or acetic acid on supported isolated rhodium catalysts *Nature* **551** 605–8

[36] Qin R, Liu K, Wu Q and Zheng N 2020 Surface coordination chemistry of atomically dispersed metal catalysts *Chem. Rev.* **120** 11810–99

[37] Ji S, Chen Y, Wang X, Zhang Z, Wang D and Li Y 2020 Chemical synthesis of single atomic site catalysts *Chem. Rev.* **120** 11900–55

[38] Wang A, Li J and Zhang T 2018 Heterogeneous single-atom catalysis *Nat. Rev. Chem.* **2** 65–81

[39] Jiao L, Yan H, Wu Y, Gu W, Zhu C, Du D and Lin Y 2020 When nanozymes meet single-atom catalysis *Angew. Chem.* **132** 2585–96

[40] Gao C, Low J, Long R, Kong T, Zhu J and Xiong Y 2020 Heterogeneous single-atom photocatalysts: fundamentals and applications *Chem. Rev.* **120** 12175–216

[41] Yi D, Lu F, Zhang F, Liu S, Zhou B, Gao D, Wang X and Yao J 2020 Regulating charge transfer of lattice oxygen in single-atom-doped titania for hydrogen evolution *Angew. Chem. Int. Ed.* **59** 15855–9

[42] Feng H *et al* 2022 Porphyrin-based Ti-MOFs conferred with single-atom Pt for enhanced photocatalytic hydrogen evolution and NO removal *Chem. Eng. J.* **428** 132045

[43] Apte S K, Garaje S N, Naik S D, Waichal R P and Kale B B 2013 Environmentally benign enhanced H_2 production from abundant copious waste H_2S using size tuneable cubic bismuth (Bi0) quantum dots-GeO_2 glass photocatalyst under solar light *Green Chem.* **15** 3459–67

[44] Zhao C, Li W, Liang Y, Tian Y and Zhang Q 2016 Synthesis of BiOBr/carbon quantum dots microspheres with enhanced photoactivity and photostability under visible light irradiation *Appl. Catal. Gen.* **527** 127–36

[45] Murphy C J 2002 Peer reviewed: optical sensing with quantum dots *Anal. Chem.* **74** 520A–6A

[46] Kandi D, Martha S and Parida K M 2017 Quantum dots as enhancer in photocatalytic hydrogen evolution: a review *Int. J. Hydrogen Energy* **42** 9467–81

[47] Das N and Kandimalla S 2017 Application of perovskites towards remediation of environmental pollutants: an overview *Int. J. Environ. Sci. Technol.* **147** 1559–72

[48] Mahmoudi F, Saravanakumar K, Maheskumar V, Njaramba L K, Yoon Y and Park C M 2022 Application of perovskite oxides and their composites for degrading organic pollutants from wastewater using advanced oxidation processes: review of the recent progress *J. Hazard. Mater.* **436** 129074

[49] Xu Q L, Zhang L Y, Cheng B, Fan J J and Yu J G 2020 S-Scheme Heterojunction Photocatalyst *Chem.* **6** 1543–59

[50] Liu Y, Hao X Q, Hu H Q and Jin Z L 2021 High Efficiency Electron Transfer Realized over NiS_2/$MoSe_2$ S-Scheme Heterojunction in Photocatalytic Hydrogen Evolution *Acta Phys.-Chim. Sin.* **37** 2008030

[51] Jia X M, Hu C, Sun H Y, Cao J, Lin H L, Li X Y and Chen S F 2023 A dual defect co-modified S-scheme heterojunction for boosting photocatalytic CO_2 reduction coupled with tetracycline oxidation *Appl. Catal., B* **324** 122232

[52] Zhang X, Chen Z, Li X M, Wu Y, Zheng J F, Li Y Q, Wang D B, Yang Q, Duan A B and Fan Y C 2023 Promoted electron transfer in Fe^{2+}/Fe^{3+} co-doped $BiVO_4$/Ag_3PO_4 S-scheme heterojunction for efficient photo-Fenton oxidation of antibiotics *Sep. Purif. Technol.* **310** 123116

[53] Li Y F, Xia Z L, Yang Q, Wang L X and Xing Y 2022 Review on g-C_3N_4-based S-scheme heterojunction photocatalysts *J. Mater. Sci. Technol.* **125** 128–44

[54] Zhang X D, Yu J G, Macyk W, Wageh S, Al-Ghamdi A A and Wang L X 2023 C_3N_4/PDA S-Scheme Heterojunction with Enhanced Photocatalytic H_2O_2 Production Performance and Its Mechanism *Adv. Sustain. Syst.* **7** 2200113

[55] Chu H, Lee C and Tai N 2016 Green preparation using black soybeans extract for graphene-based porous electrodes and their applications in supercapacitors *J. Power Sources* **322** 31–9

[56] Tarcan R, Todor-Boer O, Petrovai I, Leordean C, Astilean S and Botiz I 2020 Reduced graphene oxide today *J. Mater. Chem. C* **8** 1198–224

[57] Witjaksono G *et al* 2021 Effect of nitrogen doping on the optical bandgap and electrical conductivity of nitrogen-doped reduced graphene oxide *Molecules* **26** 6424

[58] Ngidi N P D, Ollengo M A and Nyamori V O 2020 Tuning the properties of boron-doped reduced graphene oxide by altering the boron content *New J. Chem.* **44** 16864–76

[59] Hu C, Lu T, Chen F and Zhang R 2013 A brief review of graphene-metal oxide composites synthesis and applications in photocatalysis *J. Chin. Adv. Mater. Soc.* **1** 21–39

[60] Fujishima A and Honda K 1972 Electrochemical photolysis of water at a semiconductor electrode *Nature* **238** 37–8

[61] Yoshida M, Yamakata A, Takanabe K, Kubota J, Osawa M and Domen K 2009 ATR–SEIRAS investigation of the Fermi level of Pt cocatalyst on a GaN photocatalyst for hydrogen evolution under irradiation *J. Am. Chem. Soc.* **131** 13218

[62] Mittal M, Sharma M and Pandey O P 2016 Fast and quick degradation properties of doped and capped ZnO nanoparticles under UV–visible light radiations *Sol. Energy* **125** 51–64

[63] Cho S, Jang J W, Kim J, Lee J S, Choi W and Lee K H 2011 Three-dimensional type II ZnO/ZnSe heterostructures and their visible light photocatalytic activities *Langmuir* **27** 10243–50

[64] Varghese Alex K, Tamil Pavai P, Rugmini R, Shiva Prasad M, Kamakshi K and Sekhar K C 2020 Green synthesized Ag nanoparticles for bio-sensing and photocatalytic applications *ACS Omega* **5** 13123–9

[65] Akter S, Lee S Y, Siddiqi M Z, Balusamy S R, Ashrafudoulla M, Rupa E J and Huq M A 2020 Ecofriendly synthesis of silver nanoparticles by *Terrabacter humi* sp. nov. and their antibacterial application against antibiotic-resistant pathogens *Int. J. Mol. Sci.* **21** 9746

[66] Garba Z N, Zhou W, Zhang M and Yuan Z 2020 A review on the preparation, characterization and potential application of perovskites as adsorbents for wastewater treatment *Chemosphere* **244** 125474

[67] Kafle B 2020 Introduction to nanomaterials and application of UV–visible spectroscopy for their characterization *Chemical Analysis and Material Characterization by Spectrophotometry* (Amsterdam: Elsevier) 6 147–98

[68] Huston M, DeBella M, DiBella M and Gupta A 2021 Green synthesis of nanomaterials *Nanomaterials* **11** 2130

[69] Shivashankar A, Prashantha S, Anantharaju K, Malini S, Manjunatha H, Vidya Y, Sridhar K and Munirathnam R 2022 Rod shaped zirconium titanatenanoparticles: synthesis, comparison and systematic investigation of structural, photoluminescence, electrochemical sensing and supercapacitor properties *Ceram. Int.* **48** 35676

[70] Shan A Y, Ghazi T I M and Rashid S A 2010 Immobilisation of titanium dioxide onto supporting materials in heterogeneous photocatalysis: a review *Appl. Catal. A Gen.* **389** 1–8

[71] Mavengere S and Kim J-S 2018 UV–visible light photocatalytic properties of $NaYF_4$:(Gd, Si)/TiO_2 composites *Appl. Surf. Sci.* **444** 491–6

[72] Zheng Y, Zheng L, Zhan Y, Lin X, Zheng Q and Wei K 2007 Ag/ZnO heterostructure nanocrystals: synthesis, characterization, and photocatalysis *Inorg. Chem.* **46** 6980–6

[73] Wu H B, Hng H H and Lou X W D 2012 Direct synthesis of anatase TiO_2 nanowires with enhanced photocatalytic activity *Adv. Mater.* **24** 2567–71

[74] Liu Z, Zhang X, Nishimoto S, Jin M, Tryk D A, Murakami T and Fujishima A 2007 Highly ordered TiO2 nanotube arrays with controllable length for photoelectrocatalytic degradation of phenol *J. Phys. Chem.* C **112** 253–9

[75] Fan T, Han T, Chow S K and Zhang D 2010 Biogenic N–P-codoped TiO_2: synthesis, characterization and photocatalytic properties *Bioresour. Technol.* **101** 6829–35

[76] Bora T, Sathe P, Laxman K, Dobrestov S and Dutta J 2017 Defect engineered visible light active ZnO nanorods for photocatalytic treatment of water *Catal. Today* **284** 11–8

[77] Jiao Y, Zhou L, Ma F, Gao G, Kou L, Bell J, Sanvito S and Du A 2016 Predicting Single-Layer Technetium Dichalcogenides (TcX_2, X = S, Se) with Promising Applications in Photovoltaics and Photocatalysis *ACS Appl. Mater. Interfaces* **8** 5385–92

[78] Fujishima A and Honda K 1972 Electrochemical Photolysis of Water at a Semiconductor Electrode *Nature* **238** 37–8

[79] Zhuang H L and Hennig R G 2013 Single-Layer Group-III Monochalcogenide Photocatalysts for Water Splitting *Chem. Mater.* **25** 3232–8

[80] Cao Y, Gao Q, Li Q, Jing X, Wang S and Wang W 2017 Synthesis of 3D porous MoS_2/g-C_3N_4 heterojunction as a high efficiency photocatalyst for boosting H_2 evolution activity *RSC Adv.* **7** 40727–33

[81] Gordon T R, Cargnello M, Paik T, Mangolini F, Weber R T, Fornasiero P and Murray C B 2012 Nonaqueous synthesis of TiO_2 nanocrystals using TiF_4 to engineer morphology, oxygen vacancy concentration, and photocatalytic activity *J. Am. Chem. Soc.* **134** 6751–61

[82] Yang Y, Wang S-Q, Wen H, Ye T, Chen J, Li C-P and Du M 2019 Nanoporous gold embedded ZIF composite for enhanced electrochemical nitrogen fixation *Angew. Chem. Int. Ed.* **58** 15362–6

[83] Zhang L, Huang J, Hu Z, Li X, Ding T, Hou X, Chen Z, Ye Z and Luo R 2022 Ni $(NO_3)_2$-induced high electrocatalytic hydrogen evolution performance of self-supported fold-like WC coating on carbon fiber paper prepared through molten salt method *Electrochim, Acta* **422** 140553

[84] Zhao C, Xi M, Huo J and He C 2021 B-doped 2D-InSe as a bifunctional catalyst for CO_2/CH_4 separation under the regulation of an external electric field *Phys. Chem. Chem. Phys.* **23** 23219–24

[85] Zhang L, Hu Z, Huang J, Chen Z, Li X, Feng Z, Yang H, Huang S and Luo R 2022 Experimental and DFT studies of flower-like Ni-doped Mo_2C on carbon fiber paper: a highly efficient and robust HER electrocatalyst modulated by $Ni(NO_3)_2$ concentration *J. Adv. Ceram.* **11** 1294–306

[86] Wang D, Wang X-X, Jin M L, He P and Zhang S 2022 Molecular level manipulation of charge density for solid–liquid TENG system by proton irradiation *Nano Energy* **103** 107819

[87] Yang B, Liu T, Guo H, Xiao S and Zhou L 2019 High-performance meta-devices based on multilayer meta-atoms: interplay between the number of layers and phase coverage *Sci. Bull.* **64** 823–35

[88] Yakout S M 2019 Inclusion of cobalt reinforced Ag doped SnO_2 properties: electrical, dielectric constant, magnetic and photocatalytic insights *J. Mater. Sci., Mater. Electron.* **30** 17053–65

[89] Entradas T, Cabrita J F, Dalui S, Nunes M R, Monteiro O C and Silvestre A J 2014 Synthesis of sub-5 nm Co-doped SnO_2 nanoparticles and their structural, microstructural, optical and photocatalytic properties *Mater. Chem. Phys.* **147** 563–71

[90] Nihal S, Rattan M, Anjali H, Kumar S, Sharma M, Tripathi S K and Goswamy J K 2021 Synthesis and characterization of Ag metal doped SnO_2, WO_3 and WO_3–SnO_2 for propan-2-ol sensing *Results Mater.* **9** 100127

[91] Kuspanov Z, Bakbolat B, Baimenov A, Issadykov A, Yeleuov M and Daulbayev C 2023 Photocatalysts for a sustainable future: innovations in large-scale environmental and energy applications *Sci. Total Environ.* **885** 163914–24

[92] Wang J, Wang Z, Dai K and Zhang J 2023 Review on inorganic–organic S-scheme photocatalysts *J. Mater. Sci. Technol.* **165** 187–218

[93] Sahai A, Ikram S, Rai R, Shrivastav S, Dass V R and Satsangi 2017 Quantum dots sensitization for photoelectrochemical generation of hydrogen: a review *Renew. Sustain. Energy Rev.* **68** 19–27

[94] Xiong Z, Lei Z, Ma S, Chen X, Gong B and Zhao Y 2017 Photocatalytic CO_2 reduction over V and W codoped TiO_2 catalyst in an internal-illuminated honeycomb photoreactor under simulated sunlight irradiation *Appl Catal Environ* **219** 412–24

[95] Liu X, Chen X, Wang S, Yan L, Yan J and Guo H 2022 Promoting the photocatalytic H_2 evolution activity of $CdLa_2S_4$ nanocrystalline using few-layered WS_2 nanosheet as a co-catalyst *Int. J. Hydrogen Energy* **47** 2327–37

[96] Zhu Y, Wang T, Xu T, Li Y and Wang C 2019 Size effect of Pt co-catalyst on photocatalytic efficiency of g-C_3N_4 for hydrogen evolution *Appl. Surf. Sci.* **464** 36–42

[97] Sun Q, Yu Z, Jiang R, Hou Y, Sun L and Qian L 2020 CoP QD anchored carbon skeleton modified CdS nanorods as a co-catalyst for photocatalytic hydrogen production *Nanoscale* **12** 19203–12

[98] García C R, Diaz-Torres L A, Salas P, Guzman M and Angeles-Chavez C 2015 Photoluminescent and photocatalytic properties of bismuth doped strontium aluminates blended with titanium dioxide *Mater. Sci. Semicond. Process.* **37** 105–11

[99] Wang Y, Xu K, Li D, Zhao H and Hu Z 2014 Persistent luminescence and photocatalytic properties of $Ga_2O_3:Cr^{3+}$, Zn^{2+} phosphors *Opt. Mater.* **36** 1798–801

[100] Sato J, Kobayashi H, Zkarashi K, Saito N, Nishiyama H and Inoue Y 2004 *J. Phys. Chem. B* **108** 4369–75

[101] Girija K, Thirumalairajan S, Avadhani G S, Mangalaraj D, Ponpandian N and Viswanathan C 2013 *Mater. Res. Bull.* **48** 2296–303

[102] Serpone N and Emeline A V 2012 Semiconductor photocatalysis-past, present, and future outlook *J. Phys. Chem. Lett.* **35** 673–7

[103] Lu D, Zelekew O A, Abay A K, Huang Q, Chen X and Zheng Y 2019 Synthesis and photocatalytic activities of a CuO/TiO_2 composite catalyst using aquatic plants with accumulated copper as a template *RSC Adv.* **9** 2018–25

[104] Rajendran S, Manoj D, Nimita Jebaranjitham J, Kumar B G, Bharath G, Banat F, Qin J, Vadivel S and Gracia F 2020 Nanosized titania-nickel mixed oxide for visible light photocatalytic activity *J. Mol. Liq.* **311** 113328

[105] Kudo A and Miseki Y 2009 Heterogeneous photocatalyst materials for water splitting *Chem. Soc. Rev.* **38** 253–78

[106] Dawood F, Anda M and Shafiullah G M 2020 Hydrogen production for energy: an overview *Int. J. Hydrogen Energy* **45** 3847–69

[107] Holladay J D, Hu J, King D L and Wang Y 2009 An overview of hydrogen production technologies *Catal. Today* **139** 244–60

[108] Li Y and Tsang S C E 2020 Recent progress and strategies for enhancing photocatalytic water splitting *Mater. Today Sustain.* **9** 100032

[109] Armaroli N and Balzani V 2010 The Hydrogen Issue *ChemSusChem.* **4** 21–36

[110] Hariharan D 2020 Enhanced photocatalysis and anticancer activity of green hydrothermal synthesized Ag@ TiO_2 nanoparticles *J. Photochem. Photobiol. B* **202** 111636

[111] Ahmed M A *et al* 2020 Rapid photocatalytic degradation of RhB dye and photocatalytic hydrogen production on novel curcumin/SnO_2 nanocomposites through direct Z-scheme mechanism *J. Mater. Sci., Mater. Electron.* **31** 19188

[112] Hamdy M S *et al* 2021 Fabrication of novel polyaniline/ZnO heterojunction for exceptional photocatalytic hydrogen production and degradation of fluorescein dye through direct Z-scheme mechanism *Opt. Mater.* **117** 111198

[113] Sayed M A *et al* 2022 Mesoporous polyaniline/SnO_2 nanospheres for enhanced photocatalytic degradation of bio-staining fluorescent dye from an aqueous environment *Inorg. Chem. Commun.* **139** 109326

IOP Publishing

Photocatalysts for Energy and Environmental Sustainability

Vijay B Pawade and Bharat A Bhanvase

Chapter 2

Rare-earth-doped metal oxide photocatalysts for wastewater treatment

G A Suganya Josephine, S Rubesh Ashok Kumar and D Vasvini Mary

Semiconductor metal oxide photocatalysts are well-known candidates for wastewater treatment. The enhancement of photocatalytic activity is often attempted using the approach of doping metals and nonmetals. In recent years, rare-earth (RE) materials have explored. These materials are characterized by the presence of f *orbitals*, which play a vital role in bandgap modification in doped materials. The *f–f transitions* contribute to electron trapping, thereby enhancing the lifetime of free holes and accelerating the photocatalytic process. This chapter explores the various rare-earth dopants and the doping process employed for nanomaterial preparation. The most important application of these nanomaterials is for the treatment of wastewater under UV, visible light, and natural solar irradiation, which are further discussed.

2.1 Introduction

Global environmental degradation results from intensified industry brought on by population growth and development. Industries pollute the environment by discharging heavy metal ions and harmful chemicals into waterways [1]. Therefore, protecting the environment from the damaging effects of discarded chemicals and heavy metal ions is crucial. Wastewater containing organic pollutants cannot be treated efficiently using traditional supplies or methods [2, 3]. In order to provide safe and clean water, the water industry needs to use low-cost, sustainable, efficient, and effective water treatment processes [4]. Numerous industrial applications and water treatment methods successfully utilize nanotechnology [5]. As one of the many nanotechnology-based remediation methods, semiconductor-mediated heterogeneous advanced oxidation technologies have become a viable option. As part of advanced oxidation processes (AOPs), extremely reactive hydroxyl radicals ($^{\bullet}$OH)

doi:10.1088/978-0-7503-5697-8ch2

are produced on-site and used to turn various organic contaminants into non-hazardous molecules [6–8].

Semiconductor metal oxides are a class of materials widely used for various applications in the field of science and technology. Almost all areas of research and technology have used semiconductor nanoparticles. Researchers from all over the world have been studying semiconductors at the nanoscale, and their use in heterogeneous photocatalysis has become increasingly important in biomedical nanotechnology. Although TiO_2 nanoparticles are superior semiconductors for photocatalytic applications, they are expensive [9]. In order to replace TiO_2, a wide range of metal oxides have been researched, including ZnO, SnO_2, Bi_2O_3, WO_3 and other metal oxides [10]. When exposed to UV/visible light, surface changes occur that prevent the recombination of electron–hole pairs, which is necessary for the generation of radical species [11].

Researchers from all over the world have tried to modify pure metal oxides by doping them with metals, nonmetals, and RE materials; by producing composites and core–shell materials; by supporting metal oxides on SiO_2, Al_2O_3, graphene oxide, or other materials; etc [12]. In the periodic table, the lanthanides are found among the RE elements. The lanthanides consist of 15 different elements, ranging from lutetium (71) to lanthanum (57), and the f *orbitals* in these compounds are mainly responsible for these materials' optical and catalytic capabilities [13]. In the years after Carl Axel Arrhenius discovered a black mineral, 'ytterbite,' in 1787, the REs gradually become known to the world. Lanthanides are mostly used as up-conversion catalysts. They are also added to metal oxides as dopants to improve their characteristics [14]. In addition to their function as doping agents, these substances are also employed to create composites and core–shell materials.

The luminescence of the metal oxide ZnO is improved when lanthanide (RE^{3+}) ions are used to modify its surface. The metal oxide with embedded lanthanide converts near-IR and visible radiation into UV wavelengths, which results in the sequential absorption of numerous infrared photons. The lanthanides' partially filled f-*orbitals* or f-subshells contribute to the *f–f transition*, which is what causes the luminescent property. Since the RE ion has higher photocatalytic activity, it appears that these transitions can be viewed as narrow emission bands that are unique to it [15]. This chapter covers: (i) types of RE material that have been employed to enhance the photocatalytic activity of metal oxides and (ii) REs employed as photocatalysts under UV light or visible light. The preparation techniques employed, the morphology of the materials, the pollutants of interest, and the percentage of degradation are detailed in the following sections.

2.2 Semiconductor metal oxides

The photocatalytic splitting of water by TiO_2 under visible light was first performed in 1972 by Fujishima and Honda. Since then, research into the preparation of an alternative semiconductor metal oxide for use in photocatalysis has been strongly encouraged [16, 17]. TiO_2 has been extensively explored by numerous research groups worldwide for a variety of applications such as photocatalytic processes for

Table 2.1. Semiconductor-based photocatalytic removal of organic dyes under visible light.

S. No.	Catalyst	Effluent	Time (min)	Degradation (%)	Reference
1	N/ZnO	Rhodamine B	20	95	[19]
2	N/TiO$_2$	Alizarin Red S	30	90	[20]
3	C/TiO$_2$	Azo dyes	60	99	[21]
4	Carbon dots/TiO$_2$	Methyl blue	60	98	[22]
5	Bi@Bi$_2$O$_3$	Rhodamine B	30	96.6	[23]
6	Fe^{3+} doped TiO$_2$	Malachite green	60	99	[24]
7	Bi$_2$O$_3$–ZnO	Acid black 1	60	97	[25]
8	Cr/ZnO	Azo dye	60	99	[26]
9	CdS/TiO$_2$	Rhodamine B	30	90	[27]
10	CuS@CuAl$_2$O$_4$	Eosin yellow	60	98	[28]

organic degradation, water splitting, solar cells, energy production, photocatalytic conversion, sensors, and so on. However, because TiO$_2$ has a large bandgap, light can only activate it [18]. Photocatalysts have been intensively investigated for many decades due to their widespread applicability in renewable energy production, organic reactions, and the abatement of environmental damage. Because of the wide range of semiconductor materials that have been synthesized, the applications of semiconductor photocatalysis have grown dramatically. As a result, attempts to find a suitable substitute have been made utilizing different metal oxides such as ZnO, MoO$_3$, CeO$_2$, ZrO$_2$, WO$_3$, α-Fe$_2$O$_3$, SnO$_2$, and metal chalcogenides (i.e. ZnS, CdS, CdSe, WS$_2$, and MoS$_2$), among others. However, the rapid recombination of charge carriers and the inability of O$_2$ to effectively trap electrons at the conduction band (CB) to produce superoxide radicals have blocked the practical use of metal oxide nanomaterials as photocatalysts. To address this problem, scientists are doping or combining pristine materials with other metal oxide materials to change the bandgap energies of the resulting composites. This reduces the recombination of charge carriers and improves the photocatalytic activity of the composites. Table 2.1 summarizes the removal of a variety of organic dyes by semiconductor metal-oxide-based photocatalysts under visible light irradiation.

The N-doped ZnO composites have demonstrated an effective photodegradation process in which approximately 95% of the RhB decomposed within 20 min. Only a ~57% decrease in RhB was seen over the same time period in control studies that used pure ZnO as the photocatalyst. Under UV light irradiation, N–ZnO nano-particles demonstrated enhanced photocatalytic performance in comparison to pure ZnO [19]. For example, when compared to pure TiO$_2$, N-TiO$_2$ rice grains demonstrated improved UV-light-assisted photocatalysis in the photodegradation of the industrially significant anthraquinone dye alizarin red S [20]. It possessed more photocatalytic activity for the breakdown of acid black 1 (AB 1) under UV light than pure ZnO. Bi$_2$O$_3$·Bi$_2$O$_3$–ZnO is a potential choice for the photocatalytic treatment of dye effluent for 60 min, reaching an efficiency of 97% at neutral pH 7 [25].

The UV–vis absorption spectra of pure ZnO and Cr-doped ZnO both showed a bandgap absorption peak at around 365 nm. In the UV–vis absorption spectra of Cr-doped ZnO, an extra wide tail developed from 400 to 750 nm compared to that of pure ZnO. Under visible light, photocatalytic tests revealed that synthesized Cr-doped ZnO nanowires were a potential photocatalyst for the restoration of water polluted by various chemically stable azo dyes [26].

2.3 Rare-earth elements

Lanthanides, a group of elements with atomic numbers 57–71, are divided into two groups based on electron configuration [29]. The light RE elements, also known as the cerium group, include lanthanum to gadolinium (atomic numbers 57–64) and the heavy RE elements include terbium to lutetium (atomic numbers 65–71). Despite being labeled as 'rare,' these elements are abundant in the Earth's crust. The most prevalent rare-earth metals, namely cerium, lanthanum, and yttrium, have abundances (atomic fractions) comparable to those of typical industrial metals such as nickel, copper, and lead. Even though thulium is the least abundant RE, its abundance is still 200 times higher than that of gold.

The similarities between the physical and chemical characteristics of REs are strongly influenced by their electrical structure. Because $5d$ and $4f$ electrons have comparable energies, lanthanide elements have two sorts of electronic configuration: $4f^n\ 6s^2$ and $4f^{n-1}\ 5d^1\ 6s^2$, with $n = 1$–14. Lanthanum, cerium, gadolinium, and lutetium have electronic configurations of the $4f^{n-1}\ 5d^1\ 6s^2$ type, while the other lanthanide elements have electronic configurations of the $4f^n\ 6s^2$ type. All REs may easily lose electrons in $ns^2\ (n-1)\ d^1$ or $4f^1$, which allows them all to become M^{3+} ions. M^{2+} and M^{4+} ions can be formed by several lanthanides. In comparison to the M^{3+} state, the M^{2+} and M^{4+} states are often less stable.

2.4 The preparation of rare-earth-doped metal oxides

Lanthanide metals (Ln(III)) can drastically alter metal oxide photocatalysts' electrical, physical, and chemical characteristics. Lanthanide metal ions are useful in modifying metal oxide surfaces for diverse photocatalytic applications. Various techniques for obtaining different particle sizes, shapes, and large surface areas are being investigated by research groups (figure 2.1); these characteristics are important factors that determine the activity of the surface.

Chemical, physical, and mechanical methods such as the coprecipitation technique [30–33], the sol–gel process [34, 35], hydrothermal synthesis [36–38], the electrospinning method [39, 40], the paralysis spray method [41], the combustion method [42, 43], the sonochemical method [44, 45], microwave-assisted synthesis [46], thermal evaporation [47], etc. may all be used to prepare RE-doped metal oxides, which may be altered in terms of their sizes, forms, and other characteristics. A few of the techniques adopted and the materials prepared are detailed in table 2.2.

A study investigated the optical characteristics of ZnO nanorods codoped with Eu (1%) and Ce (1%) [53]. This study illustrated that the optical absorption and emission of ZnO are easily modified by the addition of RE ions and that the resulting

Hydrothermal

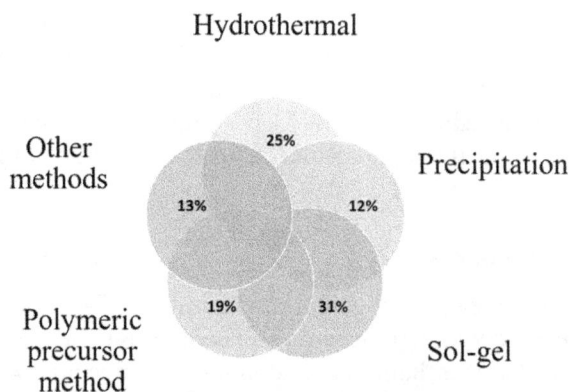

Figure 2.1. Major synthesis methods for the preparation of RE-doped metal oxides.

Table 2.2. The preparation of selected RE-doped metal oxides.

S. No.	Rare-earth-doped metal oxide	Preparation technique	Application	References
1	$LuFeO_3$	Polymeric precursor method	Photocatalysis	[48]
2	$YFeO_3$	Polymeric precursor method	Photocatalysis	[49]
3	Gd-doped CeO_2	Polymeric precursor method	Photocatalysis	[50]
4	Ce- and Y-doped SnO_2	Polymeric precursor method	Photocatalysis	[51]
5	$LaMnO_3$	Microemulsion— reverse micelle	Photocatalysis	[52]
6	Ce- and Eu-doped ZnO	Coprecipitation	Photocatalysis	[53]
7	Dy-doped ZnO	Coprecipitation	Photocatalysis	[54]
8	Ce-doped ZnO	Coprecipitation	Photocatalysis	[55]
9	Sm-doped TiO_2	Sol–gel	Photocatalysis	[56]
10	Ce-doped TiO_2	Sol–gel	Photocatalysis	[57]
11	Gd-doped SnO_2	Sol–gel	Photocatalysis	[58]
12	Eu-doped ZnO	Sol–gel	Photo- and thermoluminescence	[59]
13	Er-doped ZnO	Hydrothermal	Photoluminescence and optical properties	[60]
14	Ho-doped TiO_2	Sol–gel	Photocatalysis	[61]
15	Pr-doped molecularly imprinted TiO_2	Solvothermal	Photocatalysis	[62]
16	La-doped ZnO thin film	Dip coating	Photocatalysis	[63]

codoped material may be employed as an efficient electrode material in solar cell applications, optoelectronic devices, and photocatalysis. Under visible irradiation, Pr-doped molecularly imprinted TiO2 mesocrystals (Pr-MIP-TMCs) demonstrated higher photocatalytic activity than molecularly imprinted TiO2 mesocrystals (MIP-TMCs), which was attributed to its strong interactions between the target molecules and the imprinted cavities, its large adsorption capacity for the target molecules, its increased availability of adsorption sites, its narrow bandgap energy, and its efficient separation of photogenerated charges [62].

2.5 The photocatalytic removal of contaminants by rare-earth–metal oxides

RE-doped metal oxide photocatalysts have received a lot of interest due to their ability to split water into H_2 and O_2 when exposed to light irradiation (UV/visible/sunlight). The use of RE elements in metal oxide photocatalysts attracts a lot of interest compared to unmodified metal oxides because the *4f-levels* of lanthanide elements play a significant role in photocatalytic activities. Figure 2.2 shows the major photocatalytic applications of RE-doped metal oxides. The majority of studies have involved the photocatalytic degradation of pharmaceuticals and organic dyes. Tables 2.3–2.5 display some representative rare-earth elements used in photocatalysts for the degradation of pharmaceuticals, dyes, etc. under UV, visible, and solar irradiation.

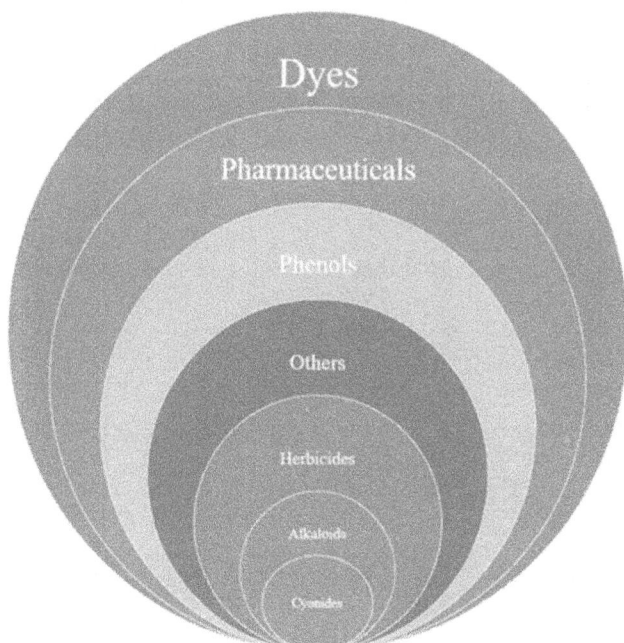

Figure 2.2. Photocatalytic applications of RE-doped metal oxides.

Table 2.3. Photocatalytic removal of endocrine disrupter.

S. No.	Catalyst	Effluent	Time (min)	Degradation (%)	References
1	Gd-doped $BiFeO_3$	Ciprofloxacin	240	80	[65]
2	FeOCl doped with Y	Ibuprofen	20	84	[66]
3	$FeCeO_x$	Diclofenac	40	83	[67]
4	$Er_2FeSbO_7/BiTiSbO_6$	Enrofloxacin	150	99.1	[68]
5	CeO_x-modified graphite felt	Carbamazepine	60	69.4	[69]
6	$LaFeO_3$/lignin biochar	Ofloxacin	75	95.6	[70]
7	CeO_2 nanosheets doped with Fe	Salicylic acid	120	96	[71]
8	$LaFeO_3$	Sulfamethoxazole	120	100	[72]
9	$LaFeO_3$/MIL-125-NH_2	Carbamazepine and caffeine	60	74 and 87	[73]
10	CeO_2/C gas diffusion electrode	Dipyrone	20	100	[74]

Table 2.4. The photocatalytic removal of organic dyes under visible light irradiation.

S. No.	Catalyst	Effluent	Time (min)	Degradation (%)	References
1	WO_3 doped with La, Gd, or Er	Methylene blue	90	98	[75]
2	$CoFe_2O_4$ doped with La or Ce	Remazol black 5, Remazol brilliant yellow	30	$\geqslant$90	[76]
3	Ce^{4+}–TiO_2 solution and nanocrystallites	Brilliant red	120	99.9	[77]
4	$PrFeO_3/CeO_2$	Methyl violet	30	80.1	[78]
5	$PrFeO_3$	Methyl violet	60		[79]
6	$LaFeO_3$ doped with Eu	Safranine-O and	20	$\geqslant$90	[80]
7	$LaFeO_3$ doped with Gd	Remazol brilliant	35		
8	$LaFeO_3$ doped with Dy	yellow	35		
9	$LaFeO_3$ doped with Nd		20		
10	$ZnFe_2O_4$ doped with Sm	Methylene blue	60	65	[81]
11	$ZnFe_2O_4$ doped with Dy	Methylene blue	45	97.3	[82]
12	$EuFeO_3$ nanoparticles	Rhodamine B	180	71	[83]
13	Eu^{2+} BiVO4	Methyl orange	180	93.6	[84]
14	Eu-N codoped TiO_2/ sepiolite nanocomposites	Orange G	540	98	[85]

(*Continued*)

Table 2.4. (*Continued*)

S. No.	Catalyst	Effluent	Time (min)	Degradation (%)	References
15	h-YbFeO$_3$/o-YbFeO$_3$	Methyl violet	180	64	[86]
16	Y-Bi$_2$WO$_6$	Rhodamine B	240	85	[87]
17	Yb/Er/Pr-Bi$_2$WO$_6$	Congo red	60	90.2	[88]
		Tetracycline	60	52.3	
		Rhodamine B	60	95.5	
18	Sm$_2$O$_3$/BiVO$_4$	Rhodamine B	210	55.1	[89]
	Eu$_2$O$_3$/BiVO$_4$			51.8	
	Tb$_2$O$_3$/BiVO$_4$			57.3	
19	Eu-doped Bi$_2$WO$_6$	Rhodamine B	60	98	[90]
20	Gd^{3+}-doped Bi$_2$MoO$_6$	Rhodamine B	10	84	[91]

Table 2.5. The photocatalytic removal of organic dyes under sunlight.

S. No.	Catalyst	Effluent	Time (min)	Degradation (%)	References
1	FeAl/Ce-Mts (montmorillonite)	Reactive blue 19	180	100	[92]
2	Gd-doped BiFeO$_3$	Methylene blue	240	96	[93]
		Rhodamine B	180	97	
3	Sm–ZnO	Methylene blue	90	96	[94]
4	FeAl/La-Mts	Reactive blue 19	180	100	[94]
5	Ce–CdS-G	Methylene blue	90	99	[95]
6	Gd/La@ZnO	Rhodamine B	100	91	[96]
7	Y/Ho-doped NiO	Methyl red	60	99	[97]
8	Eu–TiO$_2$	Methyl orange	60	98	[98]
9	NiO-CGYO (nickel-based yttria and gadolinia codoped ceria)	Methylene blue	120	90	[99]
10	BiGdYVO$_4$	Methylene blue	90	94	[100]

2.6 The photocatalytic removal of antibiotics (endocrine disrupters) by rare-earth–metal oxides

Environmental contamination has a tremendous impact on human survival and health. The overuse and misuse of antibiotics has led to a slew of issues, putting human and environmental health at risk [64]. Therefore, modern chemistry seeks excellent photocatalytic materials that can degrade antibiotics into CO_2, H_2O, and other nontoxic small molecules. Due to their particular properties, RE-doped metal oxides have emerged as next-generation photocatalysts.

Numerous approaches have been tried to remove antibiotics [65–74] from the environment using RE-doped metal oxides as catalysts. In recent years, researchers have studied the use of various RE-doped metal oxides for the degradation of antibiotics. Some degradation results are tabulated in table 2.3. One study investigated the photodegradation of the colorless organic pollutants ciprofloxacin (CIP) and levofloxacin (LFX) by a Gd-doped $BiFeO_3$ (BGFO) photocatalyst [65]. Remarkably, 240 min of sunlight was sufficient to cause deteriorations of 80% and 79% in CIP and LFX, respectively. The photocatalytic activity of BGFO may benefit from its superior shape, good crystallinity, higher optical absorption, and efficient charge carrier separation.

2.7 The photocatalytic removal of organic dye effluent by rare-earth-doped metal oxides

In photocatalysis, RE-doped metal oxide-based photocatalysts allow fast photo-generated charge carrier separation and provide abundant surface functional groups for light-harvesting materials, making their photoconversion efficiency acceptable. Therefore, numerous theoretical and experimental studies have recently shown the potential of RE-doped metal-oxide-based photocatalysts for the removal of organic dye effluents under visible light irradiation [75–91]. Several RE-doped metal-oxide-based photocatalysts have been produced and tested for their effective degradation of organic dyes (table 2.4).

WO_3 doped with La or Gd had faster and better charge separation in a composite material compared to that of pristine WO_3. The enhanced photocatalytic activity was found to occur in the following order: Gd–W > WC > Er–W > La–W > W. The findings showed excellent photocatalytic performance for 2% Gd-modified WO_3. This was attributed to the combined effects of a higher surface area, reduced charge carrier recombination, the surface hydroxyl content, the extension of photosensitivity toward the visible light absorption region, and an optimal dopant content [75].

Numerous attempts have been made to control and treat wastewater. RE-doped metal-oxide-based photocatalysts can play a significant part in the photocatalytic removal of organic dye effluents under solar irradiation [92–100]. Table 2.5 shows the photocatalytic degradation of RE-doped metal-oxide-based photocatalysts under solar irradiation. For example, methylene blue (MB) degradation under solar radiation was used to investigate the photo-induced activity of Sm-doped ZnO nanoflowers. Notably, 5% Sm–ZnO had the best degradation efficiency, achieving 96% degradation within 90 min [84].

Table 2.6 summarizes the removal of a variety of organic dyes by RE-doped metal-oxide-based photocatalysts under UV light irradiation. The photocatalytic efficiencies of ZnO nanoneedles doped with 0, 0.5, and 1% Nd were assessed by using UV light to degrade methylene blue. With 2.5 times the photocatalytic activity of undoped ZnO, the 1% Nd-doped ZnO performed best in this study [94]. As a result, this study focused on RE-doped semiconductor materials. Specifically, coprecipitation was used to synthesize cerium-doped dysprosium oxide. Under

Table 2.6.

S. No	Catalyst	Effluent	Time (Min)	Degradation %	Reference
01	ZnO-doped Nd	Methylene blue	300	92	[101]
02	Sm-doped CeO_2 nanoparticles	Rose Bengal	90	89	[102]
03	Eu^{2+}–TiO_3	Methylene blue	60	70	[103]
04	Dy-doped ZnO	Crystal violet	-	98	[104]
05	Zn–CeO_2	Methylene blue	60	97	[105]
06	Er dopant into $K_2Ta_2O_6$	Phenol	240	90	[108]
07	NiO-CGYO	Methylene blue	120	90	[99]
08	MgO- and Er^{3+}-doped MgO	Methyl Orange	250	88	[106]
09	Dy^{3+}-doped CeO_2	Orange G	120	96	[107]
10	Eu^{3+}–TiO_3	Rhodamine B	30	96	[109]

neutral pH conditions, 10 mg of photocatalyst exhibited superior activity for 10 ppm of Oregon green (OG) dye under 96% UV light irradiation [99].

2.8 The photocatalytic mechanism of rare-earth-based photocatalysts

According to all these studies carried out by diverse researchers, RE-doped metal oxides have superior photocatalytic activity for the degradation of different organic dyes when exposed to UV, visible, and solar light. The production of radical species in the reaction media provides the primary insight into the photocatalytic perform-ance [110]. In any semiconductor material, light irradiation excites electrons from the valence band (VB) into the conduction band (CB), and the holes present in the VB are free to react with water molecules to produce hydroxyl radicals ($^\bullet$OH). In addition, the superoxide radical anion (O^{2-}) is formed when oxygen in the dissolved solution reacts with the electrons in the CB. These radicals, which are powerful oxidizing species, take part in oxidation and reduction reactions. The excited electrons are held in the f subshell of the RE element(s) in the case of RE-doped metal oxides [55]. This trapping process delays the recombination process, which leads to the continuous production of reactive free radicals, which is responsible for the enhanced photocatalytic performance compared to that of the pristine metal oxide. Hence, RE-doped metal oxides show better photocatalytic activity under UV and visible light irradiation. Figure 2.3 shows the photocatalytic mechanism of RE-doped ZnO under visible light irradiation; it has been found that the photo-catalyst's absorption region shifts from the UV region to the visible region upon the addition of ZnO to the Dy_2O_3 matrix. In Dy_2O_3, it has been observed that before the trapped electron returns to the ground state, h^+ reacts with the H_2O molecule to form free $^\bullet$OH radicals, thereby initiating the degradation process.

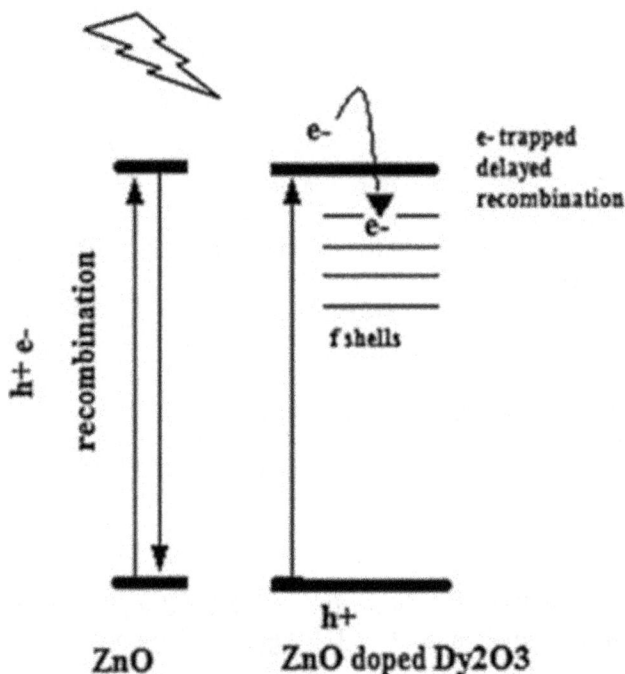

Figure 2.3. The photocatalytic mechanism of an RE-based photocatalyst. Reprinted from [55], Copyright (2014), with permission from Elsevier.

2.9 Conclusions

In brief, we conclude that RE elements are excellent dopants for metal oxides, as they enhance their properties drastically. This chapter has introduced the simple methods of synthesis employed in the research literature for the doping metal oxides with RE materials. Techniques such as the polymeric precursor method, the sol–gel method, coprecipitation, the hydrothermal method, and the solvothermal method account for the majority of the materials prepared. Since RE-doped metal oxides exhibit the characteristic *f–f transitions* associated with *F-block* elements, their properties are further enhanced. These materials contain exceptional photocatalytic, optical, and luminescent properties. Their enhanced photocatalytic property is attributed to the trapping of excited electrons in the f-*subshells*, which delays the recombination of electron–hole pairs. This further lengthens the time in which free holes can react with water molecules to generate $^{\bullet}OH$ radicals. Hence, RE-doped metal oxides serve as better nanomaterials for the degradation of organics present in wastewater under UV, visible, and natural solar irradiation. Persistent organic pollutants such as dyes, phenolics, antibiotics, and phthalates can easily be degraded. Hence, this chapter concludes that rare-earth doping of metal oxides provides better nanomaterials for environmental remediation for the treatment of wastewater under natural solar irradiation through a photocatalytic approach.

References and further reading

[1] Xiao J, Xie Y and Cao H 2015 Organic pollutants removal in wastewater by heterogeneous photocatalytic ozonation *Chemosphere* **121** 1–17

[2] 2020 *Nanoagronomy* ed S Javad (Cham: Springer Naure)

[3] Lu H, Wang J, Wang T, Wang N, Bao Y and Hao H 2017 Crystallization techniques in wastewater treatment: an overview of applications *Chemosphere* **173** 474–84

[4] Amin M T, Alazba A A and Manzoor U 2014 A review of removal of pollutants from water/wastewater using different types of nanomaterials *Adv. Mater. Sci. Eng.* **2014** 1–25

[5] Tambe Patil B B 2015 Wastewater treatment using nanoparticles *J. Adv. Chem. Eng.* **5** 10–4172

[6] Kansal S, Singh M and Sud D 2008 Studies on TiO_2/ZnO photocatalyzed degradation of lignin *J. Hazard. Mater.* **153** 412–7

[7] Li Y, Xie W, Hu X, Shen G, Zhou X, Xiang Y, Zha X and Fang P 2010 Comparison of dye photodegradation and its coupling with light-to-electricity conversion over TiO_2 and ZnO *Langmuir* **26** 591–7

[8] Tian C, Zhang Q, Wu A, Jiang M, Liang Z, Jiang B and Fu H 2012 Cost-effective large-scale synthesis of ZnO photocatalyst with excellent performance for dye photodegradation *Chem. Commun.* **48** 2858–60

[9] Fujishima A and Honda K 1972 Electrochemical photolysis of water at a semiconductor electrode *Nature* **238** 37–8

[10] Suganya G A, Josephine A and Sivasamy 2014 Nanocrystalline ZnO doped on lanthanide oxide Dy_2O_3: a novel and UV light active photocatalyst for environmental remediation *Environ. Sci. Technol. Lett.* **1** 172–8

[11] Sakthivel S, Neppolian B, Shankar M, Arabindoo B, Palanichamy M and Murugesan V 2003 Solar photocatalytic degradation of azo dye: comparison of photocatalytic efficiency of ZnO and TiO_2 *Sol. Energy Mater. Sol. Cells* **77** 65

[12] Qi K, Cheng B, Yu J and Ho W 2017 Review on the improvement of the photocatalytic and antibacterial activities of ZnO *J. Alloys Compd.* **727** 792–820

[13] Zhan W, Guo Y, Gong X, Guo Y, Wang Y and Lu G 2014 Current status and perspectives of rare earth catalytic materials and catalysis *Chinese J. Catal.* **35** 1238–50

[14] Mazierski P, Mikolajczyk A, Bajorowicz B, Malankowska A, Zaleska-Medynska A and Nadolna J 2010 The role of lanthanides in TiO_2-based photocatalysis: a review *Appl. Catal. B* **233** 301–17

[15] Yan Y and Wei S H 2008 Doping asymmetry in wide-bandgap semiconductors: origins and solutions *Phys. Status Solidi (B)* **245** 641–52

[16] Georgieva J, Valova E, Armyanov S, Philippidis N, Poulios I and Sotiropoulos S 2012 Bi-component semiconductor oxide photoanodes for the photo electrocatalytic oxidation of organic solutes and vapours: a short review with emphasis to TiO_2–WO_3 photoanodes *J. Hazard. Mater.* **211–2** 30–46

[17] Levi A, Verbitsky L, Waiskopf N and Banin U 2021 Sulfide ligands in hybrid semiconductor–metal nanocrystal photocatalysts: improved hole extraction and altered catalysis *ACS Appl. Mater. Interfaces* **14** 647–53

[18] Jiang R, Li B, Fang C and Wang J 2014 Metal/semiconductor hybrid nanostructures for plasmon-enhanced applications *Adv. Mater.* **26** 5274–309

[19] Zhang D E, Gong J Y, Ma J J, Han G Q and Tong Z W 2013 A facile method for synthesis of N-doped ZnO mesoporous nanospheres and enhanced photocatalytic activity *Dalton Trans.* **42** 16556–61

[20] Babu V J, Nair A S, Peining Z and Ramakrishna S M L 2011 Synthesis and characterization of rice grains like nitrogen-doped TiO_2 nanostructures by electrospinning–photocatalysis *Mater. Lett.* **65** 3064–8

[21] Costa E, Zamora P P and Zarbin A J G J 2011 Novel TiO_2/C nanocomposites: synthesis, characterization, and application as a photocatalyst for the degradation of organic pollutants *Colloid Interface Sci.* **368** 121–7

[22] Li F, Tian F, Liu C, Wang Z, Du Z, Li R and Zhang L 2015 One-step synthesis of nanohybrid carbon dots and TiO_2 composites with enhanced ultraviolet light active photocatalysis *RSC Adv.* **5** 8389–96

[23] Liu X, Cao H and Yin J 2011 Generation and photocatalytic activities of Bi@Bi2O3 microspheres *Nano Res.* **4** 470–82

[24] Asiltürk M, Sayılkan F and Arpaç E 2009 effect of Fe^{3+} ion doping to TiO_2 on the photocatalytic degradation of malachite green dye under UV and vis-irradiation A *J. Photochem. Photobiol. A Chem.* **203** 64–71

[25] Balachandran S and Swaminathan M 2012 Facile fabrication of heterostructured Bi_2O_3–ZnO photocatalyst and its enhanced photocatalytic activity *Phys. Chem. C* **116** 26306–12

[26] Wu C, Shen L, Zhang Y C and Huang Q 2011 Solvothermal synthesis of Cr-doped ZnO nanowires with visible light-driven photocatalytic activity *Mater. Lett.* **65** 1794–6

[27] Jiang B, Yang X, Li X, Zhang D, Zhu J and Li G J 2013 Core–shell structure CdS/TiO_2 for enhanced visible-light-driven photocatalytic organic pollutants degradation *Sol-Gel. Sci. Technol.* **66** 504–11

[28] Bellal B, Trari M and Afalfiz A 2015 Synthesis and characterization of $CdS/CuAl_2O_4$ core–shell: application to photocatalytic eosin degradation *Appl. Nanosci.* **5** 673–80

[29] Kumari M, Devi L, Maia G, Chen T, Al-Zaqri N and Ali M 2022 Mechanochemical synthesis of ternary heterojunctions $TiO_2(A)/TiO_2(R)/ZnO$ and $TiO_2(A)/TiO_2(R)/SnO_2$ for effective charge separation in semiconductor photocatalysis: a comparative study *Environ. Res.* **203** 111841

[30] Pantazis D and Neese F 2009 All-electron scalar relativistic basis sets for the lanthanides *J. Chem. Theory Comput.* **5** 2229–38

[31] Tsayn C Y and Wang M C 2013 Structural and optical studies on sol–gel derived ZnO thin films by excimer laser annealing *Ceram. Int* **39** 469

[32] Raghvendra S and Pandey A 2009 and Sanjay S optical properties of europium doped bunches of ZnO nanowires synthesized by co-precipitation methods *Chalcogenide Lett.* **6** 233–9

[33] Sin J C, Lam S M, Lee K T and Mohamed A R 2013 Fabrication of erbium-doped spherical-like ZnO hierarchical nanostructures with enhanced visible light-driven photocatalytic activity *Mater. Lett.* **91** 1–4

[34] Jia T, Fu F L Z and Zhang Q 2009 Synthesis, characterization and luminescence properties of Y-doped and Tb-doped ZnO nanocrystals *Mater. Sci. Eng. B* **162** 179–84

[35] Shahroosvand H and Ghorbani-Asl M 2013 Solution-based synthetic strategies for Eu doped ZnO nanoparticle with enhanced red photoluminescence *J. Lumin.* **144** 223

[36] Wang M, Huang C, Huang Z, Guo W, Huang J, He H, Wang H, Cao Y, Liu Q and Liang J 2009 Synthesis and photoluminescence of Eu-doped ZnO microrods prepared by hydrothermal method *Opt. Mater.* **31** 1502

[37] Aneesh P M and Jayaraj M K 2010 Red luminescence from hydrothermally synthesized Eu-doped ZnO nanoparticles under visible excitation *Bull. Mater. Sci.* **33** 227

[38] Devi L S K, Kumar K S and Balakrishnan 2011 Rapid synthesis of pure and narrowly distributed Eu doped ZnO nanoparticles by solution combustion method *Mater. Lett.* **65** 35–7

[39] Dan W D, Hai Y J, Jian C, Hui L J, Ming G and Yan L X 2011 A mini-review on rare earth metal doped ZnO nanomaterials for photocatalytic remediation of waste water *Chem., Res.* **27** 174

[40] Kumar S and Sahare P D 2014 Gd^{3+} incorporated ZnO nanoparticles: a versatile material *Mater. Res. Bull.* **51** 217

[41] Liau c c and Chao L C 2010 Growth and characterization of Er doped ZnO prepared by reactive ion beam sputtering *3rd Int. Nanoelectronics Conf. (INEC) (Hong Kong)*

[42] Reddy A J, Kokila M K, Nagabhushana H, Kumara K S, Chakradhar R P S, Nagabhushana B M and Krishna R H 2014 Luminescence studies and EPR investigation of solution combustion derived Eu doped ZnO *Spectrochim. Acta, Part A* **132** 305

[43] Pessoni H V S, Maia L J Q and Franco A 2015 Eu-doped ZnO nanoparticles prepared by the combustion reaction method: Structural, photoluminescence and dielectric characterization *Mater. Sci. Semicond. Process.* **30** 135

[44] Liu J, Huang X, Li Y, Sulieman K M, Sun F and He X 2006 Selective growth and properties of zinc oxide nanostructures *Scr. Mater.* **55** 795

[45] Murmua P P, Kennedy J, Ruck B J, Markwitz A, Williams G V M and Rubanov S 2012 Structural and magnetic properties of low-energy Gd implanted ZnO single crystals *Nucl. Instrum. Methods Phys. Res. B* **272** 100

[46] Korake P V, Kadam A N and Garadkar K M 2014 Photocatalytic activity of Eu3+-doped ZnO nanorods synthesized via microwave assisted technique *J. Rare Earths* **32** 306

[47] Ishizumi A, Fujita S and Yanagi H 2011 Influence of atmosphere on photoluminescence properties of Eu-doped ZnO nanocrystals *Opt. Mater.* **33** 1116

[48] Ahmad T and Lone I H 2017 Development of multifunctional lutetium ferrite nanoparticles: structural characterization and properties *Mater. Chem. Phys.* **202** 50–5

[49] Ahmad T, Lone I H, Ansari S G, Ahmed J, Ahamad T and Alshehri S M 2017 Multifunctional properties and applications of yttrium ferrite nanoparticles prepared by citrate precursor route *Mater. Des.* **126** 331–8

[50] Martínez J M G, Meneses R A M and da Silva C R M 2014 Synthesis of gadolinium doped ceria ceramic powder by polymeric precursor method (Pechini) *Mater. Sci. Forum* **798** 182–8

[51] Fajardo H V, Longo E, Probst L, Valentini A, Carreño N, Nunes M R, Maciel A P and Leite E R 2008 Influence of rare earth doping on the structural and catalytic properties of nanostructured tin oxide *Nanoscale Res. Lett.* **3** 194–9

[52] Ahmad T, Ramanujachary K V, Lofland S E and Ganguli A K 2006 Reverse micellar synthesis and properties of nanocrystalline GMR Ramifications of size considerations *J. Chem. Sci.* **118** 513–8

[53] Murugadoss G, Jayavel R and Rajesh Kumar M 2015 Structural and optical properties of highly crystalline Ce, Eu and co-doped ZnO nanorods *Superlattices Microstruct.* **82** 538–50

[54] Suwarnkar G V K M B and Garadkar N L G K M 2016 Solgel microwave assisted synthesis of Sm-doped TiO$_2$ nanoparticles and their photocatalytic activity for the degradation of Methyl Orange under sunlight *J. Mater. Sci., Mater. Electron.* **27** 6425–32

[55] Suganya Josephine G A and Sivasamy A 2014 Nanocrystalline ZnO doped Dy$_2$O$_3$ a highly active visible photocatalyst: the role of characteristic f orbital's of lanthanides for visible photoactivity *Appl. Catal.* B **150– 151** 288–97

[56] Suganya Josephine G A, Jayaprakash K, Meenakshi G, Sivasamy A, Nirmala Devi G and Viswanath R N 2021 Photocatalytically active ZnO flaky nanoflowers for environmental remediation under solar light irradiation: effect of morphology on photocatalytic activity *Bull. Mater. Sci.* **44** 247

[57] Choudhury B, Borah B and Choudhury A 2012 Extending photocatalytic activity of TiO$_2$ nanoparticles to visible region of illumination by doping of cerium *Photochem. Photobiol.* **88** 257–64

[58] Al-hamdi A M, Sillanpää M and Dutta J 2015 Gadolinium doped tin dioxide nanoparticles: an efficient visible light active photocatalyst *J. Rare Earths* **33** 1275–83

[59] Upadhyay P K, Sharma N and Sharma S *et al* 2021 Photo and thermoluminescence of Eu doped ZnO nanophosphors *J. Mater. Sci.: Mater. Electron.* **32** 17080–93

[60] Achehboune M, Khenfouch M, Boukhoubza I, Derkaoui I, Leontie L, Carlescu A, Mothudi B M, Zorkani I and Jorio A 2022 Optimization of the luminescence and structural properties of Er-doped ZnO nanostructures: effect of dopant concentration and excitation wavelength *J. Lumin.* **246** 118843

[61] CAI H, LIU G, Lü W, LI X, YU L and LI D 2008 Effect of Ho-doping on photocatalytic activity of nanosized TiO$_2$ catalyst *J. Rare Earths* **26** 71–5

[62] Qi H-P and Wang H-L 2020 Facile synthesis of Pr-doped molecularly imprinted TiO$_2$ mesocrystals with high preferential photocatalytic degradation performance *Appl. Surf. Sci.* **511** 145607

[63] Maache A, Chergui A, Djouadi D, Benhaoua B, Chelouche A and Boudissa M 2019 Effect of La doping on ZnO thin films physical properties: correlation between strain and morphology *Optik* **180** 1018–26

[64] Shen J, Lu Y, Liu J K and Yang X H 2016 Design and preparation of easily recycled Ag$_2$WO$_4$@ZnO@Fe$_3$O$_4$ ternary nanocomposites and their highly efficient degradation of antibiotics *J. Mater. Sci.* **51** 7793–802

[65] Sharmin F and Basith M 2022 Highly efficient photocatalytic degradation of hazardous industrial and pharmaceutical pollutants using gadolinium doped BiFeO$_3$ nanoparticles. *J. Alloys Compd.* **901** 163604

[66] Fida H, Zhang G, Guo S and Naeem A 2017 Heterogeneous Fenton degradation of organic dyes in batch and fixed bed using La-Fe montmorillonite as catalyst *J. Colloid Interface Sci.* **490** 859–68

[67] Chong S, Zhang G, Zhang N, Liu Y, Zhu J, Huang T and Fang S 2016 Preparation of FeCeO$_x$ by ultrasonic impregnation method for heterogeneous Fenton degradation of diclofenac *Ultrason. Sonochem.* **32** 231–40

[68] Luan J, Liu W, Yao Y, Ma B, Niu B, Yang G and Wei Z 2022 Synthesis and property examination of Er$_2$FeSbO$_7$/BiTiSbO$_6$ heterojunction composite catalyst and light-catalyzed retrogradation of enrofloxacin in pharmaceutical wastewater under visible light irradiation *Materials* **15** 5906

[69] Wang X, Jin Y, Chen W, Zou R, Xie J, Tang Y, Li X and Li L 2021 Electro-catalytic activity of CeO_x modified graphite felt for carbamazepine degradation via E-peroxone process *Front. Environ. Sci. Eng.* **15** 122

[70] Chen X, Zhang M, Qin H, Zhou J, Shen Q, Wang K, Chen W, Liu M and Li N 2022 Synergy effect between adsorption and heterogeneous photo-Fenton-like catalysis on $LaFeO_3$/lignin-biochar composites for high efficiency degradation of ofloxacin under visible light *Sep. Purif. Technol.* **280** 119751

[71] Wang W, Zhu Q, Qin F, Dai Q and Wang X 2018 Fe doped CeO_2 nanosheets as Fenton-like heterogeneous catalysts for degradation of salicylic acid *Chem. Eng. J.* **333** 226–39

[72] Nie Y, Zhang L, Li Y-Y and Hu C 2015 Enhanced Fenton-like degradation of refractory organic compounds by surface complex formation of $LaFeO_3$ and H_2O_2 *J. Hazard. Mater.* **294** 195–200

[73] Younes H A, Taha M, Khaled R, Mahmoud H M and Abdelhameed R M 2023 Perovskite/metal-organic framework photocatalyst: a novel nominee for eco-friendly uptake of pharmaceuticals from wastewater *J. Alloys Compd.* **930** 167322

[74] Assumpção M H M T, Moraes A, De Souza R, Reis R M, Rocha R S, Gaubeur I, Calegaro M L, Hammer P, Lanza M R d V and Santos M C D 2013 Degradation of dipyrone via advanced oxidation processes using a cerium nanostructured electrocatalyst material *Appl. Catal. A Gen.* **462** 256–61

[75] Tahir M B and Sagir M 2019 Carbon nanodots and rare metals (RM = La, Gd, Er) doped tungsten oxide nanostructures for photocatalytic dyes degradation and hydrogen production *Sep. Purif. Technol.* **209** 94–102

[76] Sharma R, Bansal S and Singhal S 2016 Augmenting the catalytic activity of $CoFe_2O_4$ by substituting rare earth cations into the spinel structure *RSC Adv.* **6** 71676–91

[77] Xie Y and Yuan C 2003 Visible-light responsive cerium ion modified titania sol and nanocrystallites for X-3B dye photodegradation *Appl. Catal. B Environ.* **46** 251–9

[78] Seroglazova A S, Chebanenko M I, Nevedomskyi V N and Popkov V I 2023 Solution combustion synthesis of novel $PrFeO_3$/CeO_2 nanocomposite with enhanced photo-Fenton activity under visible light *Ceram. Int.* **49** 15468–79

[79] Seroglazova A S and Popkov V 2022 Synthesis of highly active and visible-light-driven $PrFeO_3$ photocatalyst using solution combustion approach and succinic acid as fuel *Nanosyst. Phys. Chem. Math.* **13** 649–54

[80] Dhiman M and Singhal S 2019 Effect of doping of different rare earth (europium, gadolinium, dysprosium and neodymium) metal ions on structural, optical and photocatalytic properties of $LaFeO_3$ perovskites *J. Rare Earths* **37** 1279–87

[81] Keerthana S, Yuvakkumar R, Kumar P S, Ravi G and Velauthapillai D 2021 Rare earth metal (Sm) doped zinc ferrite ($ZnFe_2O_4$) for improved photocatalytic elimination of toxic dye from aquatic system *Environ. Res.* **197** 111047

[82] Ismail M, Akhtar K, Khan M, Kamal T, Khan M A, Asiri M, Seo A and Khan J 2019 S.B. Pollution, toxicity and carcinogenicity of organic dyes and their catalytic bio-remediation *Curr. Pharm. Des.* **25** 3645–63

[83] Ju L, Chen Z, Fang L, Dong W, Zheng F and Shen M 2011 Sol–gel synthesis and photo-Fenton-like catalytic activity of $EuFeO_3$ nanoparticles *J. Am. Ceram. Soc.* **94** 3418–24

[84] Zhang A and Zhang J 2010 Effects of europium doping on the photocatalytic behavior of $BiVO_4$ *J. Hazard. Mater.* **173** 265–72

[85] Zhou F, Wang H, Zhou S, Liu Y and Yan C 2020 Fabrication of europium-nitrogen co-doped TiO_2/Sepiolite nanocomposites and its improved photocatalytic activity in real wastewater treatment *Appl. Clay Sci.* **197** 105791

[86] Tikhanova S M, Lebedev L A, Martinson K D, Chebanenko M I, Buryanenko I V, Semenov V G, Nevedomskiy V N and Popkov V I 2021 The synthesis of novel heterojunction h-$YbFeO_3$/o-$YbFeO_3$ photocatalyst with enhanced Fenton-like activity under visible-light *New J. Chem.* **45** 1541–50

[87] Cao R, Huang H, Tian N, Zhang Y, Guo Y and Zhang T 2015 Novel Y doped Bi_2WO_6 photocatalyst: hydrothermal fabrication, characterization and enhanced visible-light-driven photocatalytic activity for Rhodamine B degradation and photocurrent generation *Mater. Charact.* **101** 166–72

[88] Li X, Li W, Liu X, Geng L, Fan H, Ma X, Dong M and Qiu H 2022 The construction of Yb/Er/Pr triple-doped Bi_2WO_6 superior photocatalyst and the regulation of superoxide and hydroxyl radicals *Appl. Surf. Sci.* **592** 153311

[89] Gu S, Li W, Bian Y, Wang F, Li H and Liu X 2016 Highly-visible-light photocatalytic performance derived from a lanthanide self-redox cycle in Ln_2O_3/$BiVO_4$ (Ln: Sm, Eu, Tb) redox heterojunction *J. Phys. Chem.* C **120** 19242–51

[90] Xu X, Ge Y, Li B, Fan F and Wang F 2014 Shape evolution of Eu-doped Bi_2WO_6 and their photocatalytic properties *Mater. Res. Bull.* **59** 329–36

[91] Yu C, Wu Z, Liu R, He H, Fan W and Xue S 2016 The effects of Gd^{3+} doping on the physical structure and photocatalytic performance of Bi_2MoO_6 nanoplate crystals *J. Phys. Chem. Solids* **93** 7–13

[92] Huang Z, Wu P, Li H, Li W, Zhu Y and Zhu N 2014 Synthesis and catalytic properties of La or Ce doped hydroxy-FeAl intercalated montmorillonite used as heterogeneous photo Fenton catalysts under sunlight irradiation *RSC Adv.* **4** 6500–7

[93] Sharmin F and Basith M 2022 Highly efficient photocatalytic degradation of hazardous industrial and pharmaceutical pollutants using gadolinium doped $BiFeO_3$ nanoparticles *J. Alloys Compd.* **901** 163604

[94] Saravanan S, Patrick D S, Vangari G A, Krishna Mohan M, Ponnusamy S and Muthamizchelvan C 2022 Facile synthesis of Sm doped ZnO nanoflowers by Co-precipitation method for enhanced photocatalytic degradation of MB dye under sunlight irradiation *Ceram. Int.* **48** 29049–58

[95] Devendran P, Selvakumar D, Ramadoss G, Sivaramakrishnan R, Alagesan T, Jayavel R and Pandian K 2022 A novel visible light active rare earth doped CdS nanoparticles decorated reduced graphene oxide sheets for the degradation of cationic dye from waste-water *Chemosphere* **287** 132091

[96] Palanivel B, Macadangdang R R, Hossain M S, Alharthi F A, Kumar M, Chang J-H and Gedi S 2023 Rare earth (Gd, La) co-doped ZnO nanoflowers for direct sunlight driven photocatalytic activity *J. Rare Earths* **41** 77–84

[97] Munawar T, Fatima S and Nadeem M S *et al* 2023 Tunability of physical properties of NiO by the introduction of rare earth metal (Y, Ho) dual doping for natural sunlight-driven photocatalysis *J. Mater. Sci.: Mater. Electron.* **34** 687

[98] Khade G, Gavade N L and Suwarnkar M B *et al* 2017 Enhanced photocatalytic activity of europium doped TiO_2 under sunlight for the degradation of methyl orange *J. Mater. Sci.: Mater. Electron.* **28** 11002–11

[99] Karthik K, Radhika D, Raghava Reddy K, Raghu A V, Sadasivuni K K, Palani G and Gurushankar K 2021 Gd^{3+} and Y^{3+} co-doped mixed metal oxide nanohybrids for photocatalytic and antibacterial applications *Nano Express* **2** 010014

[100] Noor M, Sharmin F, Mamun M A A, Hasan S, Hakim M A and Basith M A 2022 Effect of Gd and Y co-doping in $BiVO_4$ photocatalyst for enhanced degradation of methylene blue dye *J. Alloys Compd.* **895** 162639

[101] Yayapao O, Thongtem T, Phuruangrat A and Thongtem S 2013 Ultrasonic-assisted synthesis of Nd-doped ZnO for photocatalysis *Mater. Lett.* **90** 83–6

[102] Keerthana S, Yuvakkumar R, Kumar P S, Ravi G and Velauthapillai D 2021 Rare earth metal (Sm) doped zinc ferrite ($ZnFe_2O_4$) forimproved photocatalytic elimination of toxic dye from aquatic system *Environ. Res.* **197** 111047

[103] Hernández J, Coste S, Murillo A G, Romo F C and Kassiba A 2017 Effects of metal doping (Cu, Ag, Eu) on the electronic and opticalbehavior of nanostructured TiO_2 *J. Alloys Compd.* **710** 355–63

[104] Dhiman P, Rana G, Kumar A, Dawi E A and Sharma G 2023 Rare Earth Doped ZnO Nanoparticles as Spintronics and Photo Catalyst for Degradation of Pollutants. *Molecules.* **28** 2838

[105] Efendi A F and Nurhasanah. I 2015 UV-light absorption and photocatalytic properties of Zn-doped CeO_2 nanopowders prepared by ultrasound irradiation *Mater. Sci. Forum* **827** 56–61

[106] Vijaya Shanthi R, Kayalvizhi R and John Abel M *et al* 2022 MgO nanoparticles with altered structural and optical properties by doping (Er^{3+}) rare earth element for improved photocatalytic activity *Appl. Phys.* A **128** 133

[107] Prathap Kumar M, Suganya Josephine G A and Sivasamy A 2017 Oxidation of organic dye using nanocrystalline rare earth metal ion doped CeO_2 under UV and visible light irradiations *J. Mol. Liq.* **242** 789–97

[108] Krukowska A, Winiarski M J, Strychalska-Nowak J, Klimczuk T, Lisowski W, Mikolajczyk A, Pinto H P, Puzyn T, Grzyb T and Zaleska-Medynska A 2018 Rare earth ions doped $K_2Ta_2O_6$ photocatalysts with enhanced UV–vis light activity *Appl. Catal.* B **224** 451–68

[109] Zhang Y, Zhang H, Xu Y and Wang Y 2003 Europium doped nanocrystalline titanium dioxide: preparation, phase transformation and photocatalytic properties *J. Mater. Chem.* **13** 2261–5

[110] Suganya Josephine G A and Sivasamy A 2014 Nanocrystalline ZnO doped on lanthanide oxide Dy_2O_3: a novel and UV light active photocatalyst for environmental remediation *Environ. Sci. Technol. Lett.* **1** 172–8

[111] 2014 *Photocatalytic Semiconductors: Synthesis, Characterization and Environmental Applications* ed A Hernández-Ramírez and I Medina-Ramírez (New York: Springer)

Chapter 3

Composite nanostructures based on metal oxide/ upconversion phosphors with improved photocatalytic activity

Kamila Zhumanova, Darya Goponenko and Timur Sh Atabaev

TiO_2, ZnO, and some other metal-oxide-based semiconductor nanostructures are traditionally employed in photocatalytic reactions such as organic pollutant degradation and hydrogen evolution. However, because of their large bandgaps, the photocatalytic activity of these nanostructures is limited; typically, these materials require highly energetic UV photons for activation. Several strategies, including doping, sensitization with plasmonic nanoparticles, light conversion with upconversion nanostructures, and control of surface porosity, have been proposed to extend the absorption range of metal-oxide-based nanostructures. In this chapter, we review recent advances in the synthesis and application of composite metal oxide/ upconversion phosphor nanostructures for photocatalytic activity enhancement. Special emphasis is placed on composites that are chemically stable, made from naturally abundant materials, reusable, and capable of being operated under solar light illumination.

3.1 Introduction

3.1.1 Metal oxide semiconductors

The electronic structure of semiconductor material is typically characterized by a filled valence band (VB) and an empty conduction band (CB) separated by a region known as the bandgap (E_g). In general, the bandgap value largely defines the optical properties of the semiconductor, specifically its absorption range. When a semiconductor is exposed to light photons with energies greater than or at least equal to the bandgap value, photoexcited electrons (e^-) move from the VB to the CB, resulting in the formation of holes (h^+) in the VB. It should be mentioned that the

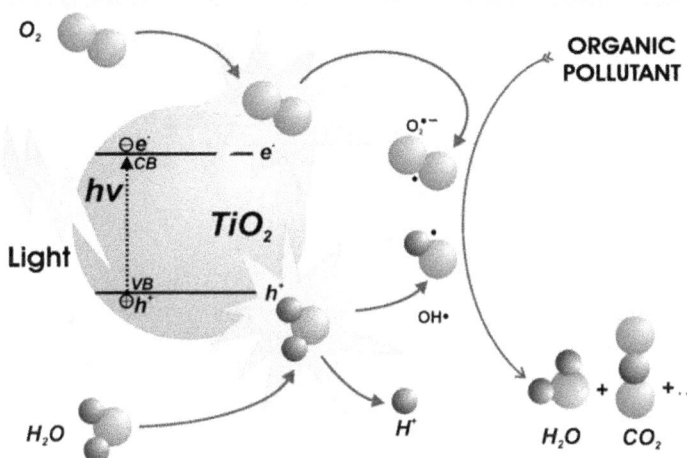

Figure 3.1. The general scheme of organic pollutant photodegradation by a TiO_2 photocatalyst. Reproduced from [3]. CC BY 4.0.

nature of the bandgap (i.e. direct or indirect bandgap) differs among various semiconductors; however, this information is outside the scope of this chapter and can be found elsewhere in the literature [1]. The photogenerated holes in the VB and the electrons in the CB can be further recombined/deactivated, react with their surroundings, or become trapped in metastable surface states. Typically, electron–hole pairs can be deactivated either radiatively in the form of light emission or non-radiatively as a release of heat. However, the deactivation process can be avoided by creating a surface defect or adding a scavenger agent that inhibits e^-/h^+ recombination. Depending on the type of scavenger, electrons or holes can react with various substances adsorbed on the semiconductor surface. For example, photogenerated holes oxidize H_2O or OH^- to produce $^\bullet OH$ radicals (equations (3.1) and (3.2)), whereas electrons produce superoxide $^\bullet O_2^-$ radicals, which are later transformed into $^\bullet OH$ radicals via a series of reactions (equations (3.3)– (3.6)). As powerful oxidizing agents, $^\bullet OH$ radicals cause the stepwise mineralization of organic compounds (equation (3.7)) and the formation of CO_2 and H_2O [2]. Figure 3.1 depicts the general scheme of organic pollutant photodegradation using TiO_2 (a commonly used photocatalytic semiconductor).

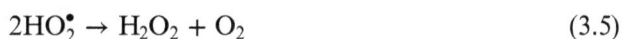

$$h^+ + H_2O \rightarrow {}^\bullet OH + H^+ \tag{3.1}$$

$$h^+ + OH^- \rightarrow {}^\bullet OH \tag{3.2}$$

$$e^- + O_2 \rightarrow {}^\bullet O_2^- \tag{3.3}$$

$$^\bullet O_2^- + H^+ \rightarrow HO_2^\bullet \tag{3.4}$$

$$2HO_2^\bullet \rightarrow H_2O_2 + O_2 \tag{3.5}$$

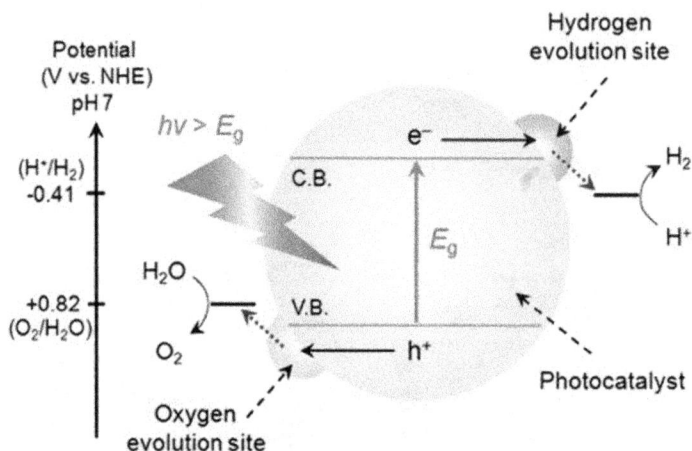

Figure 3.2. The general scheme of photochemical water splitting based on a TiO_2 photocatalyst. Reprinted with permission from [4]. Copyright (2010) American Chemical Society.

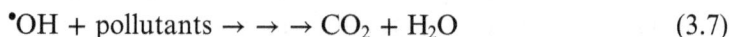

$$H_2O_2 + e^- \rightarrow {}^{\bullet}OH + OH^- \tag{3.6}$$

$${}^{\bullet}OH + \text{pollutants} \rightarrow \rightarrow \rightarrow CO_2 + H_2O \tag{3.7}$$

Photoexcitation is also required for photochemical water splitting, which is a thermodynamically unfavorable process described by two half-reactions (equations (3.8) and (3.9)) versus normal hydrogen electrode (NHE) potentials.

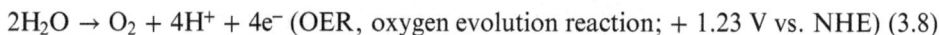

$$2H_2O \rightarrow O_2 + 4H^+ + 4e^- \text{ (OER, oxygen evolution reaction; + 1.23 V vs. NHE)} \tag{3.8}$$

$$4H^+ + 4e^- \rightarrow 2H_2 \text{ (HER, hydrogen evolution reaction; 0 V vs. NHE)} \tag{3.9}$$

Figure 3.2 shows a typical scheme for the photochemical water splitting process. In general, several conditions must be met for photochemical water splitting to happen: (a) the photon energy should be greater than or equal to the semiconductor bandgap value; (b) the bottom level of the CB should be more negative than the reduction potential of H^+/H_2, while the top level of the VB should be more positive than the oxidation potential of H_2O/O_2; (c) the ideal semiconductor should have a suitable bandgap value (slightly higher than 1.23 eV) for effective light absorption in the visible range; (d) the semiconductor should exhibit excellent chemical stability against photocorrosion, (e) it should be made from inexpensive and naturally abundant material, and (f) appropriate cocatalysts should be used for the oxygen evolution reaction (OER) and the hydrogen evolution reaction (HER). Finding a semiconductor that meets all of these requirements is challenging, so various strategies and methods have been used to overcome all of these issues.

To date, various metal oxide semiconductors such as TiO_2, ZnO, SnO_2, $SrTiO_3$, WO_3, and others have commonly been used as photocatalytic materials because they are naturally abundant, readily available on the market, and reasonably priced. Furthermore, the majority of these metal oxide semiconductors have good chemical stability and appropriate VB and CB positions, making them suitable for both organic pollutant degradation and H_2 production. However, the light absorption properties of metal oxide semiconductors are often limited because of their large bandgap values, requiring the use of highly energetic UV–blue photons for photoexcitation. Hence, the efficiency of metal oxide semiconductors is often limited because only a small portion of the solar radiation can be utilized for catalyst activation.

The formation of Z-scheme heterostructures with spatially separated oxidation/reduction sites [5–7], doping [8], plasmonic sensitization [9, 10], coupling with optical materials [11, 12], surface engineering [13], and coupling with 2D materials [14, 15] have been shown to be efficient techniques for improving the light-harvesting ability of prepared photocatalysts. Among these, optical materials that have a frequency conversion ability (upconversion (UC) materials) are considered to be the most intriguing materials and are less frequently reviewed in the literature in combination with photocatalytic materials. Hence, in this chapter, we limit ourselves to a description and analysis of the scientific literature on UC/semiconductor systems published between 2013 and 2023, i.e. the last decade's achievements. We strongly believe that the careful literature analysis and the description of various methods in this chapter will be useful in the design/construction of complex structures with improved photocatalytic properties.

3.1.2 Upconversion materials

UC materials are spectral converters that absorb one or more low-energy photons in the near-infrared (NIR) region and emit higher-energy photons, typically in the visible spectrum, though UV photons can be observed as well. To date, UC materials have been extensively employed as spectral converters in a range of applications including photocatalysis and photovoltaics [11, 12], theranostics [16], optogenetics [17], bioimaging, disease treatment, analytical science, and biosafety [18–20]. A detailed description of UC mechanisms, such as excited-state absorption (ESA), energy transfer upconversion (ETU), cooperative sensitization upconversion (CSU), and photon avalanche (PA) can be found in the literature [16, 21]. The general structure of UC materials is a host–guest system, i.e. a host matrix doped with lanthanide ions. The selection of the host matrix is crucial for the design of highly efficient UC materials. Typically, low phonon energy host matrices are utilized to maximize radiative emission while minimizing nonradiative losses [22]. The phonon energies of chlorides, bromides, and iodides are typically less than 300 cm^{-1}, but these materials are hygroscopic and thus unsuitable for the construction of UC materials because water is a known luminescence quencher [23]. Hence, the most common host materials used to date have been fluorides, oxides, phosphates, vanadates, and so on. In particular, $NaYF_4$, which has a low phonon energy of ~300–400 cm^{-1}, is known to be the most popular and widely used host matrix [24].

Trivalent lanthanide ions such as Yb, Er, Tm, Ho, and Nd are often employed as optically active centers, and their mechanism of UC emission is based on

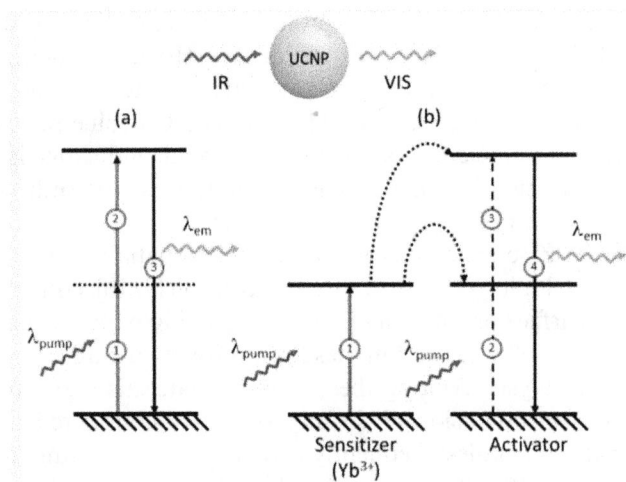

Figure 3.3. The upconversion mechanisms of the (a) ESA, and (b) ETU processes. Reproduced from [25]. CC BY 3.0.

crystal-field-induced intermixing of lanthanide electronic 4f states. It should be mentioned that the doping concentration of activators should be optimized to avoid the quenching of emission related to cross-relaxation effects. In the case of a single dopant (ESA), the ion should have a ladder-like multilevel system with intermediate levels that have relatively long lifetimes (figure 3.3(a)). Only a few elements can meet these requirements, so UC materials are frequently composed of two dopants, one of which acts as a sensitizer and the other as an optical activator (figure 3.3(b)). The ytterbium (Yb^{3+}) ion is frequently used as a sensitizer because its transition level $^2F_{7/2} \rightarrow \,^2F_{5/2}$ is resonant for many f–f transitions of other lanthanides, particularly those of Er^{3+} and Tm^{3+} ions. Moreover, the $^2F_{7/2} \rightarrow \,^2F_{5/2}$ transition of Yb^{3+} can be conveniently activated by a commercial continuous-wave (CW) 980 nm laser. Since weak f–f transitions couple with the local crystal field of the host matrix, the luminescence generated by a specific element is generally independent of the host material. It is also interesting to note that the emission color of a UC material can be tuned to achieve a specific color or cover a specific emission range by codoping the material with another activator. The most representative example is a codoped $NaYF_4$:Yb/Tm/Er system (figure 3.4), in which the concentrations of Yb, Tm, and Er can be varied depending on the desired color output [26].

In most cases, only highly energetic UV–blue photons can activate photocatalytic reactions because the majority of metal-oxide-based photocatalysts have large bandgaps that limit their effective absorption in the visible and NIR ranges. In this regard, Yb and Tm codoped UC systems are frequently selected for photocatalytic reactions due to the characteristic $^1I_6 \rightarrow \,^3F_4$ (~345 nm), $^1D_2 \rightarrow \,^3H_6$ (~360 nm), and $^1D_2 \rightarrow \,^3H_6$ (~450–460 nm) transitions of Tm^{3+} [27, 28]. In general, unabsorbed NIR photons are converted to UV–blue photons, which are then

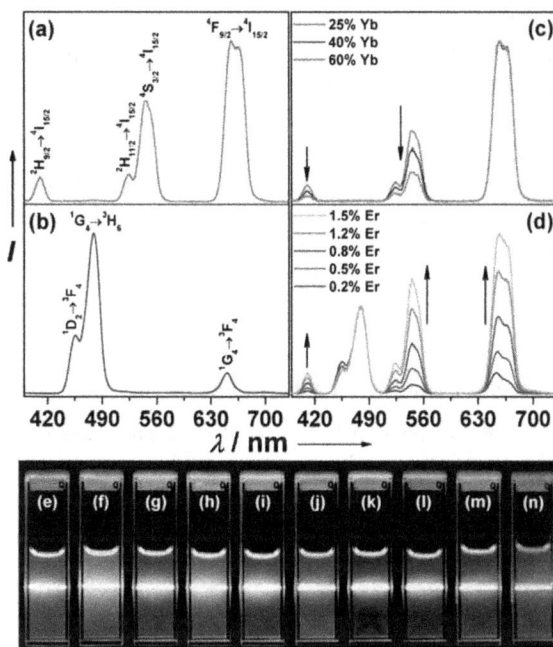

Figure 3.4. UC emissions produced by (a) $NaYF_4{:}Yb^{3+},Er^{3+}$, (b) $NaYF_4{:}Yb^{3+},Tm^{3+}$. (c) The results of varying the Yb concentration in $NaYF_4{:}Yb^{3+},Er^{3+}$ and (d) varying the Er concentration in an $NaYF_4{:}Yb/Tm/Er$ codoped system. (e–n) Digital images of colloidal suspensions of $NaYF_4{:}Yb/Tm/Er$ with various Yb, Er, and Tm concentrations under 980 nm excitation. Reprinted with permission from [26]. Copyright (2008) American Chemical Society.

further absorbed by the photocatalyst. Although such a practice has been found to be feasible to boost various photocatalytic reactions, it is important to remember that the quantum yield of UC processes is low, especially for the conversion of NIR to UV–blue photons [29]. As a result, researchers are actively involved in UC emission enhancement studies using a variety of approaches, such as plasmonic enhancement [30, 31], charge compensation/codoping [32, 33], dye sensitization, surface passivation/encapsulation with inorganic materials [34, 35], and energy management [36]. Typically, Yb^{3+} and Tm^{3+}-based UC materials with greatly improved UC luminescence are useful for improving a variety of photocatalytic reactions.

3.2 Composite nanostructures based on metal oxide/upconversion phosphors

Generally, large bandgaps, poor conductivity, and fast recombination of photo-generated electron–hole pairs significantly reduce the photocatalytic activity of metal oxides. To broaden the absorption range of metal oxide nanostructures in the visible region, various techniques such as metal/nonmetal doping, plasmonic sensitization, and coupling with other semiconductors have been extensively employed. Regardless of the methods mentioned above, there is still room for improvement in solar light capture. For example, UC materials have the potential to convert low-energy photons in the NIR region into high-energy UV–visible photons, which can then be reabsorbed by the semiconductor to generate more electron–hole pairs. Indeed, there are many research studies in the literature that report the fabrication of UC-semiconductor hybrid structures for photocatalytic reactions. Although such structures also demonstrate high potential in nanomedicine and various chemical transformations, we limit ourselves to a discussion of the UC metal-oxide-based hybrids reported in the last ten years (2013–23) for photocatalytic water treatment and green hydrogen generation.

3.2.1 Upconversion-capable photocatalytic materials for water treatment

The core–shell structure is the most common and rational design for UC-semiconductor hybrids because all the photons unabsorbed by the semiconductor reach and excite the UC structure, resulting in emission in the visible region that can be further absorbed by the semiconductor for electron–hole pair generation (figure 3.5). Close contact between the UC material and the semiconductor, as well as direct energy transfer, can efficiently extend solar light utilization and improve photocatalytic activity.

Typically, various UC-semiconductor structures have been prepared using the core–shell design. The most important parameters such as structure, pollutant model, degradation efficiency, and illumination source for some representative examples are listed in table 3.1. As can be seen, UC-semiconductor structures have been widely employed to degrade a variety of textile dyes, antibiotics, and industrial pollutants.

In some cases, a combination of two semiconductors can also be beneficial in terms of photocatalytic activity improvement. For example, Guo and coauthors [45] decorated $NaYF_4:Yb^{3+},Tm^{3+}$ (NYT) rods with CdS and TiO_2 nanoparticles to investigate the effect of the two additional semiconductors on the photocatalytic degradation of methylene blue (MB) dye. Morphological and chemical analyses

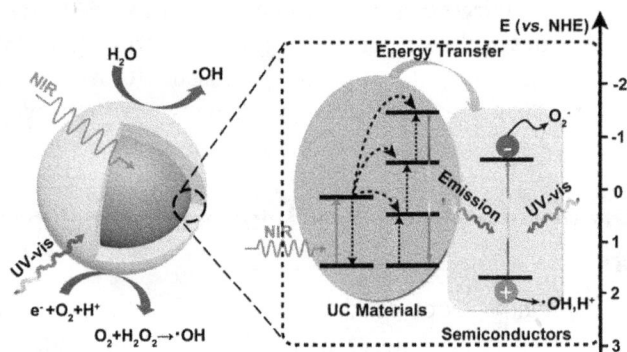

Figure 3.5. A plausible mechanism for energy transfer between a UC material and a semiconductor in a core–shell structure. Reproduced from [37] with permission from The Royal Society of Chemistry.

Table 3.1. Representative examples of UC-semiconductor structures with a core–shell design.

Structure	Pollutant (volume)	Degradation efficiency (pollutant concentration)	Illumination source	References
NaYF$_4$:Yb^{3+},Tm^{3+}@ZnO	Rhodamine B (RhB) dye (0.5 ml)	65% in 30 h (20 mg l^{-1})	980 nm diode laser (2.0 W cm^{-2})	[38]
NaYF$_4$:Yb^{3+},Tm^{3+}@TiO$_2$	Ciprofloxacin (volume information is not available)	99% in 2 h for optimal sample (1 × 10^{-5} M)	Xenon lamp (200 W)	[39]
NaYF$_4$:Yb^{3+},Tm^{3+}@TiO$_2$	Deoxynivalenol (3 ml)	95.7% in 60 min (10 μg ml^{-1})	Xenon lamp (50 W)	[40]
NaYF$_4$@CdS	Methyl orange (MO) dye, carbendazim (CZ) (25 ml for both pollutants)	52% in 6 h for MO and 64% in 6 h for CZ (5 × 10^{-5} M for both pollutants)	NIR light from solar simulator (OAI Trisol, AM 1.5, 100 mW cm^{-2})	[41]
NaYF$_4$:Yb^{3+},Tm^{3+}@ZnO	Rhodamine B (RhB) dye (50 ml)	95% in 60 min (5 mg l^{-1})	Xenon lamp (1500 mW cm^{-2})	[42]
NaYF$_4$:Yb^{3+},Tm^{3+}@ZnO	Methylene blue (MB) dye (0.5 ml)	68.7% in 10 h (15 mg l^{-1})	980 nm diode laser (10 W cm^{-2})	[43]
MoS$_2$-NaYF$_4$:Yb^{3+},Er^{3+}	Rhodamine B (RhB) dye (10 ml)	61% in 12 h (25 mg l^{-1})	980 nm diode laser (detailed information is not available)	[44]

(figure 3.6) revealed that hexagonal NYT rods with smooth surfaces were coated with a porous layer of CdS and TiO$_2$ nanoparticles. Photocatalytic experiments at 980 nm showed that the degradation rate constant for NYT/CdS/TiO$_2$ was ~3 and ~6.5 times higher than those of bare NYT/CdS and NYT/TiO$_2$, respectively. It has been shown that the synergy of two semiconductors, CdS and TiO$_2$, plays an important role in the separation of charge carriers and results in increased photo-catalytic activity. Another study [46] also found that the degradation rate constant for a NaYF$_4$:Yb,Tm@TiO$_2$/CdS composite structure was ~1.6 times higher than that of bare NaYF$_4$:Yb,Tm@TiO$_2$ when illuminated by a visible light (200–1000 nm) source.

Some research studies have focused on the development of recyclable composite structures that can be continuously reused several times. In this context, the use of a magnetic field to quickly remove photocatalytic composites can be facilitated by the co-addition of magnetic particles. As demonstrated in figure 3.7, UC structures (NaYF$_4$:Yb,Tm) can be placed on a magnetic core and then coated with a TiO$_2$ shell layer [47]. The photocatalytic activity of a prepared magnetic composite and other

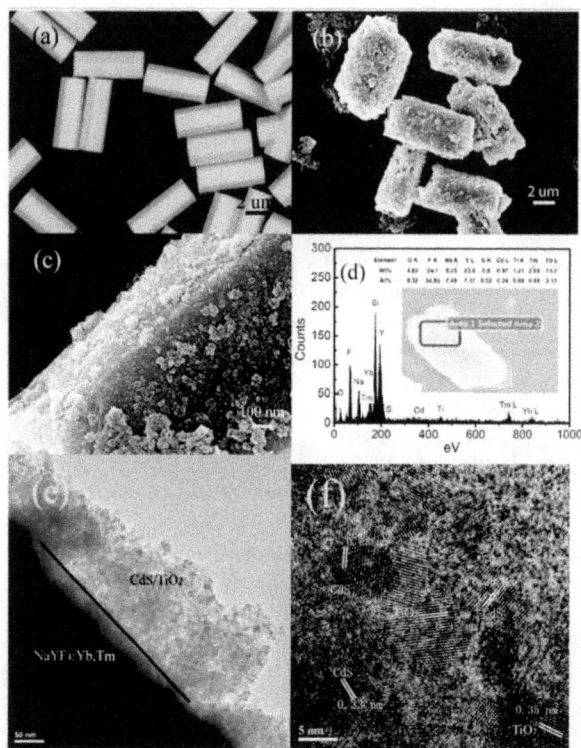

Figure 3.6. SEM images of NYT (a) and NYT/CdS/TiO$_2$ (b and c). Energy dispersive x-ray (EDX) spectroscopic analysis of NYT/CdS/TiO$_2$ (d). Transmission electron microscopy (TEM) images of NYT/ CdS/TiO$_2$ (e and f) at various magnifications. Reproduced from [45] with permission from The Royal Society of Chemistry.

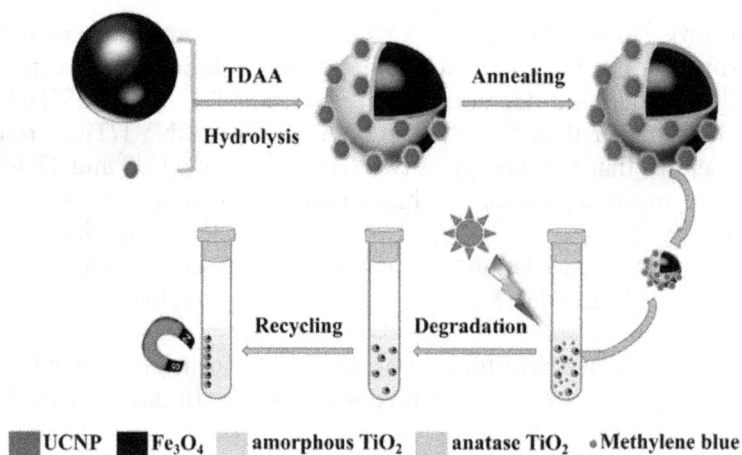

Figure 3.7. The preparation steps used for the synthesis of the magnetically recoverable (Fe$_3$O$_4$-UC)@TiO$_2$ composite. Reproduced from [47] with permission from The Royal Society of Chemistry.

reference samples was assessed by degrading MB dye under UV, UV–vis–NIR, and NIR illumination using mercury and xenon lamps with appropriate light filters. The authors showed that the magnetically recoverable (Fe$_3$O$_4$-UC)@TiO$_2$ composite exhibited higher photocatalytic activity under NIR and UV–vis–NIR illumination than commercial P25 TiO$_2$ nanoparticles. The reusability of the (Fe$_3$O$_4$-UC)@TiO$_2$ composite was tested four times and no significant loss in photocatalytic activity was found. Chen et al [48] utilized another design by decorating hexagonal prisms of NaYF$_4$:Yb,Tm with magnetic Fe$_3$O$_4$@SiO$_2$ and different amounts of P25 TiO$_2$ nanoparticles. Photocatalytic activity tests under NIR illumination revealed that the optimal sample, denoted by FS/UC/T20, was several times more active in MB, RhB, methyl orange, and phenol degradation than P25 TiO$_2$ nanoparticles. Recyclability tests, on the other hand, revealed that the photocatalytic activity dropped to ~5%–10% after the fourth cycle, which can be attributed to the partial loss of TiO$_2$ from the surface.

Plasmonic metal nanostructures are frequently coupled to semiconductor structures for efficient charge separation and light absorption in the visible region [49, 50]. In this regard, the mutual use of plasmonic and UC materials provides an undeniable advantage in terms of light absorption extension for semiconductor structures. For example, Ma and coauthors [51] tested a core–shell NaYF$_4$:Yb, Tm@TiO$_2$ composite decorated with Ag nanoparticles for R6G dye degradation under 50 W xenon lamp illumination. It was found that the degradation rate constant for NaYF$_4$:Yb,Tm@TiO$_2$/Ag was more than twice that of a NaYF$_4$:Yb, Tm@TiO$_2$ structure and ~3 times higher than that of bare TiO$_2$. Xu and colleagues [52] produced NaYF$_4$:Yb,Er,Tm microspheres that were coated with a porous TiO$_2$ layer and then decorated with varying amounts of Au nanoparticles. Figure 3.8 depicts the preparation strategy along with accompanying TEM images demonstrating the distinct morphological differences between the prepared UC structure, TiO$_2$-coated, and Au-decorated composite structures. Optimization experiments revealed that among all the tested samples, NYF@TiO$_2$–1 wt% Au had the highest degradation rate of MO dye under UV, visible, IR, and simulated solar light irradiation. In particular, the degradation rate of NYF@TiO$_2$–1 wt% Au nanoparticles under solar light was found to be nearly twice that of P25 TiO$_2$ nanoparticles. Experiments using radical scavengers revealed that both $^\bullet$OH and $^\bullet$O$_2^-$ radicals were engaged in MO photodegradation; however, the contribution of the $^\bullet$O$_2^-$ radicals was more substantial, which can be attributed to effective charge separation and the supply of additional electrons by plasmonic Au nanoparticles. A similar construction might be employed to deposit other semiconductor and metal nanoparticles on the surface of an UC structure. For example, Tian et al [53] prepared hexagonal NaYF$_4$:Yb,Tm nanoplates coated with a photocatalytic SnO$_2$ shell layer and decorated with Ag nanoparticles. The samples were subsequently evaluated for RhB dye degradation under UV, Vis, NIR, and solar radiation. The degradation rates of NaYF$_4$@SnO$_2$@Ag were reported to be ~11.45 times (Vis), ~5.25 times (NIR), and ~2.29 times higher than that of bare NaYF$_4$@SnO$_2$ under solar illumination. Finite-difference time-domain (FDTD) calculations revealed a considerable electromagnetic boost at the junction between the SnO$_2$ and the Ag,

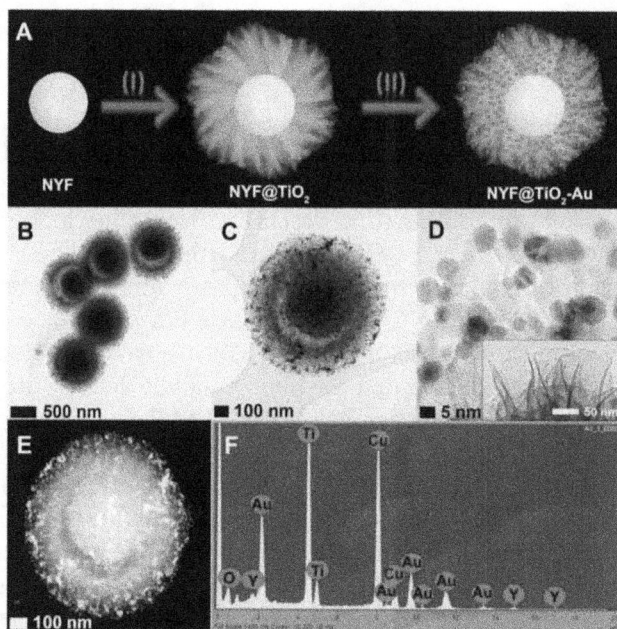

Figure 3.8. The synthesis of NaYF$_4$:Yb,Er,Tm microspheres coated with porous TiO$_2$ and decorated with Au nanoparticles (A). A TEM image of the prepared NaYF$_4$:Yb,Er,Tm microspheres (B); NaYF$_4$:Yb,Er,Tm coated with porous TiO$_2$ (C); and the final composite decorated with Au nanoparticles (D). A scanning TEM image (E) and EDX (F) analysis of the final composite. Reproduced from [52] John Wiley & Sons. © 2015 WILEY-VCH Verlag GmbH & Co. KGaA, Weinheim.

which was caused by localized surface plasmon resonance effects. As a result, the generation rate of electron–hole pairs increased dramatically, which in turn improved the photocatalytic activity of the NaYF$_4$@SnO$_2$@Ag composite.

3.2.2 Upconversion-capable photocatalytic materials for hydrogen generation

In general, UC-containing materials for hydrogen generation follow the same design principles as those of composites for wastewater treatment. The primary distinction is that these composites are often decorated with metallic cocatalysts such as Pt, Au, and Ag, which are used to lower the activation energy of H$_2$O oxidation and H$^+$ reduction; they also provide plasmonic sensitization and function as electron sinks for effective charge separation. Among them, Pt is the most active element for hydrogen evolution according to the Gibbs free energy of H-absorption (ΔG_H) calculations, and is hence mostly used for photocatalytic hydrogen generation [54]. Nevertheless, Au- and Ag-decorated structures are also frequently employed. For example, Feng and colleagues [55] examined the photocatalytic hydrogen evolution of a NaYF$_4$:Yb,Er/Au/CdS architecture with a CdS semiconductor as the outer shell layer. Controlled studies with no catalyst, NaYF$_4$:Yb,Er (NYF), NYF/Au, NYF/CdS, NYF/Au/CdS, and NYF/Au/CdS-M (mechanically mixed NYF and Au/CdS)

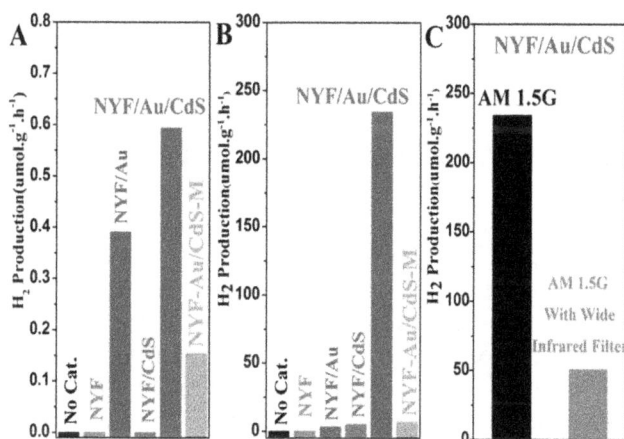

Figure 3.9. The photocatalytic activity of different samples under (A) NIR, (B) solar light, and (C) solar light without an NIR component. Reproduced from [55] with permission from The Royal Society of Chemistry.

samples were performed under NIR (A), simulated solar light (B), and solar light without NIR (C), as shown in figure 3.9. One can observe that all samples containing the NYF component are highly active under NIR illumination, emphasizing the crucial role of UC structures in photocatalytic activity improvement. Furthermore, samples containing Au also demonstrated improved catalytic performance under both NIR and solar light, indicating the necessity of cocatalyst use. The increased photocatalytic activity of the NaYF$_4$:Yb,Er/Au/CdS composite can be explained by several factors, including (a) additional photon harvesting by the UC structure, (b) plasmonic sensitization of UC and CdS structures, (c) charge carrier separation by Au, and (d) the formation of an intimate contact between NYF/Au and CdS during solvothermal synthesis.

A similar core–shell design was used in the creation of Er^{3+}:Y$_3$Al$_5$O$_{12}$/Pt–TiO$_2$ [56] and Er^{3+}:YAlO$_3$/Pt–TiO$_2$ [57] composites. Despite the fact that the resulting H$_2$ evolution rates were not as outstanding as in recent reports, photocatalytic improvement was evident in both cases with the inclusion of UC structures. For example, a recent study found that NaYF$_4$:Yb,Er particles coated with a TiO$_2$ layer and decorated with 1% Au nanoparticles generated ~350 µmol h^{-1} of hydrogen under simulated solar light irradiation [58]. This value was ~61 times and ~3.2 times greater than those of bare NaYF$_4$:Yb,Er particles covered with a TiO$_2$ layer and a reference TiO$_2$ (P25)/1% Au catalyst, respectively. Hence, the beneficial role of Au cocatalyst loading and NaYF$_4$:Yb,Er utilization is clearly apparent.

In some cases, a semiconductor may serve as the host matrix for the construction of UC structures. For example, Shang and colleagues [59] created an Er-Yb:TiO$_2$/MoO$_{3-x}$ heterostructure in which the Er-Yb:TiO$_2$ was a two-in-one UC and photocatalytic structure, while the MoO$_{3-x}$ compound acted as a light absorption material that generated 'hot electrons.' The results indicated that the Er-Yb:TiO$_2$/MoO$_{3-x}$ composite was active under both NIR and solar light, and the H$_2$

generation yield from ammonia borane under simulated sunlight was ~2.5 and ~52 times higher than those of MoO_{3-x} and bare ammonia borane, respectively. Liu and coworkers [60] proposed a Z-scheme $WO_3:Yb^{3+},Er^{3+}/Ag/Ag_3VO_4/Ag$ photocatalyst suitable both for solar antibiotic degradation and hydrogen evolution. In this case, the $WO_3:Yb^{3+},Er^{3+}$ acted as a photocatalyst as well as a UC material, supplying additional photons to plasmonic Ag nanoparticles. It was shown that such a Z-scheme configuration can yield ~489 μmol g^{-1} of H_2 in 180 min and that it can effectively degrade organic pollutants under solar illumination.

Solar hydrogen generation can also be carried out via the photoelectrochemical water splitting method, which utilizes the same charge separation process as particle-based hydrogen generation. In this scenario, a thin-film semiconductor is coupled to a metallic electrode where hydrogen is normally evolved. Nanostructuring, doping, heterojunction formation, optical/plasmonic sensitization, and control of defects are common approaches used with semiconductor thin films to improve the generated photocurrent density that is commonly measured and compared to 1.23 V vs. RHE (the minimum potential difference for water electrolysis vs. the reversible hydrogen electrode). It is interesting that even the simple placement of UC structures on the surface of semiconductor thin films can significantly improve the generated photocurrent density. For example, the photocurrent density was increased by ~16% when Pt-doped hematite (α-Fe_2O_3) nanorods were decorated with $ZrO_2:Er^{3+}$ nanocrystals [61]. In a similar manner, the photocurrent density measured at 1.23 V vs. RHE was increased to ~34% by drop-casting $NaYF_4$-based UC nanoparticles on the surface of a $ZnFe_2O_4/TiO_2$ photoanode [62]. Further investigation found that the UC-modified $ZnFe_2O_4/TiO_2$ photoanode demonstrated good activity under IR illumination, suggesting the beneficial role of UC nanostructures in photocurrent improvement. Furthermore, the idea of combining multiple materials, such as porous semiconductor films with ordered structures, with metallic and UC nanostructures appears to be promising in terms of maximum light collection in the UV, visible, and NIR regions. This concept can be realized using an inverse-opal structure, such as TiO_2, as shown in figure 3.10. Two approaches for UC and Au nanoparticle deposition were employed and various UC structures were tested, such as $NaYF_4:Yb,Er$ and $NaYF_4:Yb,Tm$ [63]. A porous structure with an ordered structure generally improves light scattering effects and, as a result, TiO_2 absorbance in the visible range. Furthermore, an increased specific area due to a porous structure can be also advantageous for photocatalytic H_2 production. It is important to note that H_2 evolution differed between the $NaYF_4:Yb,Er$ and $NaYF_4:Yb,Tm$ structures. For example, an optimized sample of $NaYF_4:Yb,Er$ yielded ~14.1 μmol cm^{-2}, while a sample of $NaYF_4:Yb,Tm$ yielded ~10.9 μmol cm^{-2} after 5 h of sunlight illumination. A detailed analysis suggested that green emission from $NaYF_4:Yb$, Er sensitizes Au nanoparticles, whereas blue emission from $NaYF_4:Yb,Tm$ primarily sensitizes TiO_2 films. Hence, this study found that the sensitization of the plasmonic structure was found to be more advantageous than the direct sensitization of the semiconductor structure.

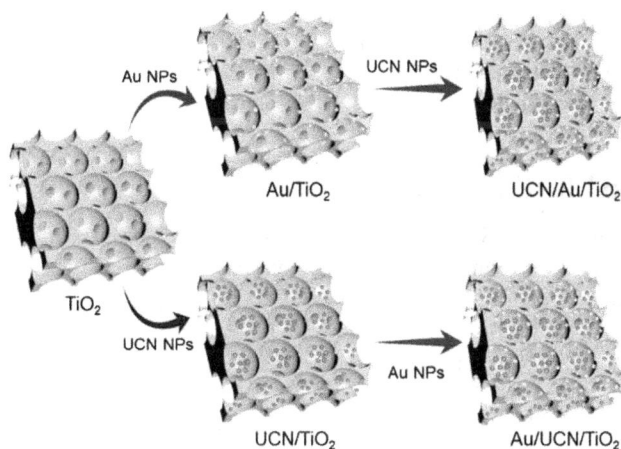

Figure 3.10. Two preparation pathways for an inverse-opal TiO$_2$ film loaded with an UC material and Au nanoparticles. Reprinted with permission from [63]. Copyright (2019) American Chemical Society.

3.3 Conclusions and future prospects

It is apparent that UC materials play an important role in various processes by supplying additional high-energy photons converted from NIR photons. Countless studies available in the research literature have demonstrated that the use of UC materials can improve any photocatalytic process. Although this feature is extremely useful in many applications, it is still uncontrollable, since photocatalytic process enhancement is primarily associated with the quality of the UC material, its emission pattern, its position in the composite structure, size, shape, and so on. Hence, it is strongly recommended that the quantum yields (QYs) of UC structures should be measured to make the methods of researchers working in different parts of the world somewhat comparable. Typically, thorough QY measurement approaches for UC materials can easily be found in the literature [64].

Another critical issue that can be addressed is the low QY of UC materials. Furthermore, weak UC luminescence can quickly be quenched by interactions with polar solvents. Hence, proper crystal engineering, surface passivation, and plasmonic sensitization could be key steps in improving the QY of UC materials. As previously mentioned, the selection of the UC material and even its placement in the composite structure may play key roles in photocatalytic processes. Therefore, careful design of the photocatalytic structures should be carried out in future studies. Moreover, these composites may be combined with 2D materials to accelerate charge carrier transfer and separation. Nevertheless, solar-harnessing devices with UC structures have the potential to be widely employed in the future green economy. Typically, they can be employed in conjunction with solar cells and photocatalytic cells for green H$_2$ generation and solar wastewater treatment.

Acknowledgments

This research was funded by the Science Committee of the Ministry of Science and Higher Education of the Republic of Kazakhstan (Grant No. AP19676347).

References and further reading

[1] Yuan L D, Deng H X, Li S S, Wei S H and Luo J W 2018 Unified theory of direct or indirect band-gap nature of conventional semiconductors *Phys. Rev.* B **98** 245203

[2] Nursam N M, Wang X and Caruso R A 2015 High-throughput synthesis and screening of titania-based photocatalysts *ACS Comb. Sci.* **17** 548–69

[3] Ibhadon A O and Fitzpatrick P 2013 Heterogeneous photocatalysis: recent advances and applications *Catalysts* **3** 189–218

[4] Maeda K and Domen K 2010 Photocatalytic water splitting: recent progress and future challenges *J. Phys. Chem. Lett.* **1** 2655–61

[5] Maeda K 2013 Z-scheme water splitting using two different semiconductor photocatalysts *ACS Catal.* **3** 1486–503

[6] Xu K, Zhang L, Yu J, Wageh S, Al-Ghamdi A A and Jaroniec M 2018 Direct Z-scheme photocatalysts: principles, synthesis, and applications *Mater. Today* **21** 1042–63

[7] Li J, Yuan H, Zhang W, Jin B, Feng Q, Huang J and Jiao Z 2022 Advances in Z-scheme semiconductor photocatalysts for the photoelectrochemical applications: a review *Carbon Energy* **4** 294–331

[8] Medhi R, Marquez M D and Lee T R 2020 Visible-light-active doped metal oxide nanoparticles: review of their synthesis, properties, and applications *ACS Appl. Nano Mater.* **3** 6156–85

[9] Atabaev T S 2018 Plasmon-enhanced solar water splitting with metal oxide nanostructures: a brief overview of recent trends *Front. Mater. Sci.* **12** 207–13

[10] Amirjani A, Amlashi N B and Ahmadiani Z S 2023 Plasmon-enhanced photocatalysis based on plasmonic nanoparticles for energy and environmental solutions: a review *ACS Appl. Nano Mater.* **6** 9085–123

[11] Yang W, Li X, Chi D, Zhang H and Liu X 2014 Lanthanide-doped upconversion materials: emerging applications for photovoltaics and photocatalysis *Nanotechnology* **25** 482001

[12] Atabaev T S and Molkenova A 2019 Upconversion optical nanomaterials applied for photocatalysis and photovoltaics: recent advances and perspectives *Front. Mater. Sci.* **13** 335–41

[13] Bai S, Jiang W, Li Z and Xiong Y 2015 Surface and interface engineering in photocatalysis *ChemNanoMat* **1** 223–39

[14] Luo B, Liu G and Wang L 2016 Recent advances in 2D materials for photocatalysis *Nanoscale* **8** 6904–20

[15] Shanker G S, Biswas A and Ogale S 2021 2D materials and their heterostructures for photocatalytic water splitting and conversion of CO_2 to value chemicals and fuels *J. Phys.: Energy* **3** 022003

[16] Chen G, Qiu H, Prasad P N and Chen X 2014 Upconversion nanoparticles: design, nanochemistry, and applications in theranostics *Chem. Rev.* **114** 5161–214

[17] Patel M, Meenu M, Pandey J K, Kumar P and Patel R 2022 Recent development in upconversion nanoparticles and their application in optogenetics: a review *J. Rare Earths* **40** 847–61

[18] Gulzar A, Xu J, Yang P, He F and Xu L 2017 Upconversion processes: versatile biological applications and biosafety *Nanoscale* **9** 12248–82

[19] Liang G, Wang H, Shi H, Wang H, Zhu M, Jing A, Li J and Li G 2020 Recent progress in the development of upconversion nanomaterials in bioimaging and disease treatment *J. Nanobiotechnol.* **18** 154

[20] Borse S, Rafique R, Murthy Z V P, Park T J and Kailasa S K 2022 Applications of upconversion nanoparticles in analytical and biomedical sciences: a review *Analyst* **147** 3155–79

[21] Auzel F 2004 Upconversion and anti-stokes processes with f and d ions in solids *Chem. Rev.* **104** 139–74

[22] Zhou J, Liu Q, Feng W, Sun Y and Li F 2015 Upconversion luminescent materials: advances and applications *Chem. Rev.* **115** 395–465

[23] Arppe R, Hyppänen I, Perälä N, Peltomaa R, Kaiser M, Würth C, Christ S, Resch-Genger U, Schäferling M and Soukka T 2015 Quenching of the upconversion luminescence of $NaYF_4:Yb^{3+},Er^{3+}$ and $NaYF_4:Yb^{3+},Tm^{3+}$ nanophosphors by water: the role of the sensitizer Yb^{3+} in non-radiative relaxation *Nanoscale* **7** 11746–57

[24] Rabouw F T, Prins P T and Norris D J 2016 Europium-doped $NaYF_4$ nanocrystals as probes for the electric and magnetic local density of optical states throughout the visible spectral range *Nano Lett.* **16** 7254–60

[25] Naccache R, Martín Rodríguez E, Bogdan N, Sanz-Rodríguez F, de la Cruz M C I, de la Fuente Á J, Vetrone F, Jaque D, Solé J G and Capobianco J A 2012 High resolution fluorescence imaging of cancers using lanthanide ion-doped upconverting nanocrystals *Cancers* **4** 1067–105

[26] Wang F and Liu X 2008 Upconversion multicolor fine-tuning: visible to near-infrared emission from lanthanide-doped $NaYF_4$ nanoparticles *J. Am. Chem. Soc.* **130** 5642–3

[27] Chen G, Qiu H, Fan R, Hao S, Tan S, Yang C and Han G 2012 Lanthanide-doped ultrasmall yttrium fluoride nanoparticles with enhanced multicolor upconversion photoluminescence *J. Mater. Chem.* **22** 20190–6

[28] Li X, Zhang F and Zhao D 2015 Lab on upconversion nanoparticles: optical properties and applications engineering via designed nanostructure *Chem. Soc. Rev.* **44** 1346–78

[29] Meijer M S *et al* 2018 Absolute upconversion quantum yields of blue-emitting $LiYF_4:Yb^{3+}$, Tm^{3+} upconverting nanoparticles *Phys. Chem. Chem. Phys.* **20** 22556–62

[30] Wu D M, Garcia-Etxarri A, Salleo A and Dionne J A 2014 Plasmon-enhanced upconversion *J. Phys. Chem. Lett.* **5** 4020–31

[31] Dong J, Gao W, Han Q, Wang Y, Qi J, Yan X and Sun M 2019 Plasmon-enhanced upconversion photoluminescence: mechanism and application *Rev. Phys.* **4** 100026

[32] Atabaev T S, Piao Z, Hwang Y H, Kim H K and Hong N H 2013 Bifunctional $Gd_2O_3:Er^{3+}$ particles with enhanced visible upconversion luminescence *J. Alloys Compd.* **572** 113–7

[33] Sinha S, Mahata M K, Swart H C, Kumar A and Kumar K 2017 Enhancement of upconversion, temperature sensing and cathodoluminescence in the K^+/Na^+ compensated $CaMoO_4:Er^{3+}/Yb^{3+}$ nanophosphor *New J. Chem.* **41** 5362–72

[34] Ti Z, Liang T, Wang Q and Liu Z 2020 Strategies for constructing upconversion luminescence nanoprobes to improve signal contrast *Small* **16** 1905084

[35] Zhu X, Zhang J, Liu J and Zhang Y 2019 Recent progress of rare-earth doped upconversion nanoparticles: synthesis, optimization, and applications *Adv. Sci.* **6** 1901358

[36] Sheng T, Xu M, Li Q, Wu Y, Zhang J, Liu J, Zhu X and Zhang Y 2021 Elucidating the role of energy management in making brighter, and more colorful upconversion nanoparticles *Mater. Today Phys.* **20** 100451

[37] Tian Q, Yao W, Wu W and Jiang C 2018 NIR light-activated upconversion semiconductor photocatalysts *Nanoscale Horiz.* **4** 10–25

[38] Guo X, Song W, Chen C, Di W and Qin W 2013 Near-infrared photocatalysis of β-NaYF$_4$: Yb^{3+},Tm^{3+}@ZnO composites *Phys. Chem. Chem. Phys.* **15** 14681–8

[39] Ma Y and Li S 2019 NaYF$_4$:Yb,Tm@TiO$_2$ core@shell structures for optimal photocatalytic degradation of ciprofloxacin in the aquatic environment *RSC Adv.* **9** 33519–24

[40] Wu S, Wang F, Li Q, Wang J, Zhou Y, Duan N, Niazi S and Wang Z 2020 Photocatalysis and degradation products identification of deoxynivalenol in wheat using upconversion nanoparticles@TiO$_2$ composite *Food Chem.* **323** 126823

[41] Balaji R, Kumar S, Reddy K L, Sharma V, Bhattacharyya K and Krishnan V 2017 Near-infrared driven photocatalytic performance of lanthanide-doped NaYF$_4$@CdS core–shell nanostructures with enhanced upconversion properties *J. Alloys Compd.* **724** 481–91

[42] Zhang F, Wang W-N, Cong H-P, Luo L-B, Zha Z-B and Qian H-S 2017 Facile synthesis of upconverting nanoparticles/zinc oxide core–shell nanostructures with large lattice mismatch for infrared triggered photocatalysis *Part. Part. Syst. Charact.* **34** 1600222

[43] Zhang J, Zhao S, Xu Z, Zhang L, Zuo P and Wu Q 2019 Near-infrared light-driven photocatalytic NaYF$_4$:Yb,Tm@ZnO core/shell nanomaterials and their performance *RSC Adv.* **9** 3688–92

[44] Chatti M, Adusumalli V N K B, Ganguli S and Mahalingam V 2016 Near-infrared light triggered superior photocatalytic activity from MoS$_2$–NaYF$_4$:Yb^{3+}/Er^{3+} nanocomposites *Dalton Trans.* **45** 12384–92

[45] Guo X, Di W, Chen C, Liu C, Wang X and Qin W 2014 Enhanced near-infrared photocatalysis of NaYF$_4$:Yb, Tm/CdS/TiO$_2$ composites *Dalton Trans.* **43** 1048–54

[46] Xu J, Shi Y, Chen Y, Wang Q, Cheng J and Li P 2018 Enhanced photocatalytic activity of TiO$_2$ in visible and infrared light through the synergistic effect of upconversion nanocrystals and quantum dots *Mater. Res. Express* **6** 025055

[47] Lv Y, Yue L, Li Q, Shao B, Zhao S, Wang H, Wu S and Wang Z 2018 Recyclable (Fe$_3$O$_4$-NaYF$_4$:Yb,Tm)@TiO$_2$ nanocomposites with near-infrared enhanced photocatalytic activity *Dalton Trans.* **47** 1666–73

[48] Chen Z and Fu M L 2018 Recyclable magnetic Fe$_3$O$_4$@SiO$_2$/β-NaYF$_4$:Yb^{3+},Tm^{3+}/TiO^2 composites with NIR enhanced photocatalytic activity *Mater. Res. Bull.* **107** 194–203

[49] Birnal P, Marco de Lucas M C, Pochard I, Herbst F, Heintz O, Saviot L, Domenichini B and Imhoff L 2023 Visible-light photocatalytic degradation of dyes by TiO$_2$–Au inverse opal films synthesized by atomic layer deposition *Appl. Surf. Sci.* **609** 155213

[50] Em S, Yedigenov M, Khamkhash L, Atabaev S, Molkenova A, Poulopoulos S G and Atabaev T S 2022 Uncovering the role of surface-attached Ag nanoparticles in photo-degradation improvement of Rhodamine B by ZnO-Ag nanorods *Nanomaterials* **12** 2882

[51] Ma Y, Liu H, Han Z, Yang L and Liu J 2015 Non-ultraviolet photocatalytic kinetics of NaYF$_4$:Yb,Tm@TiO$_2$/Ag core@comby shell nanostructures *J. Mater. Chem.* A **3** 14642–50

[52] Xu Z, Quintanilla M, Vetrone F, Govorov A O, Chaker M and Ma D 2015 Harvesting lost photons: plasmon and upconversion enhanced broadband photocatalytic activity in core@-shell microspheres based on lanthanide-doped NaYF$_4$, TiO$_2$, and Au *Adv. Funct. Mater.* **25** 2950–60

[53] Tian Q, Yao W, Wu W, Liu J, Wu Z, Liu L, Dai Z and Jiang C 2017 Efficient UV–Vis–NIR responsive upconversion and plasmonic-enhanced photocatalyst based on lanthanide-doped NaYF$_4$/SnO$_2$/Ag *ACS Sustain. Chem. Eng.* **5** 10889–99

[54] Zhu J, Hu L, Zhao P, Lee L Y S and Wong K Y 2020 Recent advances in electrocatalytic hydrogen evolution using nanoparticles *Chem. Rev.* **120** 851–918

[55] Feng W, Zhang L, Zhang Y, Yang Y, Fang Z, Wang B, Zhang S and Liu P 2017 Near-infrared-activated NaYF$_4$:Yb^{3+}, Er^{3+}/Au/CdS for H$_2$ production via photoreforming of bio-ethanol: plasmonic Au as light nanoantenna, energy relay, electron sink and co-catalyst *J. Mater. Chem.* A **5** 10311–20

[56] Li Y, Guo Y, Li S, Li Y and Wang J 2015 Efficient visible-light photocatalytic hydrogen evolution over platinum supported titanium dioxide nanocomposites coating up-conversion luminescence agent (Er^{3+}:Y$_3$Al$_5$O$_{12}$/Pt–TiO$_2$) *Int. J. Hydrogen Energy* **40** 2132–40

[57] Ma C, Li Y, Zhang H, Chen Y, Lu C and Wang J 2015 Photocatalytic hydrogen evolution with simultaneous photocatalytic reforming of biomass by Er^{3+}:YAlO$_3$/Pt–TiO$_2$ membranes under visible light driving *Chem. Eng. J.* **273** 277–85

[58] Chilkalwar A A and Rayalu S S 2018 Synergistic plasmonic and upconversion effect of the (Yb,Er)NYF-TiO$_2$/Au composite for photocatalytic hydrogen generation *J. Phys. Chem.* C **122** 26307–14

[59] Shang J, Xu X, Liu K, Bao Y, Yangyang and He M 2019 LSPR-driven upconversion enhancement and photocatalytic H$_2$ evolution for Er–Yb:TiO$_2$/MoO$_{3-x}$ nano-semiconductor heterostructure *Ceram. Int.* **45** 16625–30

[60] Liu Z, Chen L, Piao C, Tang J, Liu Y, Lin Y, Fang D and Wang J 2021 Highly active Z-scheme WO$_3$:Yb^{3+},Er^{3+}/Ag/Ag$_3$VO$_4$/Ag photocatalyst with efficient charge transfer and light utilization for enhanced levofloxacin degradation with synchronous hydrogen evolution *Appl. Catal.* A. **623** 118295

[61] Atabaev T S, Vu H H T, Ajmal M, Kim H K and Hwang Y H 2015 Dual-mode spectral convertors as a simple approach for the enhancement of hematite's solar water splitting efficiency *Appl. Phys.* A **119** 1373–7

[62] Lim Y, Lee S Y, Kim D, Han M K, Han H S, Kang S H, Kim J K, Sim U and Park Y I 2022 Expanded solar absorption spectrum to improve photoelectrochemical oxygen evolution reaction: synergistic effect of upconversion nanoparticles and ZnFe$_2$O$_4$/TiO$_2$ *Chem. Eng. J.* **438** 135503

[63] Boppella R, Mota F M, Lim J W, Kochuveedu S T, Ahn S, Lee J, Kawaguchi D, Tanaka K and Kim D H 2019 Plasmon and upconversion mediated broadband spectral response in TiO$_2$ inverse opal photocatalysts for enhanced photoelectrochemical water splitting *ACS Appl. Energy Mater.* **2** 3780–90

[64] Jones C M S, Gakamsky A and Marques-Hueso J 2021 The upconversion quantum yield (UCQY): a review to standardize the measurement methodology, improve comparability, and define efficiency standards *Sci. Technol. Adv. Mater.* **22** 810–48

[65] Würth C, Fischer S, Grauel B, Alivisatos A P and Resch-Genger U 2018 Quantum yields, surface quenching, and passivation efficiency for ultrasmall core/shell upconverting nanoparticles *J. Am. Chem. Soc.* **140** 4922–8

Chapter 4

Nanocomposites and their applications in photocatalytic degradation processes

Shubham Bonde, Gauri Kallawar, Bharat A Bhanvase and Bhaskar Sathe

4.1 Introduction

In today's world, the needs for clean water and large-scale food production are becoming increasingly important due to the growing global population and heightened industrial demands. However, the use of agrochemicals to enhance agricultural productivity has led to concerns regarding their adverse impacts on the environment and human health. Pesticides are known to have properties such as low biodegradability, high bioaccumulation, and long persistence, which can cause significant risks to soil, water, and food chains. In addition, wastewater dye effluents from various industries have emerged as major water pollutants, posing threats to aquatic ecosystems and human health. Researchers are actively exploring technological and non-technological approaches to address the urgent need for wastewater treatment. Various adsorptive and advanced oxidation processes (AOPs) such as the Fenton method [1], the photo-Fenton method [2], heterogeneous photocatalysis [3], and ozonation [4] systems have been studied for their ability to remove and degrade organic pollutants. AOPs utilize potent free radicals such as hydroxyl ($HO^{\bullet}$) and superoxide ($O_2^{-\bullet}$) radicals with strong oxidizing capabilities to break down pollutants into lower-molecular-weight intermediates and inorganic precursors. Heterogeneous photocatalysis has the potential to be used as an effective AOP wherein a semiconductor is activated by sunlight or artificial light. It utilizes water molecules and dissolved oxygen in oxidation–reduction reactions to efficiently degrade organic pollutants, including dyes, drugs, and pesticides. Commonly used metallic nano oxides such as zinc oxide [5] and titanium dioxide [6] have favorable properties such as low toxicity, chemical stability, large surface area/porosity, and photocorrosion resistance. However, they face drawbacks such as high bandgap energy, charge recombination, and agglomeration during photocatalysis [7]. To address these issues, researchers are exploring the use of nanocomposites, in which

doi:10.1088/978-0-7503-5697-8ch4

the active substance is dispersed in low concentrations (around 0.5–5 wt%) on a support. This approach mitigates the drawbacks associated with conventional nanocatalysts and enhances the overall efficiency of the photocatalytic process for environmental remediation.

This chapter delves into the catalytic applications of nanomaterials, particularly nanocomposites, in the treatment of pharmaceutical wastewater, pesticides, organic dyes, insecticides, soil and air. The high surface area of nanocomposites directly influences their catalytic roles, and nanomaterials have proven valuable in achieving high selectivity, specificity, stability, and recovery due to their nanoscale nature. Various types of nanomaterials, including nanofibers, nanoparticles, and nano-platelets, have demonstrated their unique roles in catalysis. Hybrid nanomaterials that combine the distinct properties of two components have shown great promise. These nanocomposites, ranging from organic/organic to inorganic/organic and polymer/inorganic composites, offer enhanced properties that can be leveraged to catalyze the degradation of pollutant effluents. The catalytic efficacy of nanosized semiconductor photocatalysts, such as ZnO, TiO_2, and SiO_2, has been extensively explored, providing insights into their application for wastewater treatment.

This chapter also examines how graphene and altered graphene demonstrate catalytic capabilities when paired with different inorganic compounds such as metals, metal oxides, and polymers. The following sections discuss selected studies and progress in the realm of nanocomposites, highlighting their involvement in catalyzing the breakdown of dye effluents and their contribution to ongoing initiatives aimed at addressing water pollution.

4.2 Nanocomposite materials

Nanocomposite materials, considered a crucial category of advanced materials, are solid multiphase materials in which one phase possesses at least one dimension that is less than 100 nm. This process entails integrating nanoparticles into larger matrices, such as polymers, metals, ceramics, or composites. This incorporation imparts exceptional properties and enhanced performance characteristics to nano-composites, making them valuable for various applications. At the nanoscale, they exhibit unique features, such as an increased surface-to-volume ratio, which significantly enhances catalytic efficiency by providing a higher density of active sites for adsorption—often doubling it compared to the active site densities of bulk materials [8]. As discussed in the previous section, nanocomposites address limi-tations associated with nanoparticle-based photocatalytic activity.

Nanocomposite photocatalysts fall into different types: metal-, polymer-, ceramic-, and carbon-based. Each type contributes significantly to environmental purification. Metal nanocomposite photocatalysts boost pollutant degradation by leveraging the combined impact of metal nanoparticles and semiconductors [9]. They efficiently enhance the separation of electron–hole pairs, speeding up the process of photodegradation. This makes them crucial for purifying water [10] and air [11]. Polymer nanocomposite photocatalysts are advanced materials used for environmental remediation. They consist of nanoscale photocatalytic particles embedded in polymer matrices. These materials efficiently degrade pollutants under UV or visible light, addressing critical pollution issues [12]. Ceramic nanocomposite photocatalysts employ nanostructures within ceramic materials to enhance the light-driven degradation of pollutants, making them effective for environmental cleanup. These materials possess a large surface area, enhanced photocatalytic activity, and the capability to efficiently degrade both organic and inorganic pollutants in air and water [13]. Carbon-based nanocomposites blend carbon structures with photo-catalytic nanoparticles, improving pollutant degradation under light exposure, and showing potential for environmental remediation and wastewater treatment [14].

Nanocomposites exhibit promise as materials with improved photocatalytic activity, which is attributed to their better light absorption, enhanced charge separation, and increased surface area. Various techniques are employed to comprehend their features. X-ray diffraction (XRD) and transmission electron microscopy (TEM) are utilized to analyze their crystal structure, particle size, and morphology. Meanwhile, X-ray photoelectron spectroscopy (XPS) and energy-dispersive X-ray spectroscopy (EDS) help to identify chemical composition and elemental distribution. UV–Vis spectroscopy is applied to assess optical properties, focusing on absorption and bandgap features. Brunauer–Emmett–Teller (BET) analysis is employed to measure surface area and porosity, crucial factors influencing catalytic activity. Photoluminescence (PL) and transient absorption spectroscopy techniques are used to probe photogenerated charge separation. In addition, scanning electron microscopy (SEM) aids in visualizing surface morphology and particle size, providing insights into the nanocomposite's physical structure and uniformity.

4.2.1 Types of nanocomposite materials

Nanocomposite photocatalysts are broadly categorized into metal-, carbon- polymer-, and ceramic-based nanocomposites. Metal oxides, such as titanium dioxide (TiO_2) and zinc oxide (ZnO), display excellent photocatalytic activity under UV light, which can be enhanced by combining them with other nanoscale materials. Metal sulfides such as cadmium sulfide (CdS) and zinc sulfide (ZnS) are effective in the visible light range. Carbon-based nanocomposites involve graphene, a single-layer carbon lattice, and carbon nanotubes (CNTs), which improve charge separation and transport. Polymer-based nanocomposites include stable combinations of polymers (e.g. polyvinyl alcohol) with metal oxides, which offer improved mechanical properties, and conjugated polymers (e.g. polythiophene), known for their visible light absorption and enhanced charge separation. These nanocomposites stand at the forefront of advancements in sustainable technologies. Further, ceramic-based nanocomposites are being considered because of their inherent chemical stability, expansive surface area which effectively enables pollutant adsorption, and catalytic activity, particularly in the case of advanced oxidation processes.

4.2.1.1 Metal-based nanocomposites
Metal-based nanocomposite photocatalysts have become increasingly important in environmental remediation. These materials exhibit superior photocatalytic activity and unique properties. They are produced by incorporating metal nanoparticles into

different support matrices, such as carbon nanotubes [15], metal oxides [16], or graphene [17], which work synergistically to enhance their efficiency in degrading diverse environmental pollutants. The use of metal-based nanocomposites is a viable and sustainable method with which to tackle challenges associated with air and water pollution, fostering the creation of environmentally friendly technologies for the elimination of pollutants. Precise control over their structure, morphology, and composition through rational design and synthesis allows their performance to be optimized for targeted pollutant removal, overcoming the limitations of traditional nanoparticles. For example, traditional photocatalysts such as TiO_2 nanoparticles work primarily using electron–hole recombination within particles, hindering chemical transformations. Their substantial bandgap of 3.2 eV requires the use of UV light for activation, resulting in inefficient solar energy utilization (only 5% of the spectrum) [18, 19]. Titania's nonporous, polar surface limits the absorption of nonpolar organic contaminants. Agglomeration and aggregation adversely impact TiO_2 nanoparticle photoactivity and light absorption. It is critical to address these issues to enhance the effectiveness of TiO_2 in photocatalytic applications. To overcome these constraints, nanocomposites have been utilized to augment the light absorption capacity of titanium dioxide (TiO_2) and broaden its spectral responsiveness to encompass the visible spectrum. This approach alters the photocatalyst's properties, reducing electron–hole recombination and introducing structural defects that trap electronic carriers, thereby extending their lifespans. For instance, a ZnO/TiO_2 nanocomposite photocatalyst with an 18–42 nm particle size was synthesized by Upadhyay et al [20] using the technique known as the sol–gel process. The experimental outcomes demonstrated that the introduction of 4% ZnO resulted in a notable increase in the degradation of methylene blue (MB), elevating the efficiency of the degradation process from 75% to 90%. Furthermore, the incorporation of ZnO led to enhanced absorption of visible light during the experiment. Bethi et al [21] synthesized nano photocatalysts by incorporating cerium and iron into TiO_2 using ultrasound. The application of ultrasound induced cavitation during the synthesis process. The resulting photocatalysts were then utilized in the decolorization of crystal violet dye. Notably, in the investigation of crystal violet dye decolorization, it was observed that TiO_2 doped with 0.8% iron exhibited the highest level of photocatalytic activity. An investigation conducted by Sukriti et al [22] reported the successful fabrication of nanoparticles comprising samarium-doped zinc oxide. The study demonstrated notable photocatalytic efficacy in terms of the maximum photodegradation of MB dye, particularly with the 4% samarium-doped photocatalyst, achieving an efficiency of 94.94% under visible light irradiation. The efficiency of the photocatalyst in degrading MB exhibited a substantial augmentation, reaching 6.98 times that of its undoped zinc oxide counterpart. $CdS–TiO_2$ nanocomposites prepared by Qutub et al [23] demonstrated heightened photocatalytic performance in the presence of visible light. The CdS–TiO2 nanocomposites demonstrated an 84% degradation efficiency of the organic dye AB-29, surpassing the efficiencies of CdS or TiO_2 alone. The strategy involved utilizing the UV region for TiO_2 excitation and the visible region for CdS excitation, optimizing electron–hole pair generation and trapping. In another study by Karimi-Maleh et al [24], metal oxide/Ag nanocomposites were successfully synthesized and achieved highly efficient photocatalytic degradation of organic dyes. The TiO_2/Ag nanocomposite, characterized by a bandgap of 2.78 eV and a remarkably increased surface area of 68.2 $m^2.g^{-1}$ compared to that of ZnO/Ag, exhibited superior photocatalytic performance. Both nanocomposites demonstrated efficient degradation of methyl orange (MO) and 4-chlorophenol under visible light, and TiO_2/Ag surpassed expectations by degrading 86% of MO within 180 min. In conclusion, maximizing the potential of advanced metal-based nanocomposite photocatalysts requires further research to optimize cost-effective and scalable synthesis processes, ensure long-term stability and reusability, and address environmental and health concerns. Collaboration between researchers, engineers, and policymakers is vital to transforming polluted areas into thriving ecosystems.

4.2.1.2 Polymer-based nanocomposites

Polymer-based nanocomposite photocatalysts have emerged as promising materials for environmental remediation, particularly in photocatalytic pollutant degradation. These photocatalysts, combining polymeric matrices such as polyvinyl alcohol (PVA) with nanoparticles such as titanium dioxide (TiO_2), exhibit enhanced photocatalytic performance [25]. The synergy between the polymer matrix and nanomaterials enables efficient charge separation and migration, improving pollutant degradation under light irradiation. Tunable properties, including the surface area and bandgap, allow for customized design, optimizing efficiency for specific pollutant degradation. The polymer matrix ensures stability and recyclability, making the materials durable and economically viable for repeated photocatalysis cycles. Some nanocomposites incorporate biocompatible polymer matrices, enhancing eco-friendliness for water treatment applications. These materials find use in water and air purification systems, as well as self-cleaning coatings for diverse surfaces, showcasing versatility in degrading organic dyes, pesticides, and pharmaceuticals and addressing emerging contaminants. Additionally, certain nanomaterials extend absorption to visible light, enabling efficient pollutant degradation under natural sunlight. In a recent investigation conducted by Malesic-Eleftheriadou et al [26], PET/TiO_2 nanocomposite films were fabricated with varying TiO_2 contents (10, 30, and 47 wt%) using the phase-inversion method. These films demonstrated photocatalytic activity in the degradation of a mixture of antibiotics under simulated solar irradiation. Initially, heightened TiO_2 concentrations contributed to improved photocatalytic efficiency. However, PET-30 wt% TiO_2 and PET-47 wt% TiO_2 exhibited instability in subsequent cycles, as active oxidative species generated in the initial cycle attacked PET chains, resulting in TiO_2 particle leaching. In contrast, PET-10 wt% TiO_2 maintained stability even after five cycles, effectively degrading antibiotics such as lincomycin, moxifloxacin, isoniazid, metronidazole, and norfloxacin within 2 h. Extended reaction times (6 h) were required for other compounds, which reached degradation levels of up to 98%. In another study, Feizpoor et al [27] synthesized TiO_2/$CoMoO_4$/polyaniline (PANI) ternary photocatalysts using a refluxing technique that incorporated $CoMoO_4$ and PANI into TiO_2. The resulting nanocomposites displayed augmented BET-specific surface areas: 45.8 m^2 g^{-1} for TiO_2, 50.1 m^2 g^{-1} for TiO_2/$CoMoO_4$ (30%), and 57.9 m^2 g^{-1}

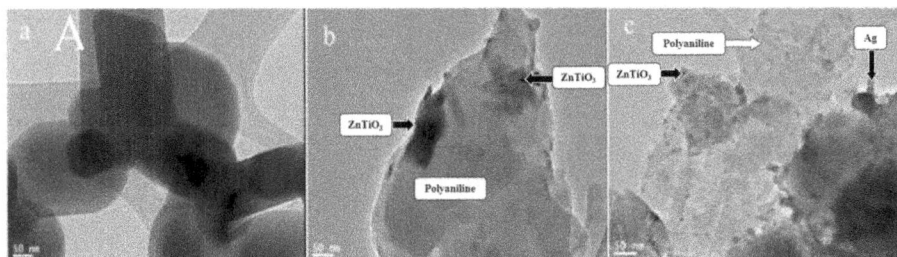

Figure 4.1. TEM images depicting (a) $ZnTiO_3$, (b) $10\%PANI/ZnTiO_3$, and (c) $1\%Ag/10\%PANI/ZnTiO_3$ nanocomposite photocatalysts. Reprinted from [28], Copyright (2021), with permission from Elsevier.

for $TiO_2/CoMoO_4/PANI$ (30%). The heightened surface area of $TiO_2/CoMoO_4/$ PANI (30%) contributed to enhanced visible-light photocatalytic performance. In comparison to TiO_2 and $TiO_2/CoMoO_4$, $TiO_2/CoMoO_4/PANI$ (30%) exhibited notable efficiency in the degradation of organic pollutants and reduction of Cr(VI). For instance, the efficiency of RhB photodegradation performed using $TiO_2/$ $CoMoO_4/PANI$ (30%) surpassed that of pure TiO_2 by 21.7 times and that of $TiO_2/CoMoO_4$ (30%) by 4.9 times. Moreover, the efficiency of the nanocomposite in Cr(VI) reduction exceeded those of TiO_2 and $TiO_2/CoMoO_4$ (30%) by 13.8 and 4.9 times, respectively. Similarly, Faisal *et al* [28] successfully synthesized Ag/PANI/ $ZnTiO_3$ nanocomposites through a cost-effective sol gel method followed by ultra-sonication. XRD analysis disclosed the presence of the rhombohedral phase in $ZnTiO_3$. Figure 4.1 shows that Ag nanoparticles (10–20 nm) were attached to $ZnTiO_3$ nanoparticles (~70 nm), along with sheet-like PANI particles (>300 nm). The ternary photocatalyst, denoted by $1\%Ag/10\%PANI/ZnTiO_3$, showcased out-standing performance, achieving a notable 95.6% removal of the target MB molecule. This degradation rate surpassed that of undoped $ZnTiO_3$ by 1.85 times. The heightened performance was ascribed to the diminutive size of the Ag nano-particles and the synergistic interactions between the Ag, PANI, and $ZnTiO_3$ constituents. Furthermore, the ternary framework exhibited notable efficacy in the visible-light degradation of the insecticide imidacloprid.

Kumar *et al* [29] successfully employed an in-situ codeposition and oxidation approach to synthesize nanocomposites featuring both rutile and anatase phases. The resulting catalysts exhibited a spherical morphology. Their bandgap energies, determined using Tauc plots, were found to be 3.2, 2.86, 2.6, and 2.3 eV for TiO_2, pure PANI, $TiO_2/PANI$ (T/P), and $TiO_2/PANI/GO$ (T/P/G), respectively. The surface areas of TiO_2, T/P, and T/P/G were measured and found to be 39.2, 78.8, and 94.4 $m^2\,g^{-1}$, respectively. Photocatalytic investigations under visible light using T/P/G, T/P, and TiO_2 at neutral pH with a catalyst dose of 1600 mg l^{-1} demonstrated significant photodegradation (85%–99%, 60%–97%, and 10%–20%) of thymol blue (TB) and rose bengal (RB) at a concentration of 25 ppm over 180 min. Reductions in total organic carbon (TOC) substantiated photodegrada-tion, and a kinetic study revealed first-order kinetics. Figure 4.2 visually depicts the enhanced nanocomposite-driven rate of dye photodegradation.

■ TiO2/PANI/GO ■ TiO2/PANI ■ TiO2

Rose Bengal
0.0224
0.0147
0.0035

Thymol Blue
0.025
0.016
0.004

0   0.005   0.01   0.015   0.02   0.025

Degradation rates

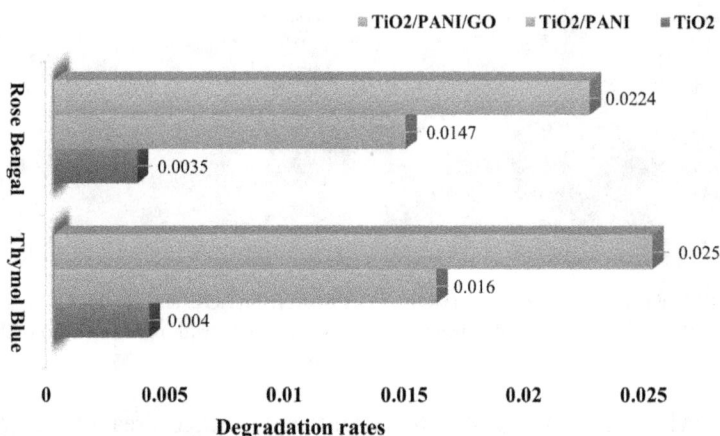

Figure 4.2. The photodegradation rates of thymol blue and rose bengal achieved using various nano-composites. Drawn based on data from [29] Copyright (2023) from Elsevier.

Nano-fillers possess the capacity to increase the overall functionality of nano-composite photocatalysts. This increment is achieved through an enhancement in the surface area, an enhancement in the pollutant adsorption capabilities, and an acceleration of reaction rates. Functionalized nano-fillers can selectively target pollutants, and they also reinforce strength, enhance stability, and increase resistance. The impact of these fillers on the photocatalytic activity of polymer-based photocatalysts depends on factors such as their type, dispersion, and functionalization [30]. When coupled with efficient photocatalytic activity, these fillers are invaluable in environmental remediation. Further research and development in this field could lead to sustainable solutions for addressing contemporary pollution challenges.

4.2.1.3 Ceramic-based nanocomposites

In the area of environmental remediation, ceramic-based nanocomposites are the preferred choice due to their inherent chemical stability, expansive surface area that facilitates effective pollutant adsorption, and catalytic activity, particularly in the context of AOPs. Their durability ensures sustained effectiveness in harsh conditions, which is crucial for long-term applications such as water treatment [31]. Notably, ceramics such as titanium dioxide exhibit photocatalytic properties and are capable of degrading pollutants when exposed to light, as demonstrated in photo-catalytic water treatment. The diverse range of ceramic materials offers flexibility in tailoring nanocomposites to target specific pollutant types and environmental conditions. Cerium oxide, known for its resistance to chemical corrosion and UV absorption in the 385–400 nm range [32] has proved to be an effective catalyst. Doping CeO_2 ceramics with suitable dopants enhances their photocatalytic performance [33]. An innovative methodology presented by Zinatloo-Ajabshir *et al* [34] employed a novel environmentally friendly reducing and stabilizing agent—Phoenix dactylifera extract. This extract played a crucial role in the synthesis of a nano-structure composed of $ZnCo_2O_4$ and Co_3O_4. This nanomaterial exhibited remarkable efficiency under visible light, achieving energy gap values of approximately 2.68 and 2.87 eV for the $ZnCo_2O_4$–Co_3O_4 product and pure $ZnCo_2O_4$, respectively, in nanoscale particles ranging from 30 to 55 nm. The photocatalytic degradation efficiencies for RB, acid black 1, and phenol red reached 100%, 95.7%, and 100%, respectively, after 75 min of exposure to visible light. The $ZnCo_2O_4$–Co_3O_4 nano-structure had excellent stability and recyclability over nine consecutive runs. In another study, Mahdavi et al [35] synthesized novel photocatalytic nanocomposites (Dy_2O_3-SiO_2) utilizing tetraethylenepentamine (Tetrene) as a basic agent through a facile and efficient sonochemical method. Morphological analyses revealed that optimal control of the sonication time (10 min) and ultrasonic power (400 W) resulted in the formation of a porous nanocomposite comprising spherical nanoparticles with sizes ranging from 20 to 60 nm. The binary Dy_2O_3-SiO_2 nanophotocatalyst exhibited an energy gap of 3.41 eV, rendering these nano-composites conducive for contaminant removal. The nanocomposite demonstrated heightened efficacy in degrading the pollutant erythrosine, achieving a degradation efficiency of 92.9%. While these advantages confirm that ceramic-based nano-composites are valuable for pollutant degradation, it is essential to consider the context, pollutant characteristics, and treatment goals when selecting nanocompo-site materials, acknowledging that other types such as polymer-based or metal-based nanocomposites may also be suitable for certain applications, depending on the specific requirements.

4.2.1.4 Carbon-based nanocomposites

Carbon-based nanocomposites that incorporate nanoparticles such as carbon nano-tubes or graphene are designed to efficiently degrade pollutants. Nanomaterials characterized by distinctive nanoscale attributes can be tailored to enhance either catalytic or adsorption functionalities for specific pollutants. In environmental remediation, these nanocomposites interact with pollutants, improving degradation or aiding removal. Leveraging the high surface area and reactivity of carbon-based nanomaterials proves advantageous in wastewater treatment, air purification, and soil remediation [36]. Graphene, along with its derivatives including graphene oxide (GO) and reduced graphene oxide (rGO), constitutes a class of carbon-based materials that act as efficient substrates for the dispersion of nanostructured metal oxides. These platforms play a crucial role in averting electron–hole pair recombination, conse-quently enhancing photocatalytic activity. GO, acting as a semiconductor photo-catalyst, undergoes gradual bandgap shifts during the catalytic process due to the loss of oxygen functionalities, resulting in enhanced conductivity. Recently, Mohamed et al [37] utilized hydrothermal synthesis to fabricate nanocomposites comprising Ag–CdSe/GO/CA, revealing heightened photocatalytic activity in the degradation of malachite green under UV–visible light conditions. The enhanced performance of the composite resulted from the multifunctional roles played by GO, silver (Ag), and cellulose acetate (CA) in providing structural support. The GO served as an electron acceptor, promoting carrier separation, while copper nanoparticles on semiconducting rGO exhibited notable degradation of organic dyes [38]. Similarly, Rosalin *et al* [39] synthesized nanocomposites of Ag-loaded ZnO-graphene (AgZG) utilizing hydrothermal techniques, resulting in notable degradation of MO under both UV radiation and sunlight. The heightened photocatalytic performance was ascribed to the judicious loading of Ag, which facilitated the enhanced production of $OH^\bullet$ radicals, augmented the absorption of visible light, and increased the efficiency of the charge transfer processes. Gu *et al* [40] synthesized graphene-doped TiO_2 using the hydrothermal method; their results revealed that the graphene played a critical function in sequestering generated electrons, inhibiting electron–hole recombination, enhancing the capacity for pollutant adsorption, and extending the range of light absorption wavelengths. Li *et al* [41] fabricated a nanocomposite photocatalyst comprising silver/bismuth oxyiodide/graphene oxide (Ag/BiOI/GO), which exhibited heightened efficiency in rhodamine B degradation. The study underscored the enhanced photocatalytic capability of the Ag/BiOI/GO, which was attributed to the presence of both Ag and GO in the composite. Fulzele *et al* [42] synthesized a high-performance photocatalyst, an $rGO/Ag_3PO_4/CeO_2$ nanocomposite, using a sonochemical method. The catalyst demonstrated remarkable photocatalytic efficacy in the degradation of MB dye under visible light, achieving close to complete degradation. This exceptional performance was ascribed to the efficient separation of electron–hole pairs, underscoring the nanocomposite's stability and effectiveness.

Carbon nanotube nanocomposites are recognized for their efficacy in pollutant degradation, which is due to their expansive surface area, elevated conductivity, and chemical stability. Some of these nanocomposites exhibit photocatalytic activity and can be customized for tailored solutions, often synergizing with other nanomaterials for enhanced performance. For instance, polyaniline/multiwalled carbon nanotube nanocomposites (PANI/MWCNT NCs) were prepared by Ardani *et al* [43] using the ultrasonic method. The outcomes demonstrated that elevating the ultrasonic power and duration resulted in a reduction in the nanocomposites' particle size, which ranged from around 190–103 nm, accompanied by a specific surface area of 25.34 $m^2\ g^{-1}$. Photodegradation investigations revealed that PANI/MWCNT NCs containing 7 wt% of MWCNTs demonstrated notable efficacy, achieving a removal efficiency of 98.39% for MO dye within 45 min. This surpassed the performance of pure PANI, which achieved a removal efficiency of 79.80%. Kinetic analysis indicated a good fit with the pseudo-second-order model. In another study conducted by Rad *et al* [44], zinc chromium layered double hydroxide (ZnCr-LDH) nanocomposites were synthesized by integrating CNTs and bamboo charcoal (BC), resulting in an increase in specific surface areas, as determined through BET analysis. The initial surface area of the ZnCr-LDH was 1.2 $m^2\ g^{-1}$, which was significantly enhancement to 5.0 and 7.0 $m^2\ g^{-1}$ upon integration with CNTs and BC, respectively. The hybridized materials manifested decreased bandgap values of 2.46, 2.22, and 1.60 eV, indicating enhanced electron–hole pair generation under ultrasonic irradiation and light sources. The hybrid materials exhibited noteworthy catalytic efficacy in the sonophotocatalytic degradation of rifampicin (RF), achieving 100% degradation efficiency within 40 min when exposed to 150 W of ultrasound and irradiation with visible light at a concentration of 0.6 g L^{-1} ZnCr-LDH/BC. The use of CNT composite photocatalysts for wastewater treatment is a promising approach, but it faces various challenges. High production costs, due to the expensive synthesis of quality CNTs, hinder its large-scale implementation in regions with budget constraints. CNTs tend to aggregate together, reducing their surface area and hindering their dispersion in the photocatalyst matrix. Concerns about the toxicity of CNTs in wastewater treatment raise questions about their environmental impact and long-term risks. Photocorrosion, which is influenced by harsh conditions, may decrease the photocatalyst's activity over time. Limited absorption of visible light affects efficiency, and scaling up production faces technical challenges. It is crucial to improve the selectivity for pollutant removal without impacting non-target compounds

Fullerenes, notably C60, are unique carbon allotropes with a spherical or ellipsoidal structure. Discovered in 1985, they blend particle and wave properties, conducting electricity and exhibiting light absorption/emission. Widely used, especially in Fullerene nanocomposite photocatalysts, they enhance light-driven catalysis in applications such as environmental remediation and energy production. A Fullerene nanocomposite photocatalyst comprises fullerene nanoparticles within a semiconductor host matrix, such as titanium dioxide (TiO_2) or zinc oxide (ZnO). Fullerenes such as C60 or C70 which have unique light-absorbing properties enhance the photocatalyst's performance. In these nanocomposites, fullerenes act as sensitizers, absorbing light to generate and efficiently separate electrons. The semiconductor matrix facilitates redox reactions upon photon absorption, which is crucial for pollutant degradation or hydrogen production. A recent study conducted by Zhang *et al* [45] synthesized a hybrid nanocomposite comprising titanium dioxide (TiO_2) and fullerene using an ultrasonication–evaporation technique. The resulting TiO_2/fullerene composite, consisting of well-dispersed nanoparticles around 25 nm in size, demonstrated high efficiency (over 96%) in degrading rhodamine B dye under visible light ($k > 400$ nm). The introduction of fullerene into TiO_2 extended the absorption spectrum of the photocatalyst into the visible light domain, concurrently increasing adsorption capacity and degradation efficiency. The observed improvement was attributed to the synergistic influence arising from the interaction between fullerene and TiO_2, making the composite a promising, efficient, and reusable photocatalyst for environmental applications [45]. Fullerene nanocomposite photocatalysts offer advantages such as tunable bandgaps and efficient charge separation, making them strong contenders for certain applications. However, their costly synthesis, limited solubility, and potential toxicity pose challenges. Comparatively, graphene and carbon nanotubes provide high surface area but struggle with charge separation. Carbon black's high cost and limited surface area restrict its performance. The optimal carbon-based photocatalyst choice depends on application-specific requirements, including factors such as cost, stability, and potential toxicity. In conclusion, carbon-based photocatalysts have a promising future in pollutant degradation. These materials exhibit exceptional catalytic properties, paving the way for sustainable environmental remediation. As research progresses, we foresee ongoing progress in the evolution of carbon-based photocatalysts characterized by enhanced efficiency and cost-effectiveness. This innovation holds the potential to revolutionize pollutant abatement technologies, offering a green and scalable solution that can address environmental challenges.

4.2.2 Methods of nanocomposite synthesis

Nanocomposite photocatalysts represent a cutting-edge advancement in materials science, particularly in the realm of environmental remediation and energy conversion. Two principal methodologies, top-down and bottom-up, are utilized for their preparation. The top-down strategy entails breaking larger materials into nanoscale components, offering control over size and composition but facing challenges in maintaining uniformity [46]. Conversely, the bottom-up method builds nanocomposites from individual molecules or nanoparticles, providing precise control over structure and composition for enhanced efficiency [46, 47]. Combining these approaches holds promise for optimizing nanocomposite photocatalyst design and fabrication, which is applicable across diverse fields such as wastewater treatment and solar energy conversion. These nanocomposites composed of nanoscale elements exhibit superior performance in photocatalysis because of their enhanced surface area, improved charge separation, and synergistic effects. Their preparation involves the integration of various nanomaterials, such as semiconductor nanoparticles and metal oxides, using a number of techniques like sol–gel synthesis, hydrothermal methods, etc. Crucial factors include the selection of the precursors and the optimization of the reaction conditions, which enable photocatalysts to be customized for specific applications such as pollutant degradation and solar energy conversion [48]. Continued research into nanocomposite photocatalyst preparation holds promise for addressing environmental sustainability and clean energy challenges. Various techniques, including precipitation, sol–gel, hydrothermal, and sonochemical methods, are utilized in the synthesis of nanocomposites. Table 4.1 examines the influences of the synthesis method and process conditions on the characteristics of the resulting photocatalyst. The choice of method is contingent upon factors such as the nature of the nanocomposite, the simplicity of the method, and the specific properties of the nanocomposite.

4.2.2.1 The precipitation method

The precipitation method is a commonly used technique for synthesizing nanocomposites. It involves mixing two or more precursor solutions that contain the desired components of the nanocomposite. When these solutions are mixed, a chemical reaction takes place, resulting in the formation of insoluble particles that precipitate out of the solution. These particles then act as building blocks for the nanocomposite. By adjusting factors such as the precursor concentration, temperature, and pH during synthesis, it is possible to regulate the dimensions, composition, and structural attributes of the produced nanoparticles. After precipitation, additional processing steps may be necessary to improve the nanocomposite's properties. These can include washing and drying to remove impurities and solvents. The procedure used for synthesis is depicted in figure 4.3, which shows a detailed

Table 4.1. The impacts of the preparation method and the process conditions on photocatalyst properties.

Photocatalyst nanocomposite	Preparation method	Reaction conditions	Properties of the nanocomposite	Results	Ref.
MgO/GO	Hydrothermal method	• 1:1 ratio mol % of GO and MgO • 1 h sonication • 4 h placed in autoclave at 80 °C	• Structure: flaky spherical • Nanoparticle size: 12 nm (average) • Bandgap: 1.7 eV	• MgO/GO photocatalyst demonstrated exceptional photocatalytic activity (98%). • Sustained high performance consistently observed over five consecutive cycles.	[61]
AgIO$_4$/CeO$_2$	Sol–gel and sonochemical route	• Ultrasonic bath: 200 W power, 1 h duration • Wet solid subjected to ultrasonic treatment • Dried in oven: 12 h	• Specific surface: 49 m^2 g^{-1} • Crystalline size: 17 nm • Bandgap energy: 1.98 eV	• RhB dye decomposition: 95% complete • Absence of secondary pollutant generation • Promising results for sustainable wastewater treatment	[62]
CuO/NiO	Coprecipitation	• Mixture pH: 10 • Dried product calcination: 400 °C, 5 h	• Nancomposite-specific surface area measured 22.28 m^2 g^{-1}. • Pore size: 3 nm • Bandgap: 4.68 eV	• Cefixime degradation: 90% • Kinetics: first-order • Rate constant (k): 7.4 × 10^{-3} min^{-1}	[63]
CuO/ZrO$_2$	Sol–gel method	• Reaction temperature: 40 °C • Calcination temperature: 500 °C for 4 h	• Surface area measured 200 m^2 g^{-1}, • Bandgap was determined to be 2.72 eV. • Average particle size: 45 nm	• Photocatalytic efficiency for thiophene oxidation: 100% in 120 min • Rate constant of CuO/ZrO$_2$ nanocomposite: 0.0172 min^{-1} • Enhancement compared to ZrO$_2$ NPs: 3.5-fold increase	[64]

Material	Method	Synthesis conditions	Properties	Observations	Reference
rGO/TiO$_2$	Hydrothermal method	rGO/TiO$_2$ precursor solution ultrasonically treated: • Power: 600 W • Room temperature • Duration: 1 h • Hydrothermal treatment: • Temperature: 1200 °C • Duration: 18 h	• Crystalline size: 15 nm • Bandgap: 2.02 eV	• Degradation rate: 13.54%	[65]
CuO/TiO$_2$	Ultrasound-assisted method	• Ultrasonic wave frequency: 20 kHz • Room temperature	• Shape: spherical • Particle size: 20 nm • Bandgap: 3.08 eV	• Highest decay constant observed: 1.38 min^{-1} • Target: sulfate-reducing bacteria • Significant enhancement in degradation efficiency	[66]
ZnO/CuO/Ag	Coprecipitation	• pH maintained at 9 • Precipitate annealed at 500 °C	• Nanocomposite photocatalyst bandgap: 2.85 eV • Spherical nanoparticles; average size: 140 nm • Aggregation of nanoparticles observed • Specific surface area: 21.3 m^2 g^{-1}	• The ZnO-CuO-Ag nanocomposite demonstrated improved efficiency in charge separation. • Achieved 100% photodegradation of MB after 90 min of visible light radiation. • Photocatalytic reaction rate constant measured at 0.0515 min^{-1}.	[67]
TiO$_2$/GO	Modified sol–gel method	• Reaction time: 1 h • Room temperature • Calcination temperature: 500 °C • Calcination duration: 5 h • Heating rate during calcination: 20 °C min^{-1}	• Specific surface area: 46.71 m^2 g^{-1} • Bandgap shift: 3.07 eV to 2.92 eV due to GO incorporation • Nanocomposite size: 30 nm	• MB dye degradation: up to 98% • pH $\approx$ 13 • Optimal degradation time: 60 min	[68]

(*Continued*)

Table 4.1. (*Continued*)

Photocatalyst nanocomposite	Preparation method	Reaction conditions	Properties of the nanocomposite	Results	Ref.
ZnO/SnO$_2$/rGO	Hydrothermal method	• GO synthesized using Hummer method • Nanocomposite precursor mixture sonicated for 30 min • Solution heated to 130 °C for 18 h	• Bandgap energies: ZnO: 2.8 eV; ZnO/rGO: 2.3 eV; SnO$_2$/rGO: 2.0 eV; ZnO/SnO$_2$/rGO: 1.9 eV • Average particle size of ZnO/SnO$_2$/rGO nanocomposites: 180 nm	• Orange dye degradation rate constant: 0.0241 min^{-1} • Nanocomposite: ZnO/SnO$_2$/rGO • Degradation efficiency: >91%	[69]
NiO/RGO	Hydrothermal method	• Teflon stainless autoclave used • Incubation at 180 °C for 10 h • Formation of black NiO/RGO nanocomposite powder • Annealed for 6 h • Annealing temperature: 350 °C	• Nanocomposite bandgap: 3.47 eV • Average crystal size: 22.5 nm.	• Photocatalytic degradation efficiency: 95.4% • Nanocomposite: NiO/RGO • Rate constant: 0.015 871 min^{-1} • NiO/rGO rate constant 1.4 times higher than that of NiO	[70]
ZnO/MgO/CaO	Biosynthesis	Initial solution: • Temperature: 80 °C. Duration: 8 h Homogeneous solution: • Calcination temperature: 700 °C • Calcination duration: 2 h	• Nanocomposite exhibits crystalline nature • Triangular shapes with a size of approximately 59 nm • Nanoparticle optical bandgap measures 5.3 eV	• MB dye degradation: 93% • Duration of UV light exposure: 120 min • Rate constant (kobs): 0.007 min^{-1}	[71]

ZnO/TiO$_2$/rGO	Laser ablation method	Laser parameters: • Pulse energy: 130 mJ • Wavelength: 355 nm • Repetition rate: 10 Hz • Pulse width: 10 ns Lens type: UV lens, focal length: 200 mm • Irradiation time: 30 min	• Nanocomposites exhibit a low bandgap of 2.73 eV. • 355 nm laser irradiation transforms spherical ZnO nanoparticles into nanorods. • Morphology becomes more compact post-irradiation.	• Rate constant: 0.149 min^{-1} • ZnO/TiO$_2$/rGO nanocomposite achieved >98.5% degradation of MB within 60 min • Flat sheet surface area increased with higher rGO content • The presence of ZnO and TiO$_2$ contributed to enhanced surface area	[72]

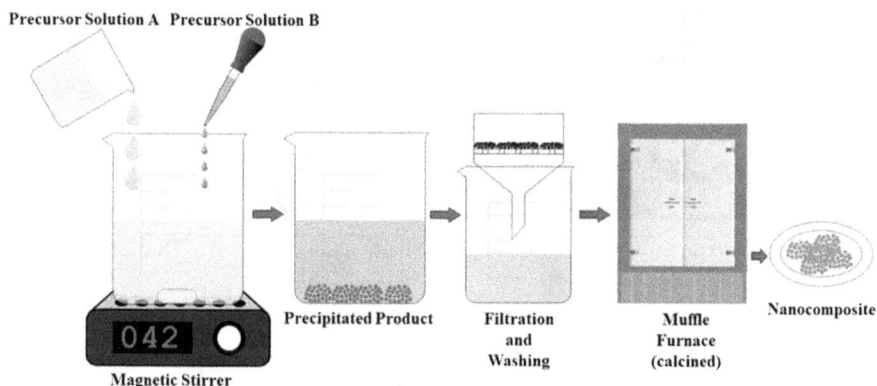

Figure 4.3. A step-by-step schematic representation of the precipitation method.

schematic representation of the precipitation method. Researchers can also treat the nanoparticles or combine them with other materials to create a composite with specific characteristics. The precipitation method is widely used because it is simple and cost-effective and gives control over particle size and composition. It is scalable, making it suitable for industrial-scale production. The homogeneous distribution of nanoparticles within the composite ensures uniform photocatalytic activity.

In a study, Wu *et al* [49] effectively produced a nanocomposite of Ag_2CO_3/ZnO using a chemical precipitation method. Zinc oxide nanowires weighing 1 g and corresponding to 12.3 millimoles were dispersed in 500 milliliters of distilled water. Following the introduction of silver nitrate ($AgNO_3$ 0.067 g, 0.39 mmol) in NH_4OH (2.3 ml, 25 wt% NH_3), the nanowire suspension was sonicated for 30 min. The resulting suspension was magnetically stirred for 30 min at room temperature, followed by the rapid addition of sodium bicarbonate (0.5 mmol) in 50 ml of deionized water, which was continuously stirred. Following 12 h of room-temperature stirring, the precipitates were filtered, sequentially washed with distilled water and ethanol, and then dried at 80 °C for 12 h. The resulting Ag_2CO_3/ZnO nanocomposite exhibited a surface area of 73.2 m^2 g^{-1} and demonstrated significant efficacy, achieving 91% degradation of rhodamine 6G (R6G) under visible light exposure in 40 min. A study carried out by Haghighatzadeh *et al* [50] employed a straightforward chemical coprecipitation technique to synthesize ZnO/TiO_2 nanocomposite catalysts featuring adjustable TiO_2 shell thickness. The resulting photocatalysts exhibited distinct crystal structures, namely wurtzite ZnO and anatase TiO_2 as the core and shell, respectively. Their bandgap energy was 3.05 eV following the incorporation of TiO_2, a decrease from the original value of 3.4 eV. The results revealed the surface morphologies of both ZnO nanoparticles (NPs) and the ZnO/TiO_2 nanocomposites. A field emission scanning electron microscopy (FESEM) image of the ZnO NPs displayed nonuniform NPs with sheet-like shapes attributed to agglomerated spheres. The particle sizes, as illustrated in figure 4.4, ranged from 18 to 50 nm. Conversely, the ZnO/TiO_2 core–shell structures exhibited relatively

Figure 4.4. Images (a) and (b): field emission scanning electron microscopy (FESEM) visualizations. (c) and (d): the particle size distributions of ZnO and the ZnO/TiO$_2$ nanocomposites, respectively. Reproduced from [50]. © IOP Publishing Ltd. All rights reserved.

uniform spherical particles with an average dimension of roughly 30 nm, indicating the effectiveness of the precipitation method in achieving homogeneous structures.

In another study, Haounati *et al* [51] successfully prepared a Fe$_3$O$_4$/Ag$_3$PO$_4$@Sep ('Sep' denotes sepiolite) magnetic nanocomposite through an eco-friendly one-step precipitation method. This nanocomposite exhibited increased photocatalytic efficacy in the degradation of malachite green (MG) dye under sunlight conditions; the most active composite completely removed MG dye within 16 min at a 10 ppm concentration. The nanocomposite exhibited degradation times that were three, five, and six times faster than those of Ag$_3$PO$_4$@Sep, Ag$_3$PO$_4$, and Fe$_3$O$_4$, respectively. The hybrid configuration exhibited a decreased bandgap energy of 2.37 eV, which improved its photocatalytic capability under visible light. Significantly, the continuous photocatalytic activity of Fe$_3$O$_4$/Ag$_3$PO$_4$@Sep continued even following three cycles of reuse. Precursor selection allowed nanoparticles to be synthesized with specific properties, enhancing the photocatalyst's effectiveness. This method is adaptable to diverse substrates, including polymers and ceramics, which provides flexibility for various applications. Factors such as ease of synthesis and scalability should be carefully considered when choosing a method for nanocomposite photocatalyst synthesis.

4.2.2.2 The hydrothermal method
The hydrothermal technique is a crucial method for synthesizing nanocomposite photocatalysts; it involves synthesis at elevated pressure and temperature in an aqueous environment. Starting with a precursor solution containing the desired nanomaterials, such as semiconductor nanoparticles or metal oxides, contents of a sealed reaction vessel undergo controlled growth and assembly under extreme conditions. This method facilitates precise control over size, morphology, and

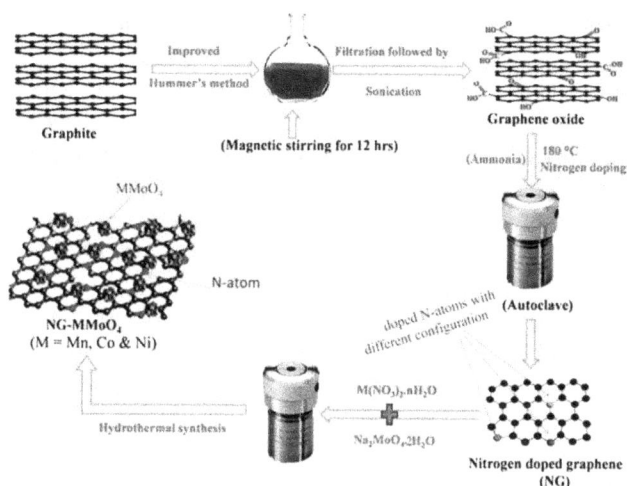

Figure 4.5. A representation of the synthesis of graphene oxide, nitrogen-doped graphene, and the nano-composite NG-MMoO$_4$. Reprinted from [52], Copyright (2017), with permission from Springer Nature.

composition, resulting in well-defined structures with enhanced photocatalytic activity. In a recent investigation conducted by Kumar *et al* [52], the hydrothermal approach was employed for the synthesis of nanocomposites consisting of nitrogen-doped graphene combined with metal molybdates (NG-MMoO$_4$), in which MMoO$_4$ represents a metal molybdate and M denotes Mn, Co, or Ni. In this process, nitrogen-doped graphene was sonicated and centrifuged to obtain a colloidal solution. Sequential additions of M(NO$_3$)$_2$·nH$_2$O (where M = Mn, Co, Ni) and Na$_2$MoO$_4$·2H$_2$O were made and the pH was adjusted to 7. The resultant suspension was transferred to a Teflon-lined stainless steel autoclave and subjected to heating at 180 °C for a period of 12 h, followed by subsequent cooling. The synthesis process is depicted in figure 4.5, which schematically represents each step in the procedure. The nanocomposites displayed nanostructures characterized by crystallite sizes ranging from 10 to 30 nm and demonstrated a remarkable 99% degradation of MB.

Similarly, another work carried out by Younas *et al* [53] successfully synthesized a WO$_3$/TiO$_2$/V$_2$O$_5$ catalyst using a hydrothermal method. Precursor solutions were combined in a stainless steel Teflon-lined autoclave and maintained at a temperature of 120 °C for 24 h. Subsequently, the resultant precipitates were washed, filtered, centrifuged, and then calcinated at 620 °C for 4 h within a muffle furnace to produce the nanocomposite. The results revealed varied structural morphologies with dimensions varying between 50 and 80 nm, and the nanocomposite displayed an average particle size of 20.44 nm. It demonstrated notable photocatalytic efficacy, yielding degradation percentages of 76%, 86%, and 86% for MB, MO, and orthonitrophenol, respectively, highlighting its excellent potential. Similarly, a study carried out by Parasuraman *et al* [54] synthesized FeWO$_4$/FeS$_2$ nanocomposites via the hydrothermal method and revealed that the 2:1 wt% FeWO$_4$/FeS$_2$ nanohybrid established a heterojunction characterized by the lowest recombination rate of electron–hole pairs and minimal electron transfer resistance. A Tauc plot analysis determined a bandgap value of 2.76 eV for $FeS_2/FeWO_4$. The nanohybrid demonstrated superior degradation activity, attaining an efficiency of 89.5% in 120 min under UV–visible irradiation. This outcome was ascribed to synergistic effects, enhanced light absorption, and proficient separation of charge carriers. Also, a novel hybrid nanocomposite comprising multiwalled carbon nanotubes (MWCNTs) with copper oxide, as well as pristine copper oxide, were synthesized via hydrothermal techniques by Shibu *et al* [55]. The MWCNTs (50 wt%) were dispersed in a solution of copper nitrate at a concentration of 0.5 M and a volume of 100 ml under continuous stirring. The pH was adjusted to 13 by adding 0.2 M NaOH. The resultant blend was subjected to hydrothermal processing at 120 °C for 15 h, subsequently undergoing centrifugation, rinsing, and heating at 150 °C for 15 h to yield the final MWCNT/copper oxide nanocomposite. A comparative investigation into the inclusion of MWCNTs versus pristine copper oxide unveiled noteworthy alterations in the crystalline phase, lattice parameters, particle characteristics, morphology, and the optical bandgap. The nanocomposite of MWCNTs and copper oxide exhibited a measured bandgap of 1.90 eV. This hybrid material displayed heightened efficacy in the photodegradation of methyl orange (MO) dye when exposed to sunlight for 6 h. The degradation process followed pseudo first-order kinetics, resulting in the breakdown of 92% of the MO dye. The MWCNT/copper oxide nanocomposite maintained a high level of recyclability, demonstrating consistent performance for up to five cycles. With the hydrothermal method, widely applied in materials science, the future of hydrothermal synthesis for the preparation of nanocomposite photocatalysts holds exciting prospects. The hydrothermal method is advantageous for synthesizing nanocomposite photocatalysts, but faces some challenges. Control of the particle size, shape, and morphology is difficult due to factors such as temperature and pressure. High temperatures and pressures may introduce impurities, demanding extra purification steps. Energy consumption is a concern for large-scale production. Some precursors are incompatible, limiting nanocomposite diversity. Certain solvents raise environmental concerns. Despite these challenges, ongoing research focuses on optimizing hydrothermal synthesis, choosing suitable precursors, and developing eco-friendly alternatives. Combined techniques, the use of computational modeling, and innovative strategies are being explored to enhance control and efficiency in nanocomposite photocatalyst synthesis. Researchers can expect to delve deeper into optimizing conditions, designing tailored nanocomposites, scaling up production, integrating with emerging technologies, and exploring new nanomaterials to unlock the full potential of these materials for sustainable and efficient applications.

4.2.2.3 The sonochemical method

The sonochemical method for preparing nanocomposite photocatalysts includes utilizing high-frequency ultrasound waves to induce acoustic cavitation in a reaction mixture. This process generates localized intense heating, high pressures, and shockwaves, promoting the formation of nanoparticles from larger precursors. Ultrasound exposure induces the collapse of bubbles, resulting in the formation of

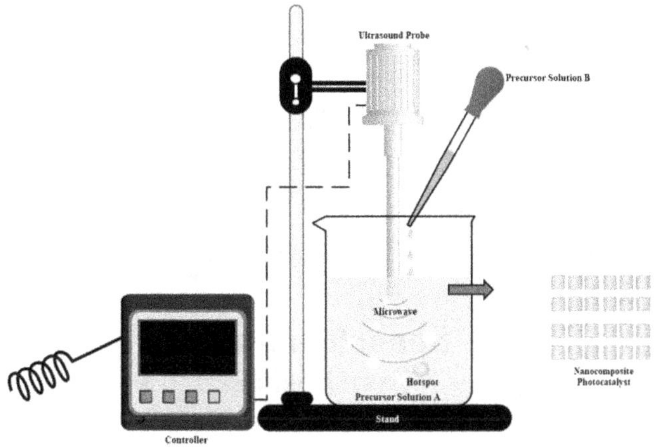

Figure 4.6. Presents a schematic illustration delineating the application of the sonochemical method in the synthesis of nanocomposites.

localized regions characterized by intense heat (5000 K) and elevated pressure (20 MPa). This non-hazardous and rapid method produces small metal particles efficiently. Figure 4.6 shows a schematic representation of the setup used in the sonochemical method for synthesizing nanocomposites. Using high-frequency waves, it is possible to precisely control the size, morphology, and composition. This approach, conducted under mild conditions, holds promise for environmentally friendly synthesis, enhancing nanocomposite properties for applications in environmental remediation and energy conversion.

In a recent study, Zinatloo-Ajabshir *et al* [56] successfully prepared a nanocomposite of $ZnCo_2O_4/Co_3O_4$ using a sonochemical process. This composite showed great potential for efficiently removing organic contaminants. According to the study, the optimal conditions with ultrasound for producing the nanocomposite with the smallest and most uniform particles were a synthesis time of 10 min and the use of cetyltrimethylammonium bromide (CTAB) as a capping agent. The outcomes revealed that the systematically optimized nanocomposite, formulated with CTAB over a duration of 10 min, exhibited notable removal efficiencies for three distinct contaminants, namely methyl violet (100%), Acid Red 14 (95.6%), and erythrosine (100%). This was achieved within just 65 min under visible radiation. The optimized sample was also found to be durable and could maintain its catalytic yield for up to eleven runs without significant loss of effectiveness. A nanocomposite of Ag_2O/CuO was synthesized via ultrasonication by Warsi *et al* [57] exhibiting crystallite sizes of 17.55 nm for Ag_2O and 17.42 nm for CuO. Tauc plots were utilized to ascertain a diminished bandgap energy of 1.51 eV for the nanocomposite. This reduction signifies favorable alterations in the electronic structure, enhancing its photocatalytic capabilities under solar irradiation when contrasted with pristine CuO (with a bandgap of 2.77 eV) and Ag_2O (with a bandgap of 1.61 eV). The nanocomposite demonstrated superior photocatalytic activity, with percentage degradation values of 97.5%, 90.5%, and 95% for MB, RB, and benzoic acid, respectively, surpassing pure metal oxides.

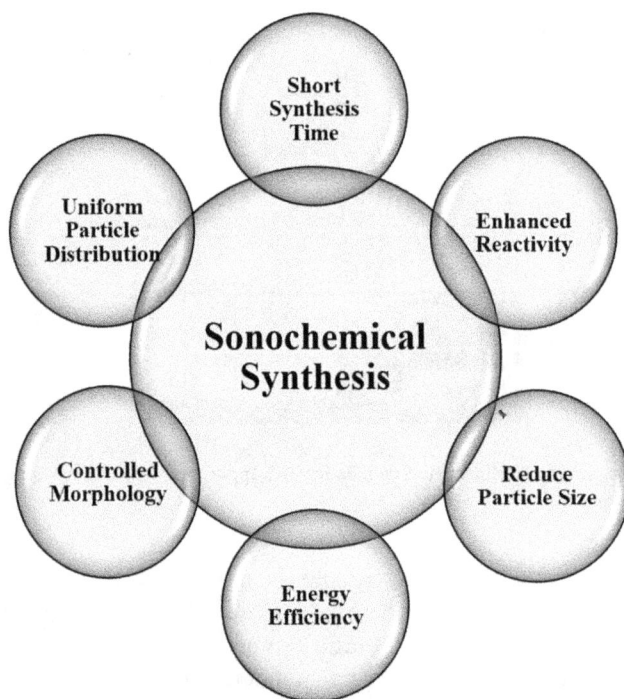

Figure 4.7. Advantages of sonochemical synthesis of nanocomposite

Sonochemical synthesis of nanocomposite photocatalysts presents several advantages shown figure 4.7, including uniform nanoparticle distribution through intense cavitation-induced mixing. This method enhances reactivity by generating high-energy conditions, fostering the formation of reactive species for improved nanocomposite structures. Sonochemistry reduces particle size, increasing surface area for better photocatalytic performance, and requires shorter synthesis times. Precise control over morphology is achieved by adjusting synthesis parameters, ensuring tailored properties for specific applications. The process is energy-efficient, operating at ambient temperatures, and exhibits versatility with various precursor materials. Sonochemical synthesis is scalable, facilitating the production of nanocomposite photocatalysts in larger quantities for practical use.

4.2.2.4 The sol–gel method

The sol–gel methodology is extensively utilized and exhibits versatility as a technique for the synthesis of nanocomposites. It involves the dissolution of metal salts or metal alkoxides in a solvent, leading to the formation of a colloidal suspension (sol), which then undergoes controlled hydrolysis and condensation reactions to transition into a solid gel phase. This process occurs at relatively low temperatures and allows for the precise dispersion of nanoscale components such as nanoparticles or nanotubes within a solid matrix. The resulting nanocomposite,

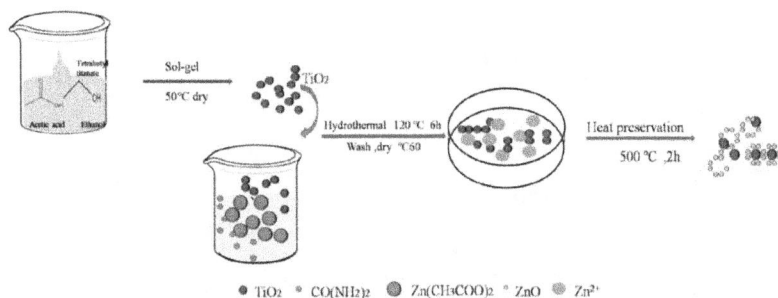

Figure 4.8. Representation of the synthesis of the TiO_2/ZnO nanocomposite through the sol–gel method. Reprinted from [59], Copyright (2021), with permission from Springer Nature.

often in gel form, offers researchers the ability to tailor its composition and structure for specific applications, such as photocatalysis. By manipulating parameters such as the precursor choice, solvent, and reaction conditions, precise control over the nanocomposite's properties can be achieved. This method's utility in creating functional materials, such as those used for photocatalysis, is evident. In an investigation conducted by Pudukudy *et al* [58], the sol–gel methodology was applied to fabricate ultrafine nanospherical composites of CeO_2/TiO_2. The introduction of 1% CeO_2 augmented the specific surface area and enhanced visible-light absorption, thereby diminishing the optical bandgap of TiO_2. The heterojunctions formed in the 4% CeO_2/TiO_2 nanocomposite substantially mitigated electron–hole pair recombination, thereby increasing photocatalytic efficiency with a tetracycline (TC) removal efficiency of ~99% after 80 min of sunlight exposure. The enhanced performance was attributed to reduced particle size, increased particle surface area, heightened absorption of visible light, reduced bandgap, and prolongation of the lifespan of photogenerated charge carriers facilitated by the formation of a heterointerface. Likewise, in a study by Bai *et al* [59], a TiO_2/ZnO nanocomposite was synthesized utilizing sol–gel methodologies. The procedural approach involved the synthesis of TiO_2/ZnO nanocomposites, as shown in figure 4.8. They diluted $Cl_6H_{36}O_4Ti$ in ethanol and added deionized water and acetic acid sequentially with constant agitation to generate a sol, which, when heated to 50 °C, transformed into a gel. The precursor was dried at 80 °C and calcinated at 550 °C for 3 h to obtain TiO_2. CH_4N_2O, $Zn (CH_3COO)_2 \cdot 2H_2O$, and the prepared TiO_2 were mixed, dissolved in deionized water, heated to 120 °C for 6 h, centrifuged, and finally calcinated at 500 °C for 2 h to yield TiO_2/ZnO. The nanocomposite displayed a decreased bandgap width of 2.89 eV and featured a surface area of 15.28 m^2 g^{-1}. Its photocatalytic degradation rate of MB was remarkable, reaching ~100% at 80 min; the first-order kinetic equation was followed, characterized by a kinetic constant of 0.037 48 min^{-1}. Notably, this the magnitude of this constant was 2.8 times higher than that of pure ZnO and 4.8 times higher than that of pure TiO_2.

Akhter *et al* [60] conducted an investigation involving the synthesis of a ZnO/TiO_2 nanocomposite through the application of the sol–gel method. The

Fig. 1. SEM images (a), (b), (c) TiO₂ nanoplates (d), (e), (f) ZnO nanoparticles (g), (h), and (i) ZnO/TiO₂ nanocomposites.

Figure 4.9. Scanning electron microscopy (SEM) images (a), (b), and (c) show TiO_2 nanoplates, while (d), (e), and (f) represent ZnO nanoparticles. Images (g), (h), and (i) illustrate the ZnO/TiO_2 composite. Reprinted from [60], Copyright (2023), with permission from Elsevier.

methodology involved the mixture of 3 g of ZnO with 10 ml of ethanol and 5 ml of distilled water, which were stirred for 15 min. Simultaneously, a mixture of 3 g of TiO_2, 10 ml of ethanol, and 5 ml of distilled water was stirred for 10 min. Five ml of concentrated HCl was mixed into this TiO_2 solution, stirred for 15 min, and gradually introduced into the ZnO solution while maintaining a pH of more than 1.5. The resultant mixture was stirred for 3 h at ambient temperature, followed by exposure to a hot water bath at 100 °C for 1 h and subsequent cooling. After 30 min of centrifugation, the nanoparticles were filtered, oven dried, dehydrated, and then subjected to calcination in a muffle furnace at 400 °C for 3 h. The synthesized ZnO/ TiO_2 nanocomposite exhibited a spherical morphology characterized by diameters ranging from 300 to 500 nm. The scanning electron microscope (SEM) images depicted in figure 4.9 elucidate the surface morphologies of the synthesized nano-particles and nanocomposites. The photocatalytic performance of ZnO/TiO_2 surpassed that of the pristine photocatalysts, resulting in a 98% degradation of MB dye, and a kinetic analysis of the photocatalysts indicated the presence of pseudo first-order kinetics.

The sol–gel methodology is commonly employed in the preparation of nano-composite photocatalysts intended for the treatment of wastewater; nevertheless, it encounters numerous challenges. The process is complex and requires critical control of the pH, temperature, and precursor concentrations. This can influence the uniformity of the ultimate product. Achieving nanoparticle homogeneity is also difficult; lack of homogeneity results in uneven distribution and reduced photo-catalytic activity. The material's pore structure and surface area may not conform to ideal specifications, which can impact reactant accessibility. The drying and gelation process can cause cracking, reducing the structural integrity of the material. Additionally, scaling up for industrial production is a challenge, and high-temperature processing is energy-intensive and thus raises concerns. Chemical residues and the high costs attributed to expensive precursors are potential drawbacks. Material compatibility issues further limit the method's applicability, despite its advantages in tailoring material properties. Ongoing research is being conducted to address these limitations.

In conclusion, the synthesis of nanocomposite photocatalysts is a dynamic and evolving field that is crucial for advancing sustainable technologies. Scientists utilize various methodologies, including sol–gel, precipitation, hydrothermal, and sono-chemical processes, each of which offers unique advantages. The sol–gel method provides precise control over composition, precipitation methods offer scalability, hydrothermal synthesis produces crystalline structures, and sonochemical methods enable rapid synthesis. These techniques, when combined, result in synergistic effects, enhancing the catalytic activity, stability, and selectivity of materials. Despite the progress made to date, ongoing research is essential to refine methods, address challenges, and explore novel possibilities. Collaborative efforts within the scientific community will drive the field forward, deepening our understanding of nanocomposite photocatalyst synthesis. Realizing the complete capabilities of this method will extend its applications in environmental remediation, energy conver-sion, and other domains, ensuring a more sustainable and promising future.

4.2.3 Properties of nanocomposite photocatalysts

Nanocomposite photocatalysts exhibit remarkable properties across multiple domains. Mechanically, their nanoscale structure provides increased surface area, enhancing mechanical strength and durability. Electrically, these nanocomposites often demonstrate improved conductivity, facilitating efficient charge transfer during photocatalytic reactions. Optically, the incorporation of nanoparticles enhances light absorption, leading to enhanced photocatalytic activity. Thermally, the nanocomposite structure promotes better heat dissipation, preventing thermal degradation and ensuring sustained performance. In summary, photocatalyst nano-composites combine enhanced mechanical strength, improved electrical conductiv-ity, superior optical absorption, and efficient thermal management, making them highly effective in various applications, particularly in harnessing light for catalytic processes.

4.2.3.1 Mechanical properties
Nanocomposite photocatalysts exhibit enhanced mechanical properties, primarily attributed to the strong interfaces formed between components, which impede crack propagation. The nanoscale size increases surface area, facilitating improved load transfer and adhesion. Synergistic effects obtained by combining materials with distinct mechanical properties enhance the overall performance [73]. Surface morphology, which influences mechanical interlocking and energy dissipation, plays a pivotal role. The material properties of individual components significantly impact nanocomposites; examples include TiO_2-SiO_2 [74] with enhanced tensile strength due to strong interfaces and CNT–polymer [75] nanocomposites exhibiting

heightened tensile strength and wear resistance. Graphene-based nanocomposites showcase exceptional mechanical attributes, due to graphene's unique 2D structure and strong covalent bonds. Applications encompass filtration membranes, structural materials, protective coatings, reinforcing agents, and energy-harvesting devices, exploiting the nanocomposites' strength, durability, and flexibility [76]. The evolving research in this field holds substantial promise, anticipating advancements. Advancements can be made by enhancing our comprehension of the underlying mechanisms and refining specialized synthesis techniques. This research stands to revolutionize industries by offering materials with unprecedented combinations of photocatalytic activity and mechanical strength.

Research into nanocomposite photocatalysts' mechanical properties is rapidly advancing. Understanding enhancements and innovating methods of synthesis can yield tailored performance, revolutionizing industries with unprecedented photocatalytic activity and mechanical strength combinations.

4.2.3.2 Electrical properties

Nanocomposite photocatalysts with unique electrical properties play a pivotal role in pollutant degradation. Key aspects include enhanced conductivity, facilitated charge separation, and transfer through components such as metal nanoparticles, leading to minimized recombination and improved efficiency [76]. Heterojunctions within the nanocomposite generate internal electric fields, directing photogenerated electrons and holes toward the surface, enhancing pollutant degradation [77]. Plasmonic effects introduced by noble-metal nanoparticles enhance light absorption and electron excitation, improving photocatalytic activity. Tuning the nanocomposite bandgap enables tailored photocatalytic activity for specific light wavelengths. Examples include TiO_2-graphene nanocomposites with improved conductivity and Ag-TiO_2 nanocomposites that display enhanced photocatalytic activity [78].

Ongoing research is focused on developing nanocomposites with optimized electrical properties for specific applications, exploring new materials and synthesis techniques, and integrating nanocomposites with other technologies for advanced environmental remediation and sustainable energy production.

4.2.3.3 Thermal properties

The effectiveness and durability of nanocomposite photocatalysts for pollutant degradation hinge on their thermal properties, influencing aspects such as light absorption and charge separation. Notably, high thermal conductivity prevents overheating during the photocatalytic process, ensuring sustained activity. Thermal stability is vital for longevity in high-temperature environments, while low specific heat capacity enables rapid heating and cooling [79]. The nanocomposites' surface area aids efficient heat transfer, averting localized overheating. These characteristics contribute to heightened light absorption, enhanced charge separation, elevated reaction rates, and improved recyclability. Examples include TiO_2-based nanocomposites for wastewater treatment, carbon-based nanocomposites for high-intensity light applications, and Bi_2O_3-based nanocomposites for improved activity at higher temperatures. Challenges include balancing activity with stability and minimizing the environmental impact of high-temperature processes. Overall, manipulating thermal properties is pivotal for developing efficient, durable, and environmentally friendly photocatalysts to combat pollution.

4.2.3.4 Optical properties

The optical properties of nanocomposite photocatalysts play a crucial role in determining their ability to degrade pollutants. These properties govern the nanocomposite's interaction with light and its capacity to produce pairs of electrons and holes, which is essential for the degradation process. The bandgap, which denotes the energy difference between the valence and conduction bands, plays a pivotal role in influencing light absorption, particularly within the visible spectrum, in turn enhancing pollutant degradation. The absorption spectrum, a measure of light absorption at various wavelengths, ensures efficient competition for incident light between the photocatalyst and pollutant [80]. Light scattering within the nanocomposite, while extending the light path length for increased interaction, must remain balanced to maintain overall efficiency. Surface plasmon resonance (SPR) in metal nanoparticles enhances light absorption, benefiting pollutant degradation by elevating electron–hole pair generation. Photoluminescence can be used to assess recombination rates; lower intensity signifies decreased recombination, which is advantageous for pollutant degradation [81]. The tailoring of these optical properties is exemplified in TiO_2-based nanocomposites, where doping narrows the bandgap for visible light absorption, and in metal–semiconductor nanocomposites, where it leverages SPR effects. In addition, graphene-based nanocomposites that have superior conductivity and extensive surface areas enhance light absorption and charge separation, contributing to improved pollutant degradation. This understanding of and control over optical properties offer promising avenues for developing efficient technologies in environmental remediation and sustainable development.

4.3 The fundamentals of photocatalysis

4.3.1 Photocatalysis

The enormous use of chemicals and their entry into bodies of air, soil, and water is leading to an alarming situation. Chemicals from different categories such as dyes, pharmaceuticals, insecticides, pesticides, and volatile organic compounds threaten the environment, even in trace amounts. Emerging contaminants, persistent organic pollutants, and endocrine-disrupting chemicals all have one thing in common, namely their potential to harm life. Across the globe, the problems associated with pollutant degradation are serious. Hence, in light of these considerations, efforts are being made to formulate effective methodologies for the degradation of these pollutants. Photocatalysis is an advanced oxidation technique that successfully degrades pollutants of organic origin. Advanced oxidation methods work by generating hydroxyl radicals and superoxide anions. Amongst the various advanced oxidation methods, such as the Fenton method, the photo-Fenton method, UV photolysis, ozonation, photocatalysis, and sonocatalysis, photocatalysis is widely accepted due to its simplicity and efficiency of operation [82].

Sunlight is a fundamental form of energy responsible for supporting many reactions which are essential for the existence of life on earth [83]. Harvesting the Sun's energy for reactions is the basis of visible-light photocatalysis. UV photocatalysis and near-infrared (NIR) photocatalysis have also been demonstrated by many researchers. Typically, photocatalysis is a photon-induced process with a wide range of applications in the fields of water treatment, light harvesting, hydrogen generation, batteries, self-cleaning materials, soil remediation, sensors, and the synthesis of novel materials. The treatment of waterborne organic pollutants using visible-light active photocatalysts has received considerable focus due to the efficacy and environmental sustainability of photocatalytic operations. The other photocatalysis operations that use UV- or NIR-activated materials have also been found to be efficient in their respective applications [84]. In a catalytic process, a material showing selective response for a particular reaction is chosen as a catalyst, and thus the process is called a catalytic process. Many industrial processes are catalyst based [85]. Photocatalysis is similar to these processes in terms of the selective action of the photoactive material in carrying out the desired process. As a competent approach for treating pollutants, photocatalysis was first applied few decades ago, and using the new nanomaterials created by various novel approaches, the field has expanded ever since [86].

Today, photocatalysis is applied for the treatment of many emerging organic contaminants and persistent pollutants [87]. Other methods involving light such as photolysis [88], the photo-Fenton method [89], photochemical oxidation [90] and photoelectrocatalysis [91] are also applied for the treatment of contaminants. Photocatalysis and photolysis are often studied in parallel in experiments. It is crucial to highlight that photocatalysis involves the utilization of a light-activated material, whereas photolysis simply involves a reaction that takes place in the presence of light. Techniques such as photoelectrocatalysis use an electrochemical cell, whereas photocatalysis does not require an electrochemical cell. Photo-Fenton oxidation involves the oxidation of organic material by Fenton's reagent. In this method, sludge formation is observed. In photochemical oxidation, an oxidant is required along with a light source. Compared to these light-based methods, there are many benefits of photocatalysis, such as the optimum use of chemicals, efficiency of degradation, the ability to harvest light energy, an absence of secondary pollution, and the inclusion of novel materials. Figure 4.10 is a schematic of photon-based methods applied for the degradation of pollutants.

In a photocatalytic process, a photon initiates an interaction with the photocatalyst, inducing electron migration from the conduction band to the valence band. This electron migration is followed by the creation of a hole. Thus, an electron–hole pair is created during photocatalysis. These electrons and holes lead to the formation of species that can degrade pollutants. Although the photon triggers the photocatalytic mechanism, the photocatalyst produces the desired outcome, which depends on the application of the photocatalyst [92].

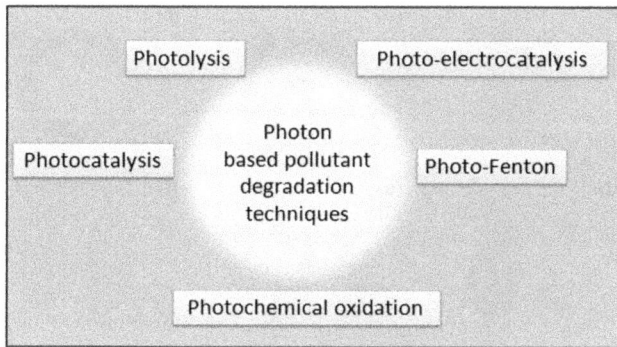

Figure 4.10. Photon-based pollutant degradation techniques

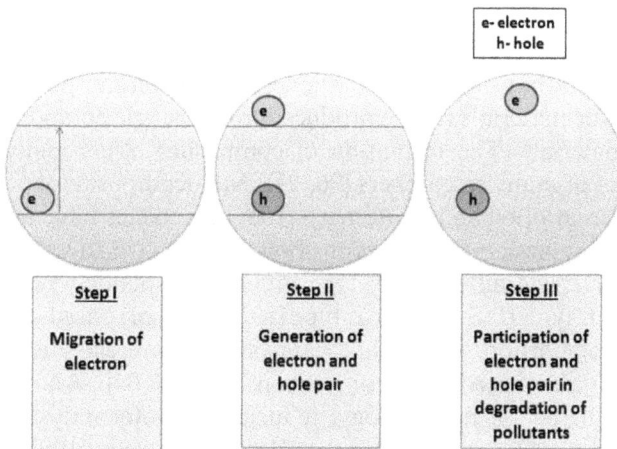

Figure 4.11. A schematic outlining the key stages in the photocatalytic degradation of pollutants.

4.3.2 What are photocatalysts?

Photocatalysts are essentially materials that are able to carry out photocatalysis. This means they are capable of producing electron–hole pairs upon exposure to photons. The three primary stages of photocatalysis are the migration of electrons from the valence band to the conduction band, the generation of electron–hole pairs, and the subsequent participation of electron–hole pairs in pollutant degradation. These steps are shown in figure 4.11.

The research literature describes different types of photocatalysts that are being applied for pollutant degradation. Thus, depending on the light source, photocatalysts may be active under UV, visible, or NIR light [93]. Photocatalysts are based on semiconductor or metal nanomaterials or a combination of these. Figure 4.12 shows the different forms of metal–semiconductor nanocomposites, such as conventional, core–shell, yolk–shell, Janus type, array type, and multi-junction structures [94].

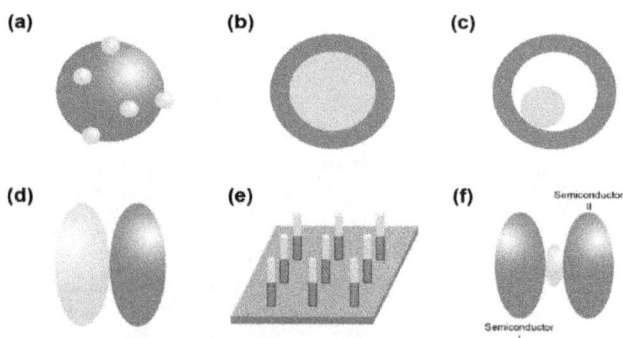

Figure 4.12. The different forms of metal–semiconductor nanocomposites: (a) the conventional structure, (b) the core–shell structure, (c) the yolk–shell structure, (d) the Janus structure, (e) the array structure, and (f) the multijunction structure. Reprinted from [94]. CC BY 4.0.

Photocatalysts have been reported to be synthesized in nanoparticle form from materials such as carbon, metals, ceramics, semiconductors, polymers, and lipids [95]. Photocatalytic materials can be produced from a single component or they may be composite materials. The formation of composites offers many merits and is therefore the aim of many researchers [96, 97]. Nanocomposite photocatalysts such as binary [98], ternary [99] and quaternary [100] composites have been synthesized. The type of heterojunction formed in a photocatalyst is also used as basis for its categorization. Depending on the type of heterojunction, photocatalysts are classified as type I, type II, S type, or Z type [101]. Further, based on the constituent forming the major portion of the photocatalyst, they are classified as metal oxide based [102], polymer based [103] or carbon based [104]. Another category of photocatalyst is based on morphology; it includes photocatalysts in the form of sheets[105], spheres [106], rods[107], cubes [108], or flowers [109]. Other morphological forms have been observed which are considered to be important for categorizing photocatalysts. Thus, there is a wide variety of photocatalytic materials. The synthesis of novel materials is improving the performance of photocatalysts in various fields of application [110].

4.3.3 Desirable properties of photocatalysts

Photocatalysts are expected to possess certain properties related to charge transfer, activity, recyclability, and efficiency, as these properties influence the performance of photocatalysts to a great extent. Photocatalytic materials possess the ability to interact with photons, facilitating the generation of electron–hole pairs. A photocatalyst's ability to interact with photons is pivotal in determining its photocatalytic behavior. The photogenerated charges further react with pollutant species, thereby degrading them. A photocatalyst is thus required to be efficient in generating electrons and holes on exposure to photons. This interaction creates photogenerated charges and is hence given a lot of attention. Once applied for degradation purposes, photocatalysts may suffer from the problem of recombination. The goal of designing a photocatalyst is to prevent the recombination of electrons and holes. Such designs involve incorporating components that enhance the photocatalytic efficacy of the nanocomposites, which is achieved through optimization methodologies [111–114].

The improvement of photocatalytic activity is also related to structural properties. Researchers have demonstrated that structures with large surface areas [115] and small particle sizes [116] are quite useful for enhancing photocatalytic activity. Photocatalytic efficacy is observed to be influenced by morphological alterations [117]. In particular, the size of the nanoparticles and their surface area have been found to affect photocatalytic operation. In this regard, the synthesis routes adopted for making photocatalysts play a crucial role. In fact, many studies have reported variations in photocatalytic performance due to variations in process parameters [118, 119]. Photocatalytic activity is also dependent on the synthesis route. For instance, compared to conventional methods, the ultrasound-assisted method leads to a more uniform distribution of nanomaterials [120]. Photocatalysts need to be stable and strong until they reach the end of their operational life. Corrosion or wear of photocatalysts reduces their efficiency of operation. Therefore, photocatalytic operations must check the recyclability of the catalyst. It is desirable for catalysts to remain stable over the maximum number of runs [121].

Among photocatalysts, TiO_2-based catalysts are extensively used due to their efficiency in photocatalysis and their nontoxic nature. The principal limitation of TiO_2 photocatalysts is their efficacy, which is maximized under UV illumination. Efforts have been made to prepare visible-light active photocatalyst materials, as this helps in utilizing the energy of sunlight or visible light [111]. An example of this approach is to layer the photocatalyst with graphene-based material. Such nanocomposites have been found to exhibit enhanced photocatalytic properties. Graphene is known to have excellent physical, electrical, mechanical, and thermal properties, as represented in figure 4.13 [111].

In conclusion, properties such as charge carrier mobility, strength, efficiency of operation, structural properties, activity in visible light, and recyclability are the pillars of photocatalyst performance. Due to these properties, photocatalysts display

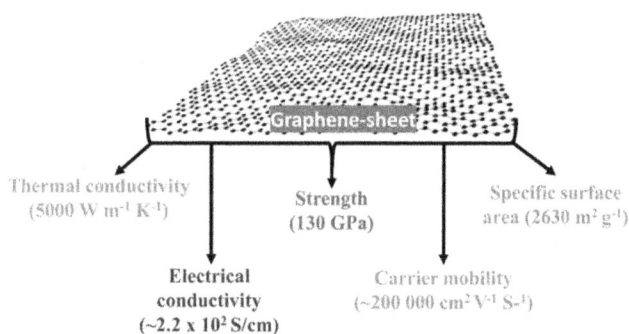

Figure 4.13. The sheet-like structure of graphene and its properties. Reproduced from [17] with permission from The Royal Society of Chemistry.

good performance in their respective applications. The benefit of photocatalytic applications is the strong degradation pathways created by room-temperature redox reactions, which are favored by photocatalysts with suitable properties.

4.3.4 How do photocatalysts work?

As mentioned in the previous section, photocatalysts generate an electron–hole pair upon interacting with a photon. The reason for the formation of this electron–hole pair is the bandgap of a material. In metals, the valence and conduction bands overlap, whereas in insulators they are separated by a large distance. Only semiconductors have a bandgap suitable for photocatalysis. In nanomaterials in which the bandgap does not fall in a range that meets the definition of a semiconductor, photocatalysis is ineffective. Therefore, the bandgap is modified by the creation of suitable composites to improve the performance of nanocomposite photocatalysts. In figure 4.14, the bandgap values of various semiconductor nano-materials are shown. It should be noted that the performance of a photocatalyst depends on the bandgap values. Photocatalysts are active in the UV or visible range, depending on these values.

For photocatalytic degradation, we are interested in the reduction and oxidation of a pollutant. Hence, a study is necessary to perform photocatalysis via these redox reactions. Four cases are possible, based on the redox potential value: (i) at a redox potential value that is below the conduction band edge of a semiconductor, reductive reactions are possible; (ii) if the redox potential of a substrate surpasses the energy level of the valence band edge of the photocatalyst, then oxidative reactions are possible; (iii) if the redox potential of the substrate is greater than that of the conduction band or lower than that of the valence band, no reaction is expected; (iv) conversely, when the redox potential of the substrate falls below the conduction band edge yet surpasses that of the valence band, either oxidative or reductive reactions may transpire [94]. Redox reactions create various pathways for the degradation of a substrate and determine the efficiency of the photocatalytic process. To understand these paths, a photocatalytic mechanism is proposed for the degradation process.

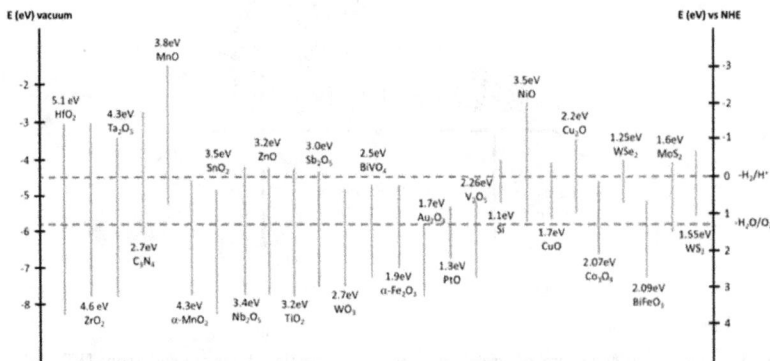

Figure 4.14. Semiconductor material bandgap values. Reprinted from [122]. CC BY 4.0.

4.3.5 Mechanisms of photocatalysis

The photocatalytic mechanism explains the stepwise pathway for the degradation of a substrate or pollutant. In general, photocatalysis involves the generation of electrons and holes which subsequently react to form other species. In particular, species with high oxidation potential values are important in photocatalysis. These powerful oxidants and their active participation are responsible for the success of photocatalysis. These oxidants mineralize pollutants into carbon dioxide and water [122]. From this study, it is clear that several species form during photocatalysis, such as electrons, holes, oxygen, oxygen anions, superoxide anions, hydroxyl radicals, and degradation products. The degradation mechanism begins with a semiconductor photocatalyst and light. It ends with degradation products. The photocatalytic mechanism thus presents the complete degradation of a pollutant in a stepwise manner, including the formation of intermediate species during the reactions.

4.3.6 Factors influencing photocatalytic activity and efficiency

In photocatalysis, photons interact with a photocatalyst, giving rise to charge carriers. The overall photocatalytic performance is therefore dependent on factors related to the source of radiation, the nature of the photocatalyst, and reactions occurring at the photocatalyst surface. As an AOP that involves light, photocatalysis is affected by many factors. The effectiveness of photocatalytic degradation for organic contaminants is contingent upon various factors, including the nature of the catalyst material, the light source, the photocatalyst's stability, the pH of the reaction, the temperature of the reaction, the surface area of the photocatalyst, and the surface morphology of the photocatalyst. These are the major factors affecting photocatalysis. As photocatalysis is a light-based operation, the light source influences photocatalysis to a great extent.

4.3.6.1 The catalyst material

Photocatalyst nanocomposites degrade pollutant species by generating electron–hole pairs. The natures and proportions of the individual component materials in a nanocomposite affect its performance. A component is added to a photocatalyst to form a nanocomposite using a doping technique. When a certain quantity of additive is included in a nanocomposite, it restricts the recombination of photo-generated charge carriers and thus improves photocatalytic operation. The selection of the dopant is a key factor in nanocomposite design. The dopant must be added in the required amount to obtain satisfactory performance from a photocatalyst. If any component material, especially the dopant, is present in excess, it does not improve photocatalytic performance [123].

Rini *et al* [109] prepared a nanocomposite photocatalyst by synthesizing zinc oxide and introducing silver doping into the zinc oxide matrix. The doping concentration of Ag was varied as follows: 0%, 2%, 4%, and 6%. An observed correlation indicated that the incremental incorporation of silver (Ag) as a dopant led to a reduction in particle size. A comparative analysis with pristine zinc oxide revealed that the Ag/ZnO nanocomposite exhibited heightened photocatalytic activity. However, at an Ag concentration of 6% in the nanocomposite, light was prevented from reaching the surface, and hence reduced photocatalytic activity was observed. This demonstrates that the optimum amount of dopant is necessary to obtain the best performance from the photocatalyst [109]. The amount of photo-catalyst added to the reaction mixture also determines the photocatalytic perform-ance. The optimum amount is decided by performing experiments with different amounts of photocatalyst. Any increase in the amount beyond the optimum value increases turbidity and reduces light penetration. Hence, it is preferable to use the optimum amount of photocatalyst [124].

4.3.6.2 The light source

The photocatalyst can be activated by a UV source, a simulated sunlight lamp, or sunlight. The photocatalytic activity is influenced by these sources. When experi-ments are carried out under natural light, variation due to the light is observed. Usually, the highest degradation efficiencies are observed when light intensity is at a maximum and vice versa. In order to overcome the problem of low-intensity operation, artificial light or simulated light is preferred [125]. Depending on the light intensity used, the photocatalytic degradation performance varies [126]. Lal *et al* [127] prepared Nd-doped TiO_2 and applied it to photodegrade MB using two different sources, namely UV and sunlight. The experiments carried out showed that the degradation efficiency was dependent on the source. The efficiency of Nd-doped TiO_2 was found to be 99.11% in ultraviolet light and 96.42% in solar light [127].

4.3.6.3 The photocatalyst's stability

The loss of the photocatalyst and its stability are major concerns for any photo-catalytic operation. A photocatalyst must remain stable until the end of the operation and for the maximum number of runs. The wear of the photocatalysts can be avoided by designing it to remain stable for the maximum number of runs. In order to recover the photocatalyst and put it back into experimental use, recovery should be easy to perform [128]. In order to ease photocatalyst recovery, a magnetically recoverable photocatalyst has been designed [129]. This reusability test is used to determine the structural integrity of a photocatalyst [112].

4.3.6.4 The reaction conditions (pH)

Parameters related to the reaction mixture, such as temperature and pH, can change the photocatalytic operation. The pH-dependent translocation of pollutants from the reaction medium to the surface of the photocatalyst is observed in specific instances, notably with regard to dyes. In the investigation of the impact of pH on photocatalysis, it is essential to take into account the point of zero charge value [124]. Another important category is the type of reaction mixture, which is decided by the type of pollutant. A given photocatalyst may show different behavior towards different pollutants. For instance, a study designed and applied CdS–Ce/MOF nanocomposites for the photoreduction of chromium and the photocatalytic degradation of crystal violet. The nanocomposite reduced 57% of the crystal violet within a 120 min timeframe, and a 95% decrease in chromium (VI) concentration was achieved within 60 min [130]. This shows that a photocatalyst can exhibit selectivity for the degradation of a pollutant.

4.3.6.5 The surface area and morphology of the photocatalyst
The surface area is one of the characteristic properties of a nanomaterial. It is desirable for a large surface area to be available for reactions. The nanocomposite's surface area thus determines its performance. For a nanocomposite with a large surface area, there are more chances for the pollutant species to reach active sites and pollutant removal is thereby improved. For example, in research carried out by Babu *et al*, it was observed that nanomaterials that had a sheet-like morphology enhanced the photocatalytic performance. Nanosheets were prepared from BiOCl and Bi_2O_3 nanofibers and it was observed that a change in morphology resulted in a change in the photocatalytic performance [117]. In another study, a novel g-C_3N_4/$PrFe_3O_4$ nanocomposite photocatalyst was synthesized. Compared to pure $PrFe_3O_4$, the g-C_3N_4/$PrFeO_3$ nanocomposite showed improved photocatalytic activity, which was attributed to its large surface area [115]. Lellala [131] reported the creation of a mesoporous material with a high surface area composed of sulfurous carbon disc graphene oxide nanosheets. The mesoporous structure was characterized by a substantial surface area of 1050 $m^2\,g^{-1}$ and a pore size of 10 nm. The nanocomposite exhibited remarkable photocatalytic power in the degradation of brilliant yellow dye [131]. However, it is necessary to use the optimum quantities of the nanocomposite components. Only at such optimum levels is the morphology improved and the desired performance obtained [113].

4.4 Nanocomposites for photocatalytic degradation

Photocatalysis was introduced decades ago. The use of TiO_2 as a photocatalyst is widely accepted due to its properties. As a conventional photocatalyst, TiO_2 faces problems due to its inactivity in visible light. This led the scientific community to think in terms of modifying the components of photocatalysts to change their properties and performance. Designing materials while keeping in mind their properties and performance is at the core of these engineered nanocomposites. The main objective of any nanocomposite is thus to enhance performance compared to the performances of the component materials. This has paved a path for many interesting novel nanomaterials to be used for photocatalytic degradation.

Photocatalytic activity is greatly improved by creating such nanocomposites. For example, many researchers have demonstrated that introducing graphene-based material into TiO_2 results in the development of an efficient nanocomposite photocatalyst. In this case of a TiO_2 and graphene-based nanocomposite, the bandgap is changed, making the nanocomposite active in visible light [111]. It is possible to design a network of species, for example, using a metal–organic framework, and use it for the photocatalytic degradation of pollutants [132]. Morphology, as a crucial element in nanocomposite performance, is the focus of many researchers. A change in surface area is observed as a result of incorporating graphene-based components [113]. Introducing the right amount of a component into a nanocomposite affects its photocatalytic performance. This opens the door to the synthesis of nanocomposites by modifying their composition [124]. Sometimes, a layered structure is synthesized to improve its photocatalytic performance. The efficiency of a photocatalyst is impeded by the challenges related to electron–hole recombination. However, by introducing a suitable component, this recombination of electrons and holes can be suppressed. This helps to enhance the photocatalytic properties of the individual materials [112].

Photocatalysts that are applied in powdered form suffer from the problem of losses during each cycle. This problem can be overcome to a great extent by creating nanocomposite photocatalysts which are recoverable at the end of each operation. Such a nanocomposite design involves creating a stable photocatalyst that is easily recovered at the end of each run. It then becomes possible to reuse this photocatalyst in the following run [121]. Photocatalytic behavior can be improved by creating a photocatalyst that increases surface plasmons. The incorporation of metal ions, such as Ag ions, results in this type of surface plasmon effect [133].

Pollutants present in wastewater that are challenging to treat by conventional routes can be efficiently treated using nanocomposite photocatalysts. So far, nanocomposite photocatalysts have been applied for the degradation of dyes, pharmaceutical products[134], insecticides [135, 136], pesticides[137–140], and heavy metals [131, 141, 142]; they have also been used to remediate air [143] and soil [144]. Previous sections have already described that an enhancement in photocatalytic behavior is observed due to the application of nanocomposites. There are many merits of the use of nanocomposites. Nanocomposites that are designed with a tailored approach exhibit improved behavior compared to those of their individual components. The incorporation of graphene-based material into nanocomposites improves charge mobility and hence the photocatalytic performance. Sometimes, surface plasmon resonance is created by the presence of certain materials. Nanocomposites are also used in the design of engineered-bandgap materials. Many nanocomposites can make materials active under visible light. A reduction in recombination is also observed in nanocomposites. It is possible to choose component materials from a vast array of materials and design novel materials by varying the synthesis route.

4.5 Challenges and future perspectives

Conventional methods such as sedimentation, aeration, activated sludge processes, etc. are available for the treatment of effluent. Although their efficiency in the removal of pollutants is questionable, they are widely accepted due to their low cost and established status. Hence, the first challenge in the application of photocatalytic operations is scalability. The research literature describes many photocatalysts that have been synthesized and applied for the degradation of organic pollutants. Some photocatalysts have also been applied to the real industrial effluents. The application of photocatalysts to real industrial effluents is a challenge due to the presence of many contaminants in addition to the target pollutant in real industrial effluents.

For instance, textile industry effluents vary significantly in terms of their pH values, colors, metallic elements, and chemical oxygen demand. Such effluents make it difficult to streamline the treatment method [145]. Photocatalysis is dependent on light as a source of photons. For this reason, visible-light-activated photocatalysis is challenging due to variations in sunlight. The photocatalyst does not function efficiently if the sunlight is of very low intensity. On the other hand, if an artificial light is used, it overcomes the problem of the variability of light but adds to the operational costs. It is important to study the lamp properties such as diameter, length, intensity, current, voltage, and lifespan [121]. There is a need to explore new light sources such as light-emitting diodes for photocatalysis. LEDs have many merits; for example, they are free from mercury, they have longer lifespans and smaller sizes, and are physically robust. Their ability to act as a light source for photocatalysis will make it easier to carry out photocatalytic operations under reduced light and help to reduce the burden of power requirements [146].

The large-scale synthesis of nanocomposites is a major challenge which affects the use of these materials. The synthesis processes that are successful on a lab scale need to be analyzed for their use in large-scale applications [147]. The properties of nanocomposites change with changes in the method. For example, in a comparative study of synthesis methods for nanocomposite ZnO/TiO_2, two methods were used for synthesis, namely the conventional route and the ultrasound-assisted route. From the study, it was clear that the nanocomposite properties were greatly changed by the method. The nanocomposite prepared using the ultrasound method showed improved properties [148]. A method which is able to prepare the nanocomposite in an eco-friendly, cost-effective, and safe way is desirable.

A nanocomposite is a tiny entity with endless possibilities. It is crucial to evaluate the components of a nanocomposite before combining them to form a nanocomposite. The individual behaviors of the components and their behaviors in the composite are important [149]. As different tailor-made nanocomposites are synthesized, it is essential to engineer the bandgaps of these materials. Several sophisticated tools are required to synthesize and confirm the preparation of these materials [150]. A vast array of nanomaterials is available for use as photocatalysts. Most studies have presented the degradation of a pollutant by a nanomaterial. Only a few studies have reported information about the degradation products and their effect on the environment. The risks associated with intermediate products need equal attention. The toxicity profiles of nanomaterials should also be studied, so that no serious harm to the environment takes place if nanoparticles are released. Further, a detailed lifecycle assessment of nanocomposites in a particular application is also essential for their wide acceptance [151, 152].

Waste generation is a problem for the environment. Therefore, efforts are being made to use such materials as starting materials for nanocomposites. One study reported the utilization of fish waste for the preparation of a nanocomposite film. In this study, corn starch, fish scale, turmeric extract, and zinc oxide nanoparticles were used to prepare a nanocomposite film. The film displayed antimicrobial properties [153]. Sheikhy *et al* [154] reported the preparation of nanocomposites from polymer wastes. In their study, waste low-density polyethylene (LDPE), high-density polyethylene (HDPE) and nano clay were processed to produce nano-composites [154]. Fatimah *et al* [155] prepared nanocomposites from iron rust waste using a hydrothermal method. $Fe_3O_4/NiFe_2O_4$ nanocomposites were pre-pared and applied for the photocatalytic degradation of methyl violet dye. The nanocomposites exhibited a degradation efficiency of more than 95% [155]. Such studies present a combined approach to the treatment of waste and the synthesis of the nanocomposite.

It is a challenging task to synthesize a nanomaterial with the desired properties using a scalable approach. In an attempt to solve this issue, Suryawanshi *et al* [156] designed a micro-reactor to produce small platinum nanoparticles capped with polyvinylpyrrolidone at a size of 5 nm. The formation of the nanoparticles was seen to be influenced by the flow of reactants [156].

Recently, Samraj *et al* [132] prepared a copper MFO–zinc tungstate heterojunc-tion and applied it for the sonophotocatalytic degradation of tetracycline. The combined application of photocatalysis and ultrasound produced high degradation results. Such a combined or hybrid approach opens possibilities for the degradation of pollutants which are not efficiently degraded by individual methods [132].

Another interesting approach is to prepare a nanocomposite mixed matrix membrane. For example, in a study, ZIF-8 decorated cellulose acetate membranes were prepared and applied for the degradation of dyes and real textile effluent. The membrane was found to be nearly 85% efficient in removing the dyes crystal violet, acid red 13, and reactive black 5. The membrane was also found to be efficient in removing chemical oxygen demand and total organic carbon, reaching efficiencies close to 70% and 80%, respectively. Heavy metals such as lead, chromium, and cadmium were also removed by the membrane [157]. The integration of nano-composites into membranes represents an effective strategy for the remediation of contaminated water. When a photocatalyst is prepared and applied in a powdered form, its loss in each stage is unavoidable. This problem can be solved by applying an immobilized photocatalyst, designing a photocatalytic membrane, or by synthe-sizing a magnetically active photocatalyst material. Effluents are rich in chemicals, and it is possible to recover this energy. A new approach is to make use of a photocatalytic fuel cell; in this way, photocatalysis and energy generation are achieved simultaneously. Such designs recover energy from wastewater while reducing its pollutant load [158].

4.6 Conclusions

The chapter provided an exhaustive examination of the evolution of nanocomposite photocatalysts in the context of degrading diverse organic pollutants. In addition, this chapter explored contemporary challenges and progresses within the domain of photocatalytic degradation. Photocatalysis is part of the family of AOPs. The successful degradation of contaminants is influenced by many factors, such as the mobility of the charge carriers, the bandgap of the photocatalyst, the process conditions, the irradiation type, and the nature of the pollutant. Therefore, the nanocomposite photocatalyst design process includes a critical study of all these factors. The effective photocatalytic decomposition of pollutants, including dyes, insecticides, pesticides, and pharmaceutical compounds has been highlighted and demonstrated by many researchers, indicating the potential of photocatalysts to degrade these substances. Some researchers have reported the degradation of real industrial effluent by combining photocatalysis with other methods. Such studies present multiple options that can be used in combination with photocatalysis for the degradation of pollutants. One such approach is to combine photocatalysis and membrane technology. Since photocatalysis is a light-dependent operation, topics such as the light source, its intensity, its penetration, and its arrangement deserve attention. A major factor in the large-scale applicability of any method is its cost. Light-efficient operation achieved using sunlight or other inexpensive light sources, such as LEDs, can reduce operational costs. While much emphasis has been given to sunlight-based operations, there is need to explore other light sources as well. In the research literature, it can be seen found that many studies have focussed on various pollutants and expanding the application of photocatalysts to a wide range of industrial contaminants. However, it is necessary to examine the details of the selective action of a photocatalyst for a particular pollutant. Such studies will open up new possibilities and degradation pathways for existing contaminants and contaminants with similar profiles. In the majority of photocatalytic processes, much emphasis has been given to degradation and this limits the application of photocatalysis to degradation alone. However, photocatalysis can degrade pollutants and simultaneously generate energy. Operations that combine photocatalysis and energy generation have many merits. Another interesting approach is to design a photocatalyst from a waste material. The waste material serves as a precursor for the synthesis. Such operations can reduce the cost of synthesis. Considering all the issues discussed in this chapter related to the synthesis of nanocomposite photocatalysts and their applications, we should anticipate the integration of cutting-edge photocatalysts with the latest methodologies in surface chemistry and materials science, leading to a new generation of catalytic entities and nano-engineering which will make photocatalysts cost-effective, scalable, and green.

References

[1] Mukherjee J, Lodh B K, Sharma R, Mahata N, Shah M P, Mandal S, Ghanta S and Bhunia B 2023 Advanced oxidation process for the treatment of industrial wastewater: a review on strategies, mechanisms, bottlenecks and prospects *Chemosphere* **345** 140473

[2] Babu Ponnusami A, Sinha S, Ashokan H, V Paul M, Hariharan S P, Arun J, Gopinath K P, Hoang Le Q and Pugazhendhi A 2023 Advanced oxidation process (AOP) combined biological process for wastewater treatment: a review on advancements, feasibility and practicability of combined techniques *Environ. Res.* **237** 116944

[3] Liu H, Li X, Zhang X, Coulon F and Wang C 2023 Harnessing the power of natural minerals: a comprehensive review of their application as heterogeneous catalysts in advanced oxidation processes for organic pollutant degradation *Chemosphere* **337** 139404

[4] Wang D, Xing Y, Li J, Dong F, Cheng H, He Z, Wang L, Giannakis S, Song S and Ma J 2023 Degradation of odor compounds in drinking water by ozone and ozone-based advanced oxidation processes: a review *ACS ES&T Water* **3** 3452–73

[5] Pattnaik A, Sahu J N, Poonia A K and Ghosh P 2023 Current perspective of nano-engineered metal oxide based photocatalysts in advanced oxidation processes for degradation of organic pollutants in wastewater *Chem. Eng. Res. Des.* **190** 667–86

[6] Navidpour A H, Abbasi S, Li D, Mojiri A and Zhou J L 2023 Investigation of advanced oxidation process in the presence of TiO_2 semiconductor as photocatalyst: property, principle, kinetic analysis, and photocatalytic activity *Catalysts* **13** 232

[7] Amulya M A S, Nagaswarupa H P, Kumar M R A, Ravikumar C R, Prashantha S C and Kusuma K B 2020 Sonochemical synthesis of $NiFe_2O_4$ nanoparticles: characterization and their photocatalytic and electrochemical applications *Appl. Surf. Sci. Adv.* **1** 100023

[8] Zhang Q, Uchaker E, Candelaria S L and Cao G 2013 Nanomaterials for energy conversion and storage *Chem. Soc. Rev.* **42** 3127

[9] Chauhan H A, Rafatullah M, Ahmed Ali K, Siddiqui M R, Khan M A and Alshareef S A 2021 Metal-based nanocomposite materials for efficient photocatalytic degradation of phenanthrene from aqueous solutions *Polymers (Basel)* **13** 2374

[10] Boruah P K, Borthakur P and Das M R 2019 Magnetic metal/metal oxide nanoparticles and nanocomposite materials for water purification *Nanoscale Materials in Water Purification* (Amsterdam: Elsevier) 473–503

[11] Mavrikos A *et al* 2022 Synthesis of Zn/Cu metal ion modified natural palygorskite clay–TiO_2 nanocomposites for the photocatalytic outdoor and indoor air purification *J. Photochem. Photobiol. A Chem.* **423** 113568

[12] Pandey N, Shukla S K and Singh N B 2017 Water purification by polymer nanocomposites: an overview *Nanocomposites* **3** 47–66

[13] Xia C, Li X, Wu Y, Suharti S, Unpaprom Y and Pugazhendhi A 2023 A review on pollutants remediation competence of nanocomposites on contaminated water *Environ. Res.* **222** 115318

[14] Kuvarega A T and Mamba B B 2017 TiO_2-based photocatalysis: toward visible light-responsive photocatalysts through doping and fabrication of carbon-based nanocomposites *Crit. Rev. Solid State Mater. Sci.* **42** 295–346

[15] Gopannagari M, Kumar D P, Park H, Kim E H, Bhavani P, Reddy D A and Kim T K 2018 Influence of surface-functionalized multi-walled carbon nanotubes on CdS nano-hybrids for effective photocatalytic hydrogen production *Appl. Catal.* B **236** 294–303

[16] Bathla A, Vikrant K, Kukkar D and Kim K-H 2022 Photocatalytic degradation of gaseous benzene using metal oxide nanocomposites *Adv. Colloid Interface Sci.* **305** 102696

[17] Khan M E 2021 State-of-the-art developments in carbon-based metal nanocomposites as a catalyst: photocatalysis *Nanoscale Adv.* **3** 1887–900

[18] Akakuru O U, Iqbal Z M and Wu A 2020 TiO2 Nanoparticles: Properties and Applications *TiO2 Nanoparticles: Applications in Nanobiotechnology and Nanomedicine* (New York: Wiley) 1 1–66

[19] Xu F 2018 Review of analytical studies on TiO_2 nanoparticles and particle aggregation, coagulation, flocculation, sedimentation, stabilization *Chemosphere* **212** 662–77

[20] Upadhyay G K, Rajput J K, Pathak T K, Kumar V and Purohit L P 2019 Synthesis of ZnO:TiO$_2$ nanocomposites for photocatalyst application in visible light *Vacuum* **160** 154–63

[21] Bethi B, Sonawane S H, Rohit G S, Holkar C R, Pinjari D V, Bhanvase B A and Pandit A B 2016 Investigation of TiO$_2$ photocatalyst performance for decolorization in the presence of hydrodynamic cavitation as hybrid AOP *Ultrason. Sonochem.* **28** 150–60

[22] Chand P and Singh V 2020 Enhanced visible-light photocatalytic activity of samarium-doped zinc oxide nanostructures *J. Rare Earths* **38** 29–38

[23] Qutub N, Singh P, Sabir S, Sagadevan S and Oh W-C 2022 Enhanced photocatalytic degradation of acid blue dye using CdS/TiO$_2$ nanocomposite *Sci. Rep.* **12** 5759

[24] Karimi-Maleh H, Kumar B G, Rajendran S, Qin J, Vadivel S, Durgalakshmi D, Gracia F, Soto-Moscoso M, Orooji Y and Karimi F 2020 Tuning of metal oxides photocatalytic performance using Ag nanoparticles integration *J. Mol. Liq.* **314** 113588

[25] Azizi-Lalabadi M, Alizadeh-Sani M, Divband B, Ehsani A and McClements D J 2020 Nanocomposite films consisting of functional nanoparticles (TiO$_2$ and ZnO) embedded in 4A-Zeolite and mixed polymer matrices (gelatin and polyvinyl alcohol) *Food Res. Int.* **137** 109716

[26] Malesic-Eleftheriadou N, Evgenidou E, Kyzas G Z, Bikiaris D N and Lambropoulou D A 2019 Removal of antibiotics in aqueous media by using new synthesized bio-based poly (ethylene terephthalate)–TiO$_2$ photocatalysts *Chemosphere* **234** 746–55

[27] Feizpoor S, Habibi-Yangjeh A, Yubuta K and Vadivel S 2019 Fabrication of TiO$_2$/CoMoO$_4$/PANI nanocomposites with enhanced photocatalytic performances for removal of organic and inorganic pollutants under visible light *Mater. Chem. Phys.* **224** 10–21

[28] Faisal M, Jalalah M, Harraz F A, El-Toni A M, Labis J P and Al-Assiri M S 2021 A novel Ag/PANI/ZnTiO$_3$ ternary nanocomposite as a highly efficient visible-light-driven photo-catalyst *Sep. Purif. Technol.* **256** 117847

[29] Kumar A, Raorane C J, Syed A, Bahkali A H, Elgorban A M, Raj V and Kim S C 2023 Synthesis of TiO$_2$, TiO$_2$/PAni, TiO$_2$/PAni/GO nanocomposites and photodegradation of anionic dyes Rose Bengal and thymol blue in visible light *Environ. Res.* **216** 114741

[30] Munyengabe A, Ndibewu P P, Sibali L L and Ngobeni P 2022 Polymeric nanocomposite materials for photocatalytic detoxification of polycyclic aromatic hydrocarbons in aquatic environments-a review *Res. Eng.* **15** 100530

[31] Ningthoujam R, Singh Y D, Babu P J, Tirkey A, Pradhan S and Sarma M 2022 Nanocatalyst in remediating environmental pollutants *Chem. Phys. Impact* **4** 100064

[32] Khakhal H R, Kumar S, Patidar D, Kumar S, Vats V S, Dalela B, Alvi P A, Leel N S and Dalela S 2023 Correlation of oxygen defects, oxide-ion conductivity and dielectric relaxation to electronic structure and room temperature ferromagnetic properties of Yb^{3+} doped CeO$_2$ nanoparticles *Mater. Sci. Eng.: B* **297** 116675

[33] Bakkiyaraj R, Bharath G, Hasini Ramsait K, Abdel-Wahab A, Alsharaeh E H, Chen S-M and Balakrishnan M 2016 Solution combustion synthesis and physico-chemical properties of ultrafine CeO$_2$ nanoparticles and their photocatalytic activity *RSC Adv.* **6** 51238–45

[34] Zinatloo-Ajabshir S, Heidari-Asil S A and Salavati-Niasari M 2021 Simple and eco-friendly synthesis of recoverable zinc cobalt oxide-based ceramic nanostructure as high-performance photocatalyst for enhanced photocatalytic removal of organic contamination under solar light *Sep. Purif. Technol.* **267** 118667

[35] Mahdavi K, Zinatloo-Ajabshir S, Yousif Q A and Salavati-Niasari M 2022 Enhanced photocatalytic degradation of toxic contaminants using Dy_2O_3–SiO_2 ceramic nanostructured materials fabricated by a new, simple and rapid sonochemical approach *Ultrason. Sonochem.* **82** 105892

[36] Kumar K, Kumar R, Kaushal S, Thakur N, Umar A, Akbar S, Ibrahim A A and Baskoutas S 2023 Biomass waste-derived carbon materials for sustainable remediation of polluted environment: a comprehensive review *Chemosphere* **345** 140419

[37] Ahmed M K, Shalan A E, Afifi M, El-Desoky M M and Lanceros-Méndez S 2021 Silver-doped cadmium selenide/graphene oxide-filled cellulose acetate nanocomposites for photocatalytic degradation of malachite green toward wastewater treatment *ACS Omega* **6** 23129–38

[38] Aragaw B A and Dagnaw A 1970 Copper/reduced graphene oxide nanocomposite for high performance photocatalytic methylene blue dye degradation *Ethiop. J. Sci. Technol.* **12** 125–37

[39] Beura R, Pachaiappan R and Paramasivam T 2021 Photocatalytic degradation studies of organic dyes over novel Ag-loaded ZnO-graphene hybrid nanocomposites *J. Phys. Chem. Solids* **148** 109689

[40] Gu Y, Xing M and Zhang J 2014 Synthesis and photocatalytic activity of graphene based doped TiO_2 nanocomposites *Appl. Surf. Sci.* **319** 8–15

[41] Li C, Wang B, Zhang F, Song N, Liu G, Wang C and Zhong S 2020 Performance of Ag/BiOBr/GO composite photocatalyst for visible-light-driven dye pollutants degradation *J. Mater. Res. Technol.* **9** 610–21

[42] Fulzele N N, Bhanvase B A and Pandharipande S L 2022 Sonochemically prepared rGO/Ag_3PO_4/CeO_2 nanocomposite photocatalyst for effective visible light photocatalytic degradation of methylene dye and its prediction with ANN modeling *Mater. Chem. Phys.* **292** 126809

[43] Ardani M R, Pang A L, Pal U, Zheng R, Arsad A, Hamzah A A and Ahmadipour M 2022 Ultrasonic-assisted polyaniline-multiwall carbon nanotube photocatalyst for efficient photodegradation of organic pollutants *J. Water Process. Eng.* **46** 102557

[44] Sadeghi Rad T, Khataee A, Sadeghi Rad S, Arefi-Oskoui S, Gengec E, Kobya M and Yoon Y 2022 Zinc-chromium layered double hydroxides anchored on carbon nanotube and biochar for ultrasound-assisted photocatalysis of rifampicin *Ultrason. Sonochem.* **82** 105875

[45] Zhang X, Wang Q, Zou L-H and You J-W 2016 Facile fabrication of titanium dioxide/fullerene nanocomposite and its enhanced visible photocatalytic activity *J. Colloid Interface Sci.* **466** 56–61

[46] Abid N, Khan A M, Shujait S, Chaudhary K, Ikram M, Imran M, Haider J, Khan M, Khan Q and Maqbool M 2022 Synthesis of nanomaterials using various top-down and bottom-up approaches, influencing factors, advantages, and disadvantages: a review *Adv. Colloid Interface Sci.* **300** 102597

[47] Thambiliyagodage C, Mirihana S, Wijesekera R, Madusanka D S, Kandanapitiye M and Bakker M 2021 Fabrication of Fe_2TiO_5/TiO_2 binary nanocomposite from natural ilmenite and their photocatalytic activity under solar energy *Curr. Res. Green Sustain. Chem.* **4** 100156

[48] Pathak D, Sharma A, Sharma D P and Kumar V 2023 A review on electrospun nanofibers for photocatalysis: upcoming technology for energy and environmental remediation applications *Appl. Surf. Sci. Adv.* **18** 100471

[49] Wu C 2014 Synthesis of Ag 2 CO 3 /ZnO nanocomposite with visible light-driven photocatalytic activity *Mater. Lett.* **136** 262–4

[50] Haghighatzadeh A, Hosseini M, Mazinani B and Shokouhimehr M 2019 Improved photocatalytic activity of ZnO-TiO 2 nanocomposite catalysts by modulating TiO 2 thickness *Mater. Res. Express* **6** 115060

[51] Haounati R *et al* 2021 Design of direct Z-scheme superb magnetic nanocomposite photocatalyst Fe_3O_4/Ag_3PO_4@Sep for hazardous dye degradation *Sep. Purif. Technol.* **277** 119399

[52] Kumar M, Singh R, Khajuria H and Sheikh H N 2017 Facile hydrothermal synthesis of nanocomposites of nitrogen doped graphene with metal molybdates (NG-MMoO$_4$) (M = Mn, Co, and Ni) for enhanced photodegradation of methylene blue *J. Mater. Sci., Mater. Electron.* **28** 9423–34

[53] Younas U, Ahmad A, Islam A, Ali F, Pervaiz M, Saleem A, Waseem M, Muteb Aljuwayid A, Habila M A and Raza Naqvi S 2023 Fabrication of a novel nanocomposite (TiO$_2$/WO$_3$/V$_2$O$_5$) by hydrothermal method as catalyst for hazardous waste treatment *Fuel* **349** 128668

[54] Parasuraman B, Kandasamy B, Murugan I, Alsalhi M S, Asemi N, Thangavelu P and Perumal S 2023 Designing the heterostructured FeWO$_4$/FeS$_2$ nanocomposites for an enhanced photocatalytic organic dye degradation *Chemosphere* **334** 138979

[55] Shibu M C, Benoy M D, Kumar G S, Duraimurugan J, Vasudevan V, Shkir M and AL-Otaibi O 2023 Hydrothermal-assisted synthesis and characterization of MWCNT/ copper oxide nanocomposite for the photodegradation of methyl orange under direct sunlight *Diam. Relat. Mater.* **134** 109778

[56] Zinatloo-Ajabshir S, Heidari-Asil S A and Salavati-Niasari M 2021 Recyclable magnetic ZnCo$_2$O$_4$-based ceramic nanostructure materials fabricated by simple sonochemical route for effective sunlight-driven photocatalytic degradation of organic pollution *Ceram. Int.* **47** 8959–72

[57] Warsi A-Z, Hussien O K, Iftikhar A, Aziz F, Alhashmialameer D, Mahmoud S F, Warsi M F and Saleh D I 2022 Co-precipitation assisted preparation of Ag$_2$O, CuO and Ag$_2$O/CuO nanocomposite: characterization and improved solar irradiated degradation of colored and colourless organic effluents *Ceram. Int.* **48** 19056–67

[58] Pudukudy M, Jia Q, Yuan J, Megala S, Rajendran R and Shan S 2020 Influence of CeO$_2$ loading on the structural, textural, optical and photocatalytic properties of single-pot sol–gel derived ultrafine CeO$_2$/TiO$_2$ nanocomposites for the efficient degradation of tetracycline under visible light irradiation *Mater. Sci. Semicond. Process.* **108** 104891

[59] Bai N, Liu X, Li Z, Ke X, Zhang K and Wu Q 2021 High-efficiency TiO$_2$/ZnO nanocomposites photocatalysts by sol–gel and hydrothermal methods *J. Solgel Sci. Technol.* **99** 92–100

[60] Akhter P, Nawaz S, Shafiq I, Nazir A, Shafique S, Jamil F, Park Y-K and Hussain M 2023 Efficient visible light assisted photocatalysis using ZnO/TiO$_2$ nanocomposites *Mol. Catal.* **535** 112896

[61] Khilji M-U-N, Nahyoon N A, Mehdi M, Thebo K H, Mahar N, Memon A A, Memon N and Hussain N 2023 Synthesis of novel visible light driven MgO@GO nanocomposite photocatalyst for degradation of Rhodamine 6G *Opt. Mater.* **135** 113260

[62] Alsalme A, AlFawaz A, Glal A H, Abdel Messih M F, Soltan A and Ahmed M A 2023 S-scheme AgIO$_4$/CeO$_2$ heterojunction nanocomposite photocatalyst for degradation of rhodamine B dye *J. Photochem. Photobiol. A Chem.* **439** 114596

[63] Ur Rahman Z, Shah U, Alam A, Shah Z, Shaheen K, Bahadar Khan S and Ali Khan S 2023 Photocatalytic degradation of cefixime using CuO-NiO nanocomposite photocatalyst *Inorg. Chem. Commun.* **148** 110312

[64] Alajmi B M, Basaleh A S, Ismail A A and Mohamed R M 2023 Hierarchical mesoporous CuO/ZrO_2 nanocomposite photocatalyst for highly stable photoinduced desulfurization of thiophene *Surf. Interfaces* **39** 102899

[65] Xu J, Su S, Song X, Luo S, Ye S and Situ W 2023 A simple nanocomposite photocatalyst $HT-rGO/TiO_2$ for deoxynivalenol degradation in liquid food *Food Chem.* **408** 135228

[66] Baig U, Dastageer M A, Gondal M A and Khalil A B 2023 Photocatalytic deactivation of sulphate reducing bacteria using visible light active CuO/TiO_2 nanocomposite photocatalysts synthesized by ultrasonic processing *J. Photochem. Photobiol., B* **242** 112698

[67] Zare A, Saadati A and Sheibani S 2023 Modification of a Z-scheme ZnO-CuO nanocomposite by Ag loading as a highly efficient visible light photocatalyst *Mater. Res. Bull.* **158** 112048

[68] Verma N, Chundawat T S, Chandra H and Vaya D 2023 An efficient time reductive photocatalytic degradation of carcinogenic dyes by TiO_2–GO nanocomposite *Mater. Res. Bull.* **158** 112043

[69] Packialakshmi J S, Albeshr M F, Alrefaei A F, Zhang F, Liu X, Selvankumar T and Mythili R 2023 Development of $ZnO/SnO_2/rGO$ hybrid nanocomposites for effective photocatalytic degradation of toxic dye pollutants from aquatic ecosystems *Environ. Res.* **225** 115602

[70] Vivek P, Sivakumar R, Selva Esakki E and Deivanayaki S 2023 Fabrication of NiO/RGO nanocomposite for enhancing photocatalytic performance through degradation of RhB *J. Phys. Chem. Solids* **176** 111255

[71] Sabouri Z, Kazemi Oskuee R, Sabouri S, Tabrizi Hafez Moghaddas S S, Samarghandian S, Sajid Abdulabbas H and Darroudi M 2023 Phytoextract-mediated synthesis of Ag-doped ZnO–MgO–CaO nanocomposite using *Ocimum basilicum* L seeds extract as a highly efficient photocatalyst and evaluation of their biological effects *Ceram. Int.* **49** 20989–97

[72] Manda A A *et al* 2023 Fast one-pot laser-based fabrication of ZnO/TiO_2-reduced graphene oxide nanocomposite for photocatalytic applications *Opt. Laser Technol.* **160** 109105

[73] Chinnasamy C, Perumal N, Choubey A and Rajendran S 2023 Recent advancements in MXene-based nanocomposites as photocatalysts for hazardous pollutant degradation—a review *Environ. Res.* **233** 116459

[74] Purkayastha M D, Pal Majumder T, Sarkar M and Ghosh S 2022 Carrier transport and shielding properties of rod-like mesoporous TiO_2–SiO_2 nanocomposite *Radiat. Phys. Chem.* **192** 109898

[75] Tamayo-Vegas S and Lafdi K 2022 Experimental and modelling of temperature-dependent mechanical properties of CNT/polymer nanocomposites *Mater. Today Proc.* **57** 607–14

[76] Osman A, Elhakeem A, Kaytbay S and Ahmed A 2022 A comprehensive review on the thermal, electrical, and mechanical properties of graphene-based multi-functional epoxy composites *Adv. Compos. Hybrid Mater.* **5** 547–605

[77] Norizan M N, Abdullah N, Halim N A, Demon S Z N and Mohamad I S 2022 Heterojunctions of rGO/metal oxide nanocomposites as promising gas-sensing materials —a review *Nanomaterials* **12** 2278

[78] Mehta M, Chandrabose G, Krishnamurthy S, Avasthi D K and Chowdhury S 2022 Improved photoelectrochemical properties of TiO_2–graphene nanocomposites: effect of

defect induced visible light absorption and graphene conducting channel for carrier transport *Appl. Surf. Sci. Adv.* **11** 100274

[79] Williams J D and Peterson G P 2021 A review of thermal property enhancements of low-temperature nano-enhanced phase change materials *Nanomaterials* **11** 2578

[80] Alzahrani H S, Al-Sulami A I, Alsulami Q A and Rajeh A 2022 A systematic study of structural, conductivity, linear, and nonlinear optical properties of PEO/PVA-MWCNTs/ ZnO nanocomposites films for optoelectronic applications *Opt. Mater.* **133** 112900

[81] Chang C-Y, Yamakata A and Tseng W J 2022 Effect of surface plasmon resonance and the heterojunction on photoelectrochemical activity of metal-loaded TiO_2 electrodes under visible light irradiation *J. Phys. Chem.* C **126** 12450–9

[82] Bethi B, Sonawane S H, Bhanvase B A and Gumfekar S P 2016 Nanomaterials-based advanced oxidation processes for wastewater treatment: a review *Chem. Engi. Process.* **109** 178–89

[83] Zhang Z, Zhang X, Porcar-Castell A, Chen J M, Ju W, Wu L, Wu Y and Zhang Y 2022 Sun-induced chlorophyll fluorescence is more strongly related to photosynthesis with hemispherical than nadir measurements: evidence from field observations and model simulations *Remote Sens. Environ.* **279** 113118

[84] Kubacka A, Caudillo-Flores U, Barba-Nieto I and Fernández-García M 2021 Towards full-spectrum photocatalysis: successful approaches and materials *Appl. Catal. A Gen.* **610** 117966

[85] Gallagher J M, Roberts B M W, Borsley S and Leigh D A 2023 Conformational selection accelerates catalysis by an organocatalytic molecular motor *Chem.*

[86] Ohtani B 2010 Photocatalysis A to Z- what we know and what we do not know in a scientific sense *J. Photochem. Photobiol.,* C **11** 157–78

[87] Petronella F, Truppi A, Ingrosso C, Placido T, Striccoli M, Curri M L, Agostiano A and Comparelli R 2017 Nanocomposite materials for photocatalytic degradation of pollutants *Catal. Today* **281** 85–100

[88] Wu H, Yang K, Wang X, Fang N, Weng P, Duan L, Zhang C, Wang X and Liu L 2023 Xenon-lamp simulated sunlight-induced photolysis of pyriclobenzuron in water: kinetics, degradation pathways, and identification of photolysis products *Ecotoxicol. Environ. Saf.* **263** 115272

[89] Łoński S, Łukowiec D, Barbusiński K, Babilas R, Szeląg B and Radoń A 2023 Flower-like magnetite nanoparticles with unfunctionalized surface as an efficient catalyst in photo-Fenton degradation of chemical dyes *Appl. Surf. Sci.* **638** 158127

[90] Shankaraiah G, Saritha P, Bhagawan D, Himabindu V and Vidyavathi S 2017 Photochemical oxidation of antibiotic gemifloxacin in aqueous solutions—a comparative study *S. Afr. J. Chem. Eng.* **24** 8–16

[91] Torres-Pinto A, Díez A M, Silva C G, Faria J L, Sanromán M Á, Silva A M T and Pazos M 2023 Photoelectrocatalytic degradation of pharmaceuticals promoted by a metal-free g-C_3N_4 catalyst *Chem. Eng. J.* **476** 146761

[92] Wu S, Jia Q and Dai W 2017 Synthesis of RGO/TiO 2 hybrid as a high performance photocatalyst *Ceram. Int.* **43** 1530–5

[93] Sang Y, Liu H and Umar A 2015 Photocatalysis from UV/Vis to near-infrared light: towards full solar-light spectrum activity *ChemCatChem.* **7** 559–73

[94] Fu Y, Li J and Li J 2019 Metal/semiconductor nanocomposites for photocatalysis: fundamentals, structures, applications and properties *Nanomaterials* **9** 359

[95] Khan I, Saeed K and Khan I 2019 Nanoparticles: properties, applications and toxicities *Arab. J. Chem.* **12** 908–31

[96] Sharma G, Kumar A, Sharma S, Naushad M, Prakash Dwivedi R, ALOthman Z A and Mola G T 2019 Novel development of nanoparticles to bimetallic nanoparticles and their composites: a review *J King Saud Univ. Sci.* **31** 257–69

[97] Raizada P, Sudhaik A and Singh P 2019 Photocatalytic water decontamination using graphene and ZnO coupled photocatalysts: a review *Mater. Sci. Energy Technol.* **2** 509–25

[98] Bresolin B-M, Sgarbossa P, Bahnemann D W and Sillanpää M 2020 $Cs_3Bi_2I_9/g$-C_3N_4 as a new binary photocatalyst for efficient visible-light photocatalytic processes *Sep. Purif. Technol.* **251** 117320

[99] Gu J, Jia H, Ma S, Ye Z, Pan J, Dong R, Zong Y and Xue J 2020 Fe_3O_4-loaded g-C_3N_4/C-layered composite as a ternary photocatalyst for tetracycline degradation *ACS Omega* **5** 30980–8

[100] Lin Y-Y, Hung J-T, Chou Y-C, Shen S-J, Wu W-T, Liu F-Y, Lin J-H and Chen C-C 2022 Synthesis of bismuth oxybromochloroiodide/graphitic carbon nitride quaternary compo-sites (BiOxCly/BiOmBrn/BiOpIq/g-C_3N_4) enhances visible-light-driven photocatalytic activity *Catal. Commun.* **163** 106418

[101] Wang G, Lv S, Shen Y, Li W, Lin L and Li Z 2023 Advancements in heterojunction, cocatalyst, defect and morphology engineering of semiconductor oxide photocatalysts *J. Materiomics*

[102] Russo S, Muscetta M, Amato P, Venezia V, Verrillo M, Rega R, Lettieri S, Cocca M, Marotta R and Vitiello G 2024 Humic substance/metal-oxide multifunctional nanoparticles as advanced antibacterial-antimycotic agents and photocatalysts for the degradation of PLA microplastics under UVA/solar radiation *Chemosphere* **346** 140605

[103] Liu Y, Xu L, Zhang N, Wang J, Mu X and Wang Y 2022 A promoted charge separation/ transfer and surface plasmon resonance effect synergistically enhanced photocatalytic performance in Cu nanoparticles and single-atom Cu supported attapulgite/polymer carbon nitride photocatalyst *Mater. Today Chem.* **26** 101250

[104] Sarwar A, Razzaq A, Zafar M, Idrees I, Rehman F and Kim W Y 2023 Copper tungstate ($CuWO_4$)/graphene quantum dots (GQDs) composite photocatalyst for enhanced degra-dation of phenol under visible light irradiation *Results Phys.* **45** 106253

[105] Hafeez H Y, Mohammed J, Suleiman A B, Ndikilar C E, Sa'id R S and Muhammad I 2023 Insights into hybrid TiO_2–g-C_3N_4 heterostructure composite decorated with rGO sheet: a highly efficient photocatalyst for boosted solar fuel (hydrogen) generation *Chem. Phys. Impact* **6** 100157

[106] Zhao Q *et al* 2020 Nonhydrolytic sol–gel in-situ synthesis of novel recoverable amorphous Fe_2TiO_5/C hollow spheres as visible-light driven photocatalysts *Mater. Des.* **194** 108928

[107] Wu P, Xu W, Gu X, Wang M, Liu B and Khan S 2020 Preparation of rod-shaped Bi_5O_7I as Bifunctional Material for Supercapacitors and Photcatalysts *Int. J. Electrochem. Sci.* **15** 11294–305

[108] Adekoya J A, Chibuokem M O, Masikane S and Revaprasadu N 2023 Heterostructures of Ag_2FeSnS_4 chalcogenide nanoparticles as potential photocatalysts *Sci. Afr.* **19** e01509

[109] Rini A S, Defti A P, Dewi R, Jasril and Rati Y 2023 Biosynthesis of nanoflower Ag-doped ZnO and its application as photocatalyst for Methylene blue degradation *Mater. Today Proc.* **87** 234–9

[110] Di Paola A, García-López E, Marcì G and Palmisano L 2012 A survey of photocatalytic materials for environmental remediation *J. Hazard. Mater.* **211–212** 3–29

[111] Tayebi M, Kolaei M, Tayyebi A, Masoumi Z, Belbasi Z and Lee B-K 2019 Reduced graphene oxide (RGO) on TiO_2 for an improved photoelectrochemical (PEC) and photocatalytic activity *Sol. Energy* **190** 185–94

[112] Gupta S and Subramanian V R 2014 Encapsulating $Bi_2Ti_2O_7$ (BTO) with reduced graphene oxide (RGO): an effective strategy to enhance photocatalytic and photoelectrocatalytic activity of BTO *ACS Appl. Mater. Interfaces* **6** 18597–608

[113] Moztahida M, Jang J, Nawaz M, Lim S-R and Lee D S 2019 Effect of rGO loading on Fe_3O_4: a visible light assisted catalyst material for carbamazepine degradation *Sci. Total Environ.* **667** 741–50

[114] Ben Saber N, Mezni A, Alrooqi A and Altalhi T 2021 Fabrication of efficient $Au@TiO_2$/rGO heterojunction nanocomposite: boosted photocatalytic activity under ultraviolet and visible light irradiation *J. Mater. Res. Technol.* **12** 2238–46

[115] Chebanenko M I, Lebedev L A, Seroglazova A S, Lobinsky A A, Gerasimov E Y, Stovpiaga E Y and Popkov V I 2023 Novel g-C_3N_4/$PrFeO_3$ nanocomposites with Z-scheme structure and superior photocatalytic activity toward visible-light-driven removal of tetracycline antibiotics *Heliyon* **9** e22038

[116] Quan Y, YiO M H N, Li Y, Myers R J and Kafizas A 2023 Influence of Bi co-catalyst particle size on the photocatalytic activity of BiOI microflowers in Bi/BiOI junctions—a mechanistic study of charge carrier behaviour *J. Photochem. Photobiol. A Chem.* **443** 114889

[117] Babu V J, Bhavatharini R S R and Ramakrishna S 2014 Bi_2O_3 and BiOCl electrospun nanosheets and morphology-dependent photocatalytic properties *RSC Adv.* **4** 29957

[118] Pirgholi-Givi G, Farjami-Shayesteh S and Azizian-Kalandaragh Y 2019 The influence of preparation parameters on the photocatalytic performance of mixed bismuth titanate-based nanostructures *Physica* B **575** 311572

[119] Zou H, Song M, Yi F, Wang X, Bian L, Li W, Zhang J, Pan N and Zeng P 2018 Effect of sintering temperature on the photocatalytic activity of carbon–Bi_2O_3–TiO_2 composite *J. Mater. Sci., Mater. Electron.* **29** 2201–8

[120] Kallawar G A, Bhanvase B A and Sathe B R 2023 Sonochemically prepared bismuth doped titanium oxide-reduced graphene oxide ($Bi@TiO_2$–rGO) nanocomposites for effective visible light photocatalytic degradation of malachite green *Diam. Relat. Mater.* **139** 110423

[121] Grzegórska A, Ofoegbu J C, Cervera-Gabalda L, Gómez-Polo C, Sannino D and Zielińska-Jurek A 2023 Magnetically recyclable TiO_2/MXene/$MnFe_2O_4$ photocatalyst for enhanced peroxymonosulphate-assisted photocatalytic degradation of carbamazepine and ibuprofen under simulated solar light *J. Environ. Chem. Eng.* **11** 110660

[122] Porcu S, Secci F and Ricci P C 2022 Advances in hybrid composites for photocatalytic applications: a review *Molecules* **27** 6828

[123] Irfan S, Zhuanghao Z, Li F, Chen Y-X, Liang G-X, Luo J-T and Ping F 2019 Critical review: Bismuth ferrite as an emerging visible light active nanostructured photocatalyst *J. Mater. Res. Technol.* **8** 6375–89

[124] Tegenaw A B, Yimer A A and Beyene T T 2023 Boosting the photocatalytic activity of ZnO-NPs through the incorporation of C-dot and preparation of nanocomposite materials *Heliyon* **9** e20717

[125] Yang L, Guo J, Yang T, Guo C, Zhang S, Luo S, Dai W, Li B, Luo X and Li Y 2021 Self-assembly Cu_2O nanowire arrays on Cu mesh: a solid-state, highly-efficient, and stable photocatalyst for toluene degradation under sunlight *J. Hazard. Mater.* **402** 123741

[126] Meng Y, Huang X, Wu Y, Wang X and Qian Y 2002 Kinetic study and modeling on photocatalytic degradation of para-chlorobenzoate at different light intensities *Environ. Pollut.* **117** 307–13

[127] Lal M, Sharma P and Ram C 2022 Synthesis and photocatalytic potential of Nd-doped TiO_2 under UV and solar light irradiation using a sol–gel ultrasonication method *Res. Mater.* **15** 100308

[128] Yin Z-C, Yang M, Gosavi S W, Kumar Singh A, Chauhan R and Jin J-C 2021 A 3D supramolecular Ag(I)-based coordination polymer as stable photocatalyst for dye degradation *Inorg. Chem. Commun.* **131** 108805

[129] Zhang Q, Yu L, Xu C, Zhao J, Pan H, Chen M, Xu Q and Diao G 2019 Preparation of highly efficient and magnetically recyclable Fe_3O_4@C@Ru nanocomposite for the photocatalytic degradation of methylene blue in visible light *Appl. Surf. Sci.* **483** 241–51

[130] Meng J, He M, Li F, Li T, Huang Z and Cao W 2023 Combining Ce–metal–organic framework with CdS for efficient photocatalytic removals of heavy metal ion and organic pollutant under visible and solar lights *Inorganica Chim. Acta* **557** 121701

[131] Lellala K 2021 Sulfur embedded on in-situ carbon nanodisc decorated on graphene sheets for efficient photocatalytic activity and capacitive deionization method for heavy metal removal *J. Mater. Res. Technol.* **13** 1555–66

[132] Jeyaprakash J S, Rajamani M, Bianchi C L, Ashokkumar M and Neppolian B 2023 Highly efficient ultrasound-driven Cu–MOF/$ZnWO_4$ heterostructure: an efficient visible-light photocatalyst with robust stability for complete degradation of tetracycline *Ultrason. Sonochem.* **100** 106624

[133] Leong K H, Gan B L, Ibrahim S and Saravanan P 2014 Synthesis of surface plasmon resonance (SPR) triggered Ag/TiO_2 photocatalyst for degradation of endocrine disturbing compounds *Appl. Surf. Sci.* **319** 128–35

[134] Antonopoulou M, Kosma C, Albanis T and Konstantinou I 2021 An overview of homogeneous and heterogeneous photocatalysis applications for the removal of pharmaceutical compounds from real or synthetic hospital wastewaters under lab or pilot scale *Sci. Total Environ.* **765** 144163

[135] McCormick W J, McCrudden D, Skillen N and Robertson P K J 2023 Electrochemical monitoring of the photocatalytic degradation of the insecticide emamectin benzoate using TiO_2 and ZnO materials *Appl. Catal. A Gen.* **660** 119201

[136] Zelić I E, Povijač K, Gilja V, Tomašić V and Gomzi Z 2022 Photocatalytic degradation of acetamiprid in a rotating photoreactor—determination of reactive species *Catal. Commun.* **169** 106474

[137] Li S-S, Wen L, He S-W, Xu Z, Ding L, Cheng Y-H and Chen M-L 2023 Enhanced photocatalytic degradation of imidacloprid by a simple Z-type binary heterojunction composite of long afterglow with metal-organic framework *Catal. Commun.* **183** 106775

[138] Zhang Y, Cao X, Yang Y, Guan S, Wang X, Li H, Zheng X, Zhou L, Jiang Y and Gao J 2023 Visible light assisted enzyme-photocatalytic cascade degradation of organophosphorus pesticides *Green Chem. Eng.* **4** 30–8

[139] Ayodhya D and Veerabhadram G 2019 Fabrication of Schiff base coordinated ZnS nanoparticles for enhanced photocatalytic degradation of chlorpyrifos pesticide and detection of heavy metal ions *J. Materiomics* **5** 446–54

[140] Yogesh Kumar K, Prashanth M K, Shanavaz H, Parashuram L, Alharethy F, Jeon B-H and Raghu M S 2023 Ultrasound assisted fabrication of $InVO_4/In_2S_3$ heterostructure for enhanced sonophotocatalytic degradation of pesticides *Ultrason. Sonochem.* **100** 106615

[141] Fatima T, Husain S and Khanuja M 2022 Superior photocatalytic and electrochemical activity of novel WS_2/PANI nanocomposite for the degradation and detection of pollutants: antibiotic, heavy metal ions, and dyes *Chem. Eng. J. Adv.* **12** 100373

[142] Meng J, He M, Li F, Li T, Huang Z and Cao W 2023 Combining Ce-metal–organic framework with CdS for efficient photocatalytic removals of heavy metal ion and organic pollutant under visible and solar lights *Inorganica Chim. Acta* **557** 121701

[143] Wood D, Shaw S, Cawte T, Shanen E and Van Heyst B 2020 An overview of photocatalyst immobilization methods for air pollution remediation *Chem. Eng. J.* **391** 123490

[144] Wang A, Teng Y, Hu X, Wu L, Huang Y, Luo Y and Christie P 2016 Diphenylarsinic acid contaminated soil remediation by titanium dioxide (P_25) photocatalysis: degradation pathway, optimization of operating parameters and effects of soil properties *Sci. Total Environ.* **541** 348–55

[145] Srivastava A and Bandhu S 2022 Biotechnological advancements and challenges in textile effluents management for a sustainable bioeconomy: Indian case studies *Case Stud. Chem. Environ. Eng.* **5** 100186

[146] Jo W-K and Tayade R J 2014 New generation energy-efficient light source for photocatalysis: LEDs for environmental applications *Ind. Eng. Chem. Res.* **53** 2073–84

[147] Kallawar G A, Barai D P and Bhanvase B A 2021 Bismuth titanate based photocatalysts for degradation of persistent organic compounds in wastewater: a comprehensive review on synthesis methods, performance as photocatalyst and challenges *J. Clean. Prod.* **318** 128563

[148] Giram D, Das A B B and Bhanvase B 2023 Comparative study of $ZnO–TiO_2$ nanocomposites synthesized by ultrasound and conventional methods for the degradation of methylene blue dye *Indian J. Chem. Technol.* **30** 693–704

[149] Bhanvase B A, Shende T P and Sonawane S H 2017 A review on graphene–TiO_2 and doped graphene–TiO 2 nanocomposite photocatalyst for water and wastewater treatment *Environ. Technol. Rev.* **6** 1–14

[150] Mohanta D, Barman K, Jasimuddin S and Ahmaruzzaman M 2021 Encapsulating band gap engineered $CoSnO_3$ mixed metal oxide nanocomposite in rGO matrix: a novel catalyst towards LED light induced photoelectrocatalytic water oxidation at neutral pH *J. Electroanal. Chem.* **880** 114830

[151] Anjum M, Miandad R, Waqas M, Gehany F and Barakat M A 2019 Remediation of wastewater using various nano-materials *Arab. J. Chem.* **12** 4897–919

[152] Alalm M G, Djellabi R, Meroni D, Pirola C, Bianchi C L and Boffito D C 2021 Toward scaling-up photocatalytic process for multiphase environmental applications *Catalysts* **11** 562

[153] Haldar S S, Halder D, Patel G, Singhania A K, R R and Pandey A 2023 Waste fish scale for the preparation of bio-nanocomposite film with novel properties *Environ. Technol. Innov.* **32** 103386

[154] El-Sheikhy R, Al-Khuraif A and Al-Shamrani M 2023 Converting polymer-wastes to green nanocomposites in Saudi Arabia: investigation of fracture parameters (KIc, KIIc, θc, σc) for quality and sustainability evaluation *Vacuum* **217** 112513

[155] Fatimah I, Yanti I, Wijayanti H K, Ramanda G D, Sagadevan S, Tamyiz M and Doong R 2023 One-pot synthesis of Fe_3O_4/$NiFe_2O_4$ nanocomposite from iron rust waste as reusable catalyst for methyl violet oxidation *Case Stud. Chem. Environ. Eng.* **8** 100369

[156] Suryawanshi P L, Gumfekar S P, Kumar P R, Kale B B and Sonawane S H 2016 Synthesis of ultra-small platinum nanoparticles in a continuous flow microreactor *Colloid Interface Sci. Commun.* **13** 6–9

[157] Malkapuram S T, Seepana M M, Sonawane S H, Lakhera S K and Randviir E 2024 ZIF-8 decorated cellulose acetate mixed matrix membrane: an efficient approach for textile effluent treatment *Chemosphere* **349** 140836

[158] Vasseghian Y, Khataee A, Dragoi E-N, Moradi M, Nabavifard S, Oliveri Conti G and Mousavi Khaneghah A 2020 Pollutants degradation and power generation by photocatalytic fuel cells: a comprehensive review *Arab. J. Chem.* **13** 8458–80

Chapter 5

Photocatalysts for hydrogen evolution

Somnath C Dhawale and Bhaskar R Sathe

The most significant green and sustainable energy resource that can meet forth-coming energy needs is hydrogen (H_2). This is because it has the potential to reduce our dependency on fossil fuels while also serving as the ultimate antidote to environmental degradation. The most promising option is the generation of solar-induced H_2 through water splitting, which uses economical solar energy and has the potential for enhanced solar-to-hydrogen conversion efficiency. In this chapter, the most recent advances in metal-free photocatalysts for H_2 production from water are discussed. It has been established that some materials, including carbon nitride and sulfonate, possess exceptional reactivity towards visible light, a sufficient number of active sites, and appropriate H^+/H_2 reduction potential to operate as effective photocathodes. Furthermore, in order address the issue of soluble mediators and simultaneously optimize the reduction and oxidation processes, the application of a soluble redox mediator in Z-scheme water splitting systems has been studied. Solar fuel is also produced using a number of photophysical and electrocatalytic techniques.

5.1 Introduction

Hydrogen (H_2) is a promising and versatile energy carrier which plays an important role in sustainable and environmentally friendly energy systems [1]. For the production of ammonia and in the petrochemical sector, hydrogen is a crucial and widely utilized chemical raw ingredient. As a result, there is a sharp rise in the annual demand for hydrogen. The three principal industrial processes for the production of H_2 are steam reforming, the gasification of coal, and the electrolysis of H_2O. However, these processes produce H_2 through the use of fossil fuels and therefore release CO_2 as a byproduct as part of these processes. Interestingly, water splitting, especially when powered by renewable energy sources, is often regarded as a more sustainable and environmentally friendly method for H_2 production.

This creates an opening for sustainable hydrogen generation, since its primary raw material, water, is plentiful and regenerative. In addition, since the products are hydrogen and oxygen, secondary environmental contamination is avoided. Renewable energy from hydropower, the wind, and the sun can be utilized for the generation of electricity needed for water splitting [2–4] and therefore water splitting has become more significant.

However, only 3%–5% of commercial hydrogen (H_2) is currently produced by water splitting reactions. Water splitting is primarily a kinetically controlled electrochemical process that has an energy-intensive nature since it results from delayed electron transfer at an electrified interface corresponding to two half cells. These correspond to water oxidation and reduction, which are also known as the oxygen evolution reaction (OER) and the hydrogen evolution reaction (HER), respectively [5]. Unfortunately, these processes have exceptionally high energy barriers, which further hampers energy conversion. Unquestionably, the fabrication/synthesis of effective and stable catalysts for the HER and the OER is vital for advancing water splitting technologies [6]. Both electrocatalytic and photocatalytic approaches are being explored to improve the performance of water splitting systems.

Photocatalysts and electrocatalysts for the HER have been thoroughly studied and reported from a number of perspectives during the last few decades [7, 8]. Nowadays, catalysts prepared using noble metals such as Pt and Pd are known to have exceptional catalytic activity, but the cost and scarcity of noble metals can be limiting factors for their commercial application in water splitting technologies. To address this challenge, significant research efforts are being directed toward developing non-precious metal-based systems as alternatives for HER catalysts. The goal is to find materials that are more abundant and cost-effective yet still offer high catalytic activity; examples include transition-metal sulfides, nitrides, selenides, and phosphides [3, 9]. However, the majority of these experience unpredictable agglomeration/dissolution throughout the HER process as well as undesired and/or unstable alterations in their morphology/crystalline structure. Due to the leakage of heavy metals into electrolytes, this low stability may potentially have negative effects on the environment.

The United Nations (UN) claims that energy-related activities are a major contributor to climate change [10]. As a result, research into new renewable energy sources has been included as target 7 of the UN's 17 Sustainable Development Goals. In this regard, a growing focus of basic and applied research has been H_2 generation as a sustainable energy source. In general, hydrogen (H_2) may be produced via various chemical processes as shown in figure 5.1. In comparison, photocatalytic water splitting is reported to be the simplest way. It requires semiconducting particles to create hydrogen and oxygen when they are in solution and exposed to sunlight. It also exhibits thermodynamic feasibility, as shown by the following equations (5.1) and (5.2).

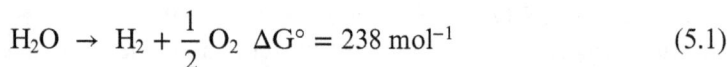

$$H_2O \; \rightarrow \; H_2 + \frac{1}{2} O_2 \; \Delta G^\circ = 238 \; mol^{-1} \tag{5.1}$$

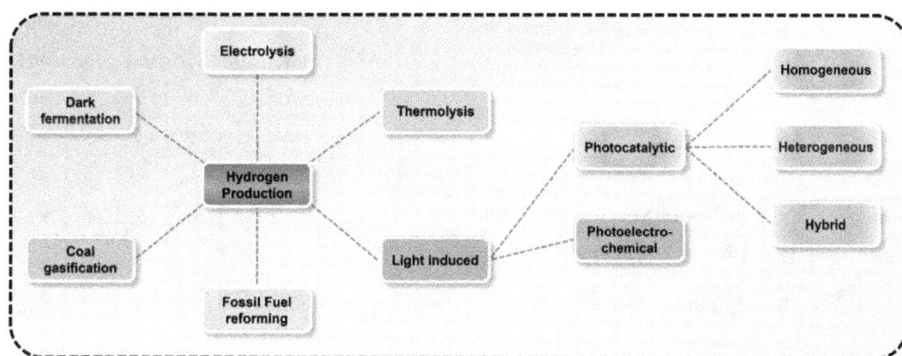

Figure 5.1. A schematic survey of the major feasible H_2-generating routes. The most promising option is solar-induced H_2 generation by water splitting, considered to be a promising and environmentally friendly approach for sustainable energy production.

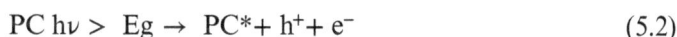

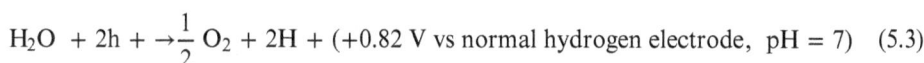

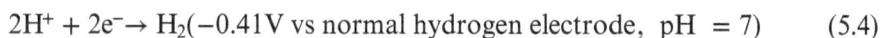

$$PC \; h\nu > \; Eg \rightarrow \; PC^* + h^+ + e^- \qquad (5.2)$$

$$H_2O \; + 2h + \rightarrow \frac{1}{2} O_2 + 2H + (+0.82 \; V \; vs \; normal \; hydrogen \; electrode, \; pH = 7) \quad (5.3)$$

$$2H^+ + 2e^- \rightarrow H_2(-0.41V \; vs \; normal \; hydrogen \; electrode, \; pH \; = 7) \qquad (5.4)$$

When water is split by means of a photocatalyst, the valence band (VB) must have a higher positive potential than the water oxidation potential required for the OER (equation (5.3)). However, in order to drive the reduction processes, a conduction band (CB) potential less negative than the normal H_2 evolution (vs. NHE) potential is needed (equation (5.4)). Ultimately, this procedure offers a pure and sustainable supply of hydrogen without generating toxic gases or other byproducts. The photocatalyst must meet energy gap requirements which are necessary to produce H_2 and O_2 simultaneously; these are considered to be challenging.

Consequently, combining catalysts to form cocatalysts is a tactic frequently used to pursue this goal. Different catalysts have been created for this purpose, including homogeneous, heterogeneous, and hybrid ones. Heterogeneous photocatalysts are well known for a variety of properties, including high chemical stability in a liquid medium, which offers resistance to degradation or corrosion during the chemical reaction. This is important for ensuring the longevity and effectiveness of the catalyst under reaction conditions, reducing operating costs, and enabling the production of specific products.

Recently, a lot of interest in semiconductor metal sulfides has been observed because of their outstanding physical and chemical characteristics. Numerous binary metal sulfides (MS), as well as their derivatives and heterostructures, are now extensively used for a variety of applications [11–16]. The most popular metals used to make sulfide compounds for photocatalytic H_2 production are listed in figure 5.2. The majority of these compounds exhibit extraordinary activity in visible light, enough reactive centers, and the right reduction potential for H_2 production to

IA																	VIIIA
H	IIA											IIIA	IVA	VA	VIA	VIIA	He
Li	Be					VIIIB						B	C	N	O	F	Ne
Na	Mg	IIIB	IVB	VB	VIB	VIIB			IB	IIB		Al	Si	P	S	Cl	Ar
K	Ca	Sc	Ti	V	Cr	Mn	Fe	Co	Ni	Cu	Zn	Ga	Ge	As	Se	Br	Kr
Rb	Sr	Y	Zr	Nb	Mo	Tc	Ru	Rh	Pd	Ag	Cd	In	Sn	Sb	Te	I	Xe
Cs	Ba	La	Hf	Ta	W	Re	Os	Ir	Pt	Au	Hg	Tl	Pb	Bi	Po	At	Rn
Fr	Ra	Ac	Rf	Db	Sg	Bh	Hs	Mt	Ds	Rg	Cn	Nh	Fl	Mc	Lv	Ts	Og

Figure 5.2. A representation of elements in the periodic table whose sulfide compounds are often employed in photocatalytic H_2 production (blue), used for dopants (green), or whose oxides or other compounds (cyan) are used to build heterostructures with MS for H_2 evolution.

function as efficient photocatalysts. In addition, new quantum size effects allow for additional tuning to obtain quicker charge transfer, prolonged excited-state durations, etc [17, 18]. Over the last ten years, a lot of work has been done on boosting solar hydrogen generation using MS semiconductor photocatalysts [19–21]. $CuInS_2$ and $CuGaS_2$ have been demonstrated to have good selectivity and a high photocatalytic H_2 generation rate compared to those of other MS materials [22, 23], particularly when combined with titanium dioxide (TiO_2). The focus of current research is on MS with reduced dimensions. Moreover, research into high H_2 generation is expanding due to the functionalization of two-dimensional (2D) MS structures [24]. One popular method of increasing photocatalytic activity is to utilize metals as dopants or other substances as cocatalysts, as shown in figure 5.2.

The evolution and further advancement of civilization depend on our capacity to harness energy beyond that provided by humans and other animals and on our ability to locate, capture, and utilize energy more and more skillfully. Among the various resources of energy, oil, gas, and coal account for more than 85% of global energy production [1]. A sustainable source equal to 1066 barrels of oil, 108 000 m^3 of natural gas, and 250 tons of coal/second would have to be generated to fulfill the world average energy consumption rate of 17.2 TW [25]. The world's energy consumption is anticipated to quadruple by 2050 due to rising population growth (1.12% per year) and industrialization [26]. An 'energy crisis' will emerge from the quick depletion of finite fossil fuel supplies, even if the present 2.1% annual rate of global primary energy use does not change. In addition, all living things and plants are harmed by the poisonous effluent emissions that result from burning fossil fuels, which also pose a hazard to humanity [27]. Climate change (according to the United Nations Framework Convention on Climate Change (UNFCCC)) poses an unstoppable threat to society. As a result, in addition to the usual choices of afforestation and bioenergy with carbon capture and storage (BECCS), the Paris Agreement asks for the growth of other net CO_2 removal technologies.

The hunt for alternative energy sources is sparked by how we respond to this situation. In order to reduce (or avoid) the anthropogenic effects produced by the combustion of fossil fuels and to address the safety of supply in the short term, it is generally acknowledged that a transformation from nonrenewable fossil fuels to renewable fuel resources is necessary. Solar energy is thought to be a feasible option for producing hydrogen from water, assuming that catalysts with the necessary activity, selectivity, and durability can be created [28]. A noteworthy aspect of H_2 fuel is that it burns cleanly with no production of CO_2 or NO_x. In addition, it can be kept in a storage unit and utilized as needed [29].

In recent years, the necessity to preserve natural resources and limit energy use has prompted the creation of energy storage devices that are renewable and sustainable. Hybrid vehicles and portable electronic gadgets have made use of unconventional energy sources such as batteries, supercapacitors, and fuel cells, which use electrochemical reactions to convert chemical energy into electrical energy [30]. However, H_2 is a regenerative and ecologically friendly energy carrier that has attracted the interest of scientists for the last three decades [31, 32]. Nowadays, H_2 is mostly generated from fossil fuels, primarily through a process known as the steam methane reforming technique, which is both ecologically and economically unfavorable [33]. H_2 is expected to be generated using renewable energy sources; a facile pathway for this purpose is photocatalytic H_2 production based on water splitting driven by solar energy technologies. This leverages solar energy for both energy storage and green H_2 production, which have garnered significant attention due to their potential to address key challenges in sustainable energy. Solar energy conversion is a sustainable solution that prevents global warming and the energy crisis [34]. Following Fujishima and Honda's development of photoelectrochemical water splitting in 1972, several research reports describing the application of various photocatalysts have been published [35]. The application of powder photocatalysts is indeed common in research related to photocatalysis, including water splitting, and environmental remediation [36–41]. Photocatalytic H_2 production is a clean technique that uses semiconductor nano-particles and abundant sunshine irradiation to manufacture environmentally acceptable fuel. The critical challenge in the photocatalytic H_2 production process is finding highly efficient and stable photocatalysts which are active under visible light and can facilitate both the HER and the OER under visible-light exposure [42]. Moreover, overcoming issues related to quantum efficiency, photostability, and visible-light absorption is critical for advancing the field and making these technologies more viable for real-world applications. In recent decades, the most widely researched photocatalysts for water splitting reactions have been TiO_2, g-C_3N_4, and CdS [34]. Some promising visible-light-driven photocatalysts such as Ta_3N_5, [43, 44], TaON [45], GaN:ZnO [13, 46], sulfides [47–49], CoO [50], and also g-C_3N_4 [51–53] have been discovered and used for H_2 production. However, the solar-to-fuel conversion efficiency is variable and still very low, which hampers practical applications.

Undoubtedly, a review of metal-free photocatalysts would contribute significantly to the development of novel materials for further energy conversion and storage applications [54–56]. Thus, the development of the same would offer a wealth of information to researchers and technologists, allowing them to acquire knowledge about the current state of the subject and to understand the problems that need to be solved in order to forge a future course. However, no review has yet been made available that has addressed a substantial collection of metal-free photocatalysts. A study of this nature is therefore highly appropriate, given the speed of scientific advancement. We anticipate that a comparison of several metal-free photocatalysts would provide information that would allow us to choose, examine, and screen the optimal metal-free photocatalyst combination for high solar-to-hydrogen conversion efficiency.

5.2 The fundamentals of photocatalytic water splitting

According to thermodynamics, the entire water-splitting reaction is an endothermic, energetically uphill reaction with a Gibbs free energy change ($\Delta G°$) of 237 kJ mol^{-1} showed in equation (5.1). Three crucial stages are involved in a photocatalytic water splitting reaction, as follows: (i) the process of producing electron–hole pairs through the photoexcitation of photocatalysts, (ii) the separation and transport of the generated electrons and holes to the surface of the photocatalyst, and (iii) the capture of the generated electrons and holes by active sites on the cocatalysts for consumption in electrode reactions for H_2O splitting reactions, as shown in figure 5.3 [37, 57, 58]. The total efficiency of solar energy conversion is calculated by equation (5.5):

$$\eta = \eta1 \cdot (\eta2 \times \eta3), \qquad (5.5)$$

where η_1 is the efficiency of light harvesting, η_2 is the efficiency of charge separation, and η_3 is the efficiency of the surface catalytic reaction.

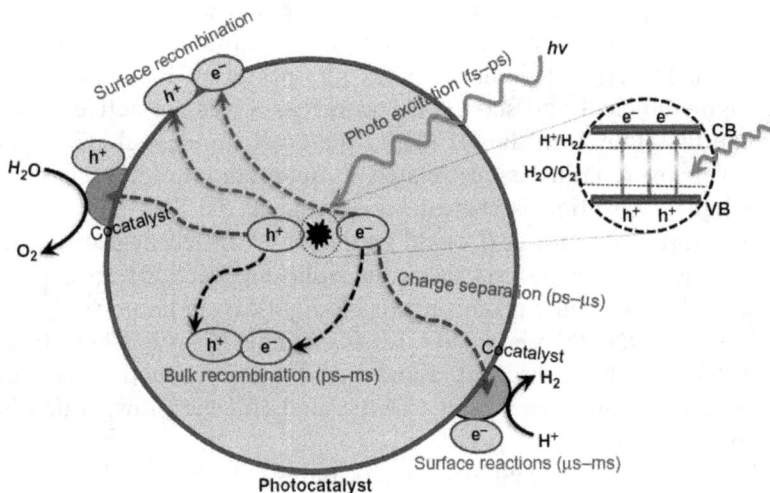

Figure 5.3. The mechanism of semiconductor-based photocatalytic water splitting. Adapted from [57], with permission from the American Chemical Society, and [59], with permission from Elsevier.

These processes typically include light absorption, charge carrier generation, migration, surface reactions, and the production of H_2 and O_2. The claimed water splitting efficiencies are often poor because recombination results in the bulk of the charges being lost.

Certain band structures are found in semiconductors, and their bandgaps are often found to lie in the energy range of 0.2–4.0 eV. If the photon energy of incident light is equivalent to or higher than the bandgap of a semiconductor material, electrons in the VB can be excited to the CB. After photoexcitation, electrons–hole pairs are released and come to the surface for the transfer processes. The holes oxidize H_2O molecules to produce protons and O_2, while the electrons convert the liberated protons to H_2. The surface reduction and oxidation processes that generate H_2 and O_2 resemble those of water electrolysis. The highest occupied molecular orbital (HOMO) and lowest unoccupied molecular orbital (LUMO) energy levels of an organic complex photocatalyst are crucial for driving redox reactions during photocatalysis. Efficient and effective photoinduced electron transfer processes can only be achieved by aligning these energy levels with the redox potentials of the involved species, such as O_2/H_2O, using the correct alignment. Hence, a photon in the optical range of 1000 nm requires a minimum Gibbs free energy of 1.23 eV for water splitting. The electrode interfacial reactions of water splitting, especially for the OER, require the photocatalysts to have bandgaps of more than than 1.23 eV due to the high overpotential.

To convert solar energy efficiently, a semiconductor-based photocatalyst must first have a bandgap suitable for both efficient light absorption and the CB and VB potentials needed for the redox reactions that split water. Second, the ideal photocatalyst should be able to separate and transmit photogenerated charges with remarkable efficiency. Charge separation and recombination are competing processes in photocatalysis. The recombination of photoexcited charges frequently occurs within a short time interval (picoseconds to microseconds). Consequently, the water splitting redox reactions on the surface must take place during the lifetimes of the photoexcited charges [41]. The efficient separation and transfer of photogenerated charges is necessary for photocatalytic water splitting. Any methods that help with charge separation should be helpful in the design and construction of extremely effective photocatalyst systems. Third, the surface-active sites should take up the electrons and holes fast, so they can join the reactions with the adsorbed species. In general, the surface response occurs over a longer period (microseconds to seconds) than the charge formation and separation processes. Therefore, methods for increasing the surface catalytic reactions are also necessary to build photocatalyst systems that work well. An efficient method for hastening the catalytic process and raising the total water splitting efficiency is to load suitable cocatalysts onto the surface of the photocatalyst to create additional active sites. It is important to note that surface catalytic reactions and charge separation are not separate processes but rather have a synergistic correlation. Effective charge separation benefits from quicker surface reactivity and vice versa.

5.3 Quantum efficiency and solar-to-H_2 efficiency

In photochemistry, the quantum yield is a measure of the efficiency of a particular process, indicating the rate at which a certain event occurs for each photon absorbed in a given unit of time. The quantum yields of light emission, product creation, reactant disappearance, and other photochemical and photophysical processes that occur in photochemical reactions are frequently measured by photochemists [60]. Comparing the activities of photocatalysts is often challenging due to the substantial dependence of their photoactivity on experimental circumstances, such as light intensity and reaction temperature [38]. It is important to note that dividing the reacted electrons by the absorbed photons yields the true quantum efficiency (the internal quantum efficiency (IQE)). However, due to light scattering and loss, determining the quantity of photons absorbed by a particular photocatalytic system is difficult. In general, when H_2 or O_2 gas is produced by a certain photocatalyst system, the number of incident photons can be measured using a thermopile. Therefore, the gain in terms of quantum efficiency is the apparent quantum efficiency (AQE) at a particular monochromatic wavelength, which may be calculated using the following equation for a given photocatalyst [61].

$$
\begin{aligned}
AQE &= \frac{\text{Number of reactive electrons}}{\text{Number of incident photons}} \times 100\% \\
&= \frac{2 \times \text{Number of evolved } H_2 \text{ molecules}}{\text{Number of incident photons}} \times 100\% \\
&= \frac{4 \times \text{Number of evolved } O_2 \text{ molecules}}{\text{Number of incident photons}} \times 100\%
\end{aligned}
\tag{5.6}
$$

$$
\begin{aligned}
IQE &= \frac{\text{Number of reactive electrons}}{\text{Number of absorbed photons}} \times 100\% \\
&= \frac{2 \times \text{Number of evolved } H_2 \text{ molecules}}{\text{Number of absorbed photons}} \times 100\% \\
&= \frac{4 \times \text{Number of evolved } O_2 \text{ molecules}}{\text{Number of absorbed photons}} \times 100\%
\end{aligned}
\tag{5.7}
$$

A realistic metric for evaluating photocatalyst performance is the solar-to-hydrogen (STH) efficiency, which may be computed using the formula below:

$$
STH = \frac{\text{Output energy as } H_2 \text{ gas}}{\text{Energy of incident solar light}} = \frac{r_{H_2} \times \Delta G}{P_{sun} \times S} \times 100\%
\tag{5.8}
$$

where R_{H_2} is the rate of H_2 production (mmol s^{-1}), P_{sun} is the solar energy flux (mW cm^{-2}), S is the reactor area (cm^2), and G is the Gibbs free energy (kJ mol^{-1}).

5.4 Overall water splitting (hydrogen evolution/oxygen evolution)

The total water splitting efficiency can be used to quantify the photocatalytic water splitting capacity in the absence of sacrificial chemicals. H_2 and O_2 are produced in a 2:1 stoichiometric ratio during the entire process (equations (5.9) and (5.10)). Sacrificial species are added to the relevant half-cell reaction to quickly control the photogenerated holes and/or electrons, which is the rate-limiting phase (figure 5.4). For instance, in the HER (equations (5.11) and (5.12)), lactic acid, methanol, or triethanolamine is commonly employed as a hole scavenger and in the OER (equations (5.13) and (5.14)), Fe^{3+}, IO_3, or Ag^+ is used. Two considerations must be addressed for half-reactions: (1) determining the thermodynamic suitability of a photocatalyst's valence or conduction bands for proton reduction or water oxidation, and (2) calculating real-world proton reduction and water oxidation kinetic reaction rates.

The overall water splitting reaction:

$$\text{Reduction: } 4H^+ + 4e^- \rightarrow 2H_2 \tag{5.9}$$

$$\text{Oxidation: } 2H_2O + 4h^+ \rightarrow 4H^+ + O_2 \tag{5.10}$$

The H_2-generating half-reaction in the presence of CH_3OH:

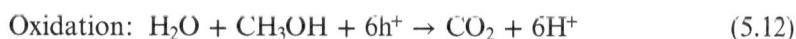

$$\text{Reduction: } 6H^+ + 6e^- \rightarrow 3H_2 \tag{5.11}$$

$$\text{Oxidation: } H_2O + CH_3OH + 6h^+ \rightarrow CO_2 + 6H^+ \tag{5.12}$$

The O_2-generating half-reaction in the presence of Ag^+:

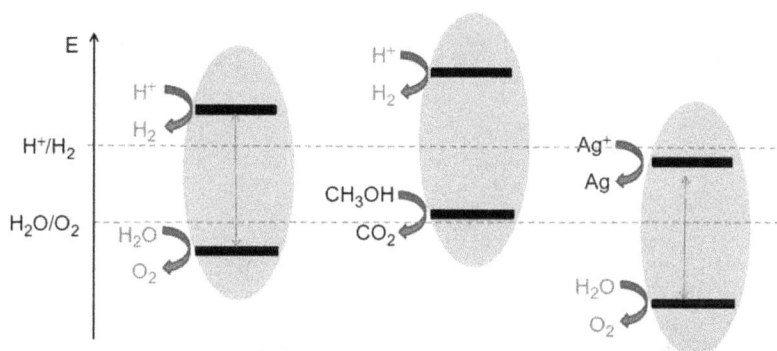

$$\text{Reduction: } 4Ag^+ + 4e^- \rightarrow 4Ag \tag{5.13}$$

$$\text{Oxidation: } 2H_2O + 4h^+ \rightarrow 4H^+ + O_2 \tag{5.14}$$

Figure 5.4. A schematic showing the half-reactions (the OER and HER) of photocatalytic water splitting in the presence of sacrificial agents. Reprinted from [59], Copyright (2017), with permission from Elsevier.

5.5 The basic concepts of photocatalysis

Thermal catalysis and photocatalysis are two subfields within the larger field of catalysis. The International Union of Pure and Applied Chemistry (IUPAC) defines photocatalysis as 'a change in the rate of a chemical reaction or its initiation in the presence of a substance—the photocatalyst—that absorbs light and is involved in the chemical transformation of the reaction partners.' At the turn of the 20th century, researchers believed that light irradiation might be utilized to stimulate a process. Between 1900 and 1920, Giacomo Ciamician was the first scientist to try to understand how light impacts chemical reactions [62]. However, the development of the field of photocatalysis began primarily in the 1970s [35, 63].

Figure 5.5(a) shows the catalytic conversion of a reactant (R) to a product (P) via active intermediate species denoted by I′ and C′. In the case of thermal catalysis, the activation of R occurs through the physical process of energy transfer, whereas in photocatalysis, it occurs when the excited state (C*) of a photocatalytic system transfers an electron, as schematically shown in figures 5.5(b) and (c). In photocatalysis, the electrical activation of R results in the production of an intermediate I (a radical or radical ion), which is theoretically also capable of being created by a thermal process.

After the electrons are transferred, C* deactivates to form C°. The intermediate I regenerates C in the final state and provides the end product P (potentially through

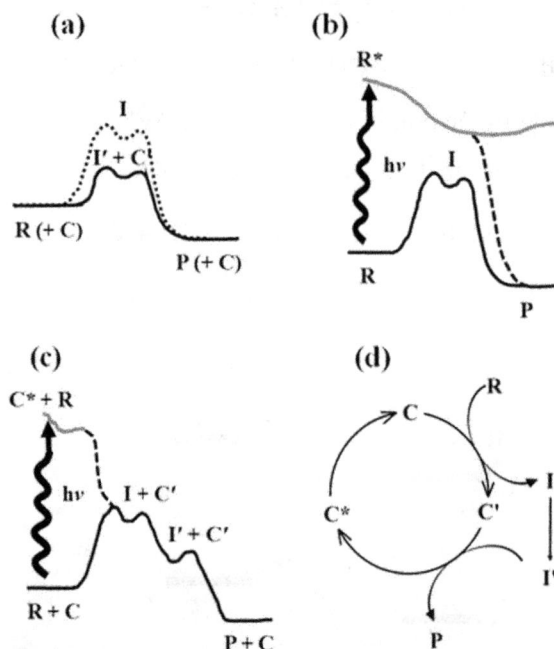

Figure 5.5. (a) A thermal reaction (R→P) that C catalyzes through the intermediate I°; (b) a photochemical reaction in which the chemical reaction starts when the reagent R* is in the excited state; (c) a photocatalyzed reaction in which R undergoes all of its chemical transformations on the ground-state surface, but the catalyst C is active only in the excited state; and (d) C serves as a photocatalyst in the R→P reaction. Reproduced from [64] with permission from the Royal Society of Chemistry.

further intermediates I'). In any reaction pathway, although the chemical route partially occurs on surface R* as depicted in figure 5.5(b), the whole chemical transformation of R to P takes place on the surface with the lowest potential energy. The process described, in which a new reaction pathway becomes available on the ground-state surface because of the absorption of light, is commonly referred to as photocatalysis [64, 65].

The shift in Gibbs energy (ΔG) is another significant characteristic that sets photocatalytic reactions apart from conventional thermal catalytic processes. ΔG may be neutral or favorable. Energy release is implied by a negative ΔG, whereas energy absorption or storage is denoted by a positive ΔG. Normal catalysts can only be used to catalyze negative ΔG reactions. This is due to the fact that a heated catalyst accelerates a chemical reaction that would continue spontaneously with a negative Gibbs energy change by minimizing the activation barrier of the transformation and altering the intermediate states. However, in photocatalytic processes, both scenarios are feasible [66]. For instance, even when ΔG is positive, an overall redox reaction can still happen, which explains why a reaction with positive ΔG cannot proceed spontaneously. Simply put, the reduction and oxidation processes must be physically or chemically distinct from one another for the reaction between the two products to occur without producing any net products. Under these conditions, the reactions of e^- and h^+ with the oxidant (ΔG_e) and the reductant (ΔG_h) must have a negative Gibbs energy change; that is, the reactions must occur spontaneously and be followed by photoexcitation, as shown in figure 5.6. In order to have a negative change in Gibbs energy for both processes, the photocatalyst material's CB bottom must be more cathodic and its VB top should be more anodic than the reference system potentials of both the oxidant and reductant species. Moreover, for both reactions, i.e. those with positive ΔG and those with negative ΔG, the reduction and oxidation must be separately performed by electrons (e^-) and holes (h^+), respectively [66].

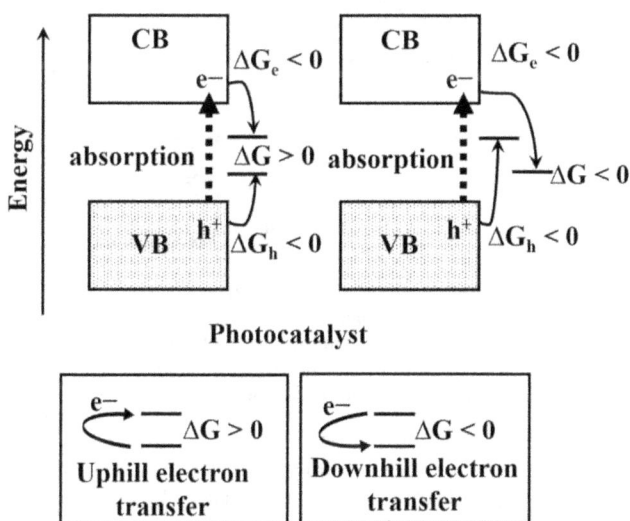

Figure 5.6. The change in Gibbs energy during photocatalytic reactions. Adopted from [66], Copyright (2010), with permission from Elsevier.

5.6 Elements of photocatalytic hydrogen generation physics and chemistry

Figure 5.7 illustrates the various photophysical and electrochemical processes that are involved in the production of photocatalytic hydrogen. The photophysical process includes the production of charge carriers and photon absorption as well as their separation and transport to the reaction sites, whereas the electrochemical process also includes redox reactions for water splitting. A photocatalyst absorbs photons of specific wavelengths when exposed to light, based on its inherent bandgap. Semiconductor photocatalysts can absorb photons that strike them at energies equal to or greater than their bandgap energy. An electron in the valence band gains energy from the absorbed photon, which enables it to cross the bandgap and move into the conduction band. A positively charged hole is left in the valence band as a result of this action. As a result, an electron–hole pair is produced, with the hole now in the valence band and the photoexcited electron now in the conduction band. Various photoinduced processes are made possible by the fact that the photoexcited electron and hole can move independently within the semiconductor material. Photon absorption results in the production of excited holes and electrons. To maintain equilibrium, excited electrons in the CB may radiatively or non-radiatively return to

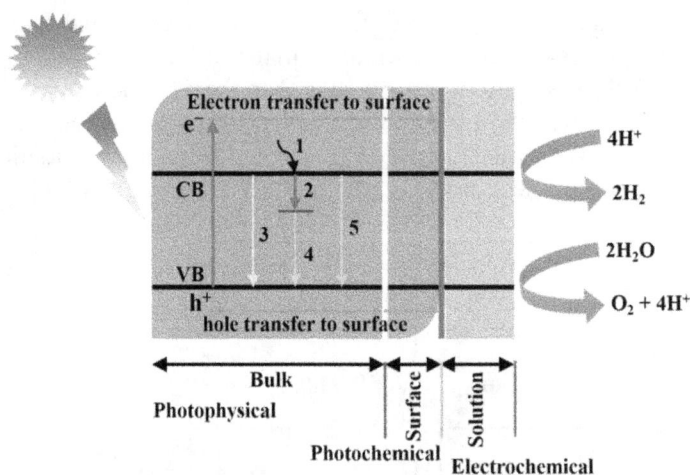

Figure 5.7. Photocatalytic water splitting involves three main dynamical processes (photophysical, photochemical, and electrochemical). The bandgap is excited by the following: (1) electron/hole relaxation at the band edges; (2) electron/hole trapping due to defects or surface states; (3) band-edge electron–hole or exciton recombination; (4) trapped electron–hole or relaxed exciton recombination; and (5) nonlinear exciton–exciton annihilation for electron/hole relaxation. Reprinted with permission from [71]. Copyright (2019) American Chemical Society.

the VB and recombine with holes. During recombination, the open electron–hole pairs are eliminated. This is an inevitable photophysical process, regardless of the type of semiconductor used as a photocatalyst [67, 68]. Recombination is an important cause of low QE in photocatalysts [69, 70].

The remaining electrons and holes must subsequently be delivered to the surface, where the HER and OER are carried out. Photoexcited carriers can be transported to the active surface sites via diffusion, or electric fields can be applied to the interface between the semiconductor and the electrolyte or between the semiconductor and the cocatalyst [70]. Excitons are produced on a timescale of less than 100 fs. Excitons generally only have lifespans of a few hundred picoseconds on average. Electrons can diffuse within a semiconductor after they are photoexcited from the VB to the CB. This diffusion involves the movement of electrons through the crystal lattice. Electron diffusion typically occurs on the order of a few picoseconds. A number of variables, including the diffusion coefficient and the distance that electrons must travel before scattering or recombining, can affect the diffusion time. On the other hand, the analogous transition time for holes is often even shorter, typically ranging from 100 to 300 femtoseconds (fs). Holes, as positive carriers, can move relatively quickly in comparison to electrons [72, 73]

5.7 Systems for producing hydrogen via photocatalysis

Water-splitting reactions can result from photocatalysis in two ways: (i) partially, in which only the reduction process is involved, or (ii) completely, in which both reduction and oxidation events are involved at the same time. Triethanolamine, alcohol, and other sacrificial electron donors and hole scavengers can be employed to produce the water splitting half-reaction. These agents help in realizing the reduction or oxidation reactions separately. When sacrificial agents are present, the photogenerated holes in the photocatalyst oxidize the sacrificial agent rather than water. The photocatalyst's electrons become more enriched as a consequence. Because this process depends on sacrificial agents, it is crucial to understand that it does not represent water splitting. The stoichiometric separation of water into H_2 and O_2 without the use of sacrificial agents is known as 'water splitting.' When sacrificial agents are present, the process should be differentiated from genuine water splitting since it entails the oxidation of the sacrificial agent. Figure 5.8 depicts examples of both one-step (single-photon) and two-step (two-photon) systems that are available for generic water splitting under visible-light irradiation [38]. The two-photon systems are also known as Z-scheme processes. For one-step complete water splitting, Z-scheme photocatalyst systems use a single semiconductor with band-edge positions that suit the H^+/H_2 and O_2/H_2O potentials [74]. A single semiconductor is used in Z-scheme systems, and its band-edge positions are chosen so that they align with the potentials for the evolution of both oxygen (O_2/H_2O) and hydrogen (H^+/H_2). This makes it possible to split water completely in one process. Separate hydrogen evolution photocatalysts (HEPs) and oxygen evolution photocatalysts (OEPs) are used in two-step (Z-scheme) water splitting systems to evolve H_2 and O_2, respectively.

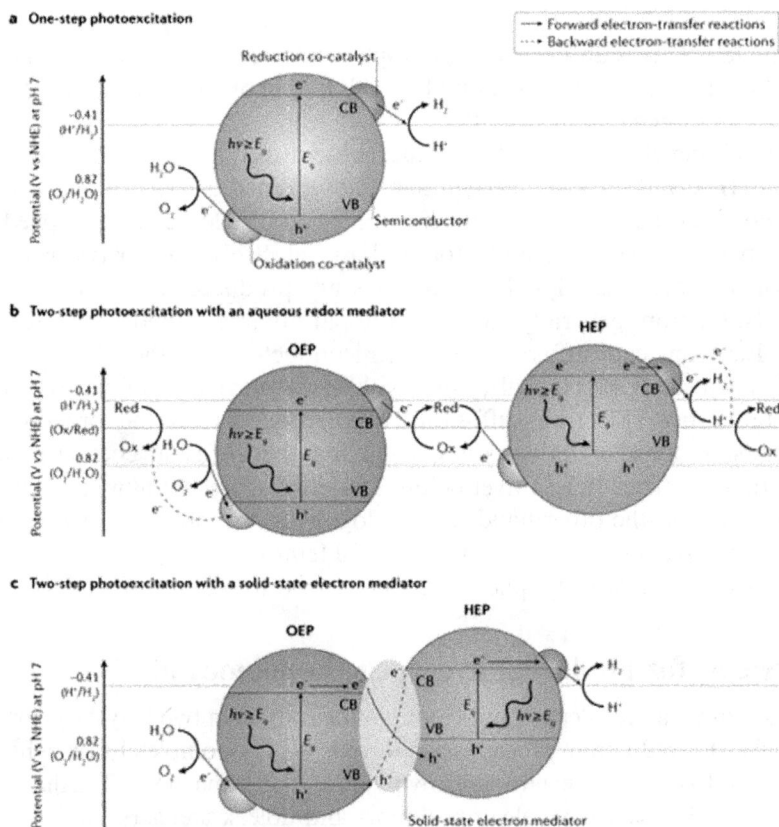

Figure 5.8. A schematic representation of several photocatalytic water splitting systems. Reprinted from [75], Copyright (2017), with permission from Nature.

It is important to note that a Z-scheme containing separate HEP and OEP photocatalysts for total water splitting is not a set rule. Semiconductors with HER or OER activity can be employed in this system. The Z-scheme thereby provides a stronger driving force by lowering the thermodynamic requirements for water splitting. The OEP and HEP in a Z-scheme system continue to have photoexcited electrons and holes, respectively. This is unlike a one-step system in which the entire photocatalytic cycle is completed in a single stage. The recombination of holes in the HEP and electrons in the OEP is necessary in the Z-scheme system to complete the photocatalytic cycle. Figures 5.8(b) and 5.8(c) illustrate the solid-state electron mediator and aqueous redox mediator that are used to facilitate this recombination. The Z-scheme system needs eight electrons to complete the cycle in two stages, as opposed to a one-step system that may require four electrons to complete the photocatalytic cycle. A one-step photoexcitation process produces twice as much H_2 and O_2 as a two-step photoexcitation system at given apparent quantum yield (AQY) and unity light absorption values [75]. Because of its susceptibility to reverse electron transfer, the Z-scheme system presents greater kinetic challenges than the

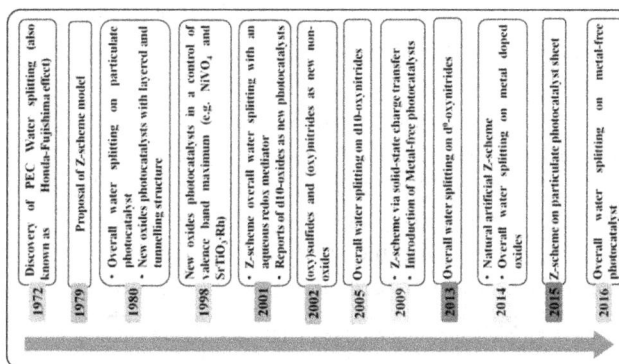

Figure 5.9. A list of significant advancements in photocatalytic water splitting over time.

one-step system. The effective functioning of a Z-scheme water splitting process depends on controlling the interparticle charge transfer between the HEP and the OEP in addition to preventing backward electron transmission.

5.8 Current advances in photocatalytic water splitting

Bard and his colleagues carried out a major push to advance photocatalytic water splitting in the late 1970s [76]. In summary, they discussed likely ways to integrate semiconductor electrodes and heterogeneous photocatalysts in PEC cells, emphasizing both the fundamental principles and practical applications of these components in the context of solar-driven processes. A significant amount of research has been done in the search for extremely reliable, robust, and affordable heterogeneous photocatalysts since 1980 [77]. Wide-bandgap metal oxides with specific cations, such as TiO_2 (titanium dioxide) with Ti^{4+}, are indeed known for their efficient photocatalytic properties under UV light. The key feature of these materials is their wide bandgap, which permits the absorption of UV light and generates electron–hole pairs, initiating photocatalytic reactions. Other metal oxides mentioned in the research literature, such as Zr^{4+}, Zn^{2+}, Mo^{6+}, Ce^{4+}, Nb^{5+}, Ge^{4+}, In^{3+}, Ta^{5+}, W^{6+}, Ga^{3+}, Sn^{4+}, Sb^{5+}, etc. may exhibit similar photocatalytic properties depending on their specific electronic configurations and bandgap characteristics. The efficiency of a photocatalyst is influenced by factors such as the bandgap width, charge carrier mobility, and recombination rate [28, 38]. It is worth noting that while TiO_2 has been widely studied and reported as an active photocatalyst under UV light because its energy levels suffice to launch the water splitting process [78]. Nitrides with d^{10} metal ions are also effective water splitting photocatalysts when exposed to UV light. In photocatalysis, non-oxide powder photocatalysts such as germanium nitride (Ge_3N_4) and cocatalyst-doped gallium nitride (GaN) have garnered interest [13, 79, 80].

With energy gaps between 4 and 3 eV, wide-bandgap photocatalysts are less effective at converting solar energy, since UV light makes up just 4% of the solar spectrum [81]. The photon absorption of wide-bandgap photocatalysts is enhanced by bandgap engineering approaches such as the use of visible-light sensitizers and metal/nonmetal ion doping [82–87]. These methods increase photon absorption, but they also affect the redox potentials, since the elements utilized to narrow the bandgap can also serve as sites of recombination for charge carriers produced by photosynthesis [84, 86]. Noble metals (Au, Ag, Pt, Rh, etc.) are effective sensitizers. Metal nanoparticles, especially Ag and Au, are used as cocatalysts and are gaining popularity as plasmonic photocatalysts [88, 89]. Large-scale uses are also limited by the expense of noble metals. Organic dyes are hindered by their instability; they break down quickly under longer-term irradiation, limiting their practicality [90]. The creation of photocatalysts that are active under visible light is essential for effective water splitting. When a sufficient electron acceptor is present, oxide photocatalysts such as WO_3 can be stable photocatalysts for O_2 evolution under visible light [91, 92].

The CB and VB positions in metal oxide photocatalysts are crucial for H_2O oxidation and reduction, respectively [93]. Strategies for making wide-bandgap metal oxides sensitive to visible light include doping them with transition metals, valence-band control using anions or p-block metal ions, and spectral sensitization [38]. In conclusion, the difficulties are in finding economical substitutes for noble metals, creating stable photocatalysts for successful water splitting that are active under visible light, and finding a balance between bandgap engineering to maximize absorption and minimizing recombination losses. The goal of ongoing research in these fields is to overcome these obstacles and open the door to useful uses for solar energy conversion.

A donor or acceptor level forms in the forbidden gap when a transition-metal cation or anion is doped into a metal oxide that has a wide bandgap; this acts as a focus for the absorption of visible light. Instead of introducing an energy band into the host material, doping introduces a distinct energy level. This may impede the fast movement of photogenerated carriers and thus affect the effectiveness of the photocatalytic process. The reason for maintaining charge balance is that dopants can function as recombination centers [94]. The production of a VB by orbitals other than those related to O2p is crucial for oxide photocatalysts to function in the presence of visible light [95, 96]. This is required to facilitate effective charge separation and allow for the absorption of visible light. Along with doping, wide-bandgap oxide semiconductor photocatalysts have been investigated in combination with visible-light sensitizers to introduce electrons into the conduction band for the formation of H_2. Small-bandgap semiconductors and organic dyes are common sensitizers. Sensitizers need to have a higher negative excited-state potential than conduction band potential of the metal oxide photocatalyst in order to effectively inject charge [97]. This ensures efficient injection of electrons into the CB, facilitating the photocatalytic reaction. In summary, while doping with transition metals introduces the levels required for visible-light absorption, it comes with challenges related to charge migration and recombination. The use of sensitizers is an alternative approach in which careful selection is needed to ensure effective charge injection into the metal oxide photocatalyst's conduction band. These considerations are essential for the fabrication of photocatalysts that are active under visible light for various applications, including the evolution of H_2. Further additions are supported oxide nanostructures for photoactivated H_2 synthesis [98], such as Ag–ZnO nanocomposites [99], Si–Co_3O_4 nanopyramids [100], CuO–ZnO nanorod arrays, etc [101].

Metal sulfides are also appealing photocatalysts for visible light. There have been several reports of metal sulfide photocatalysts [49, 102–104]. The VB of sulfide photocatalysts is made up of S3p orbitals, which are more negative than O2p orbitals. For instance, CdS is a renowned photocatalyst for visible light that has a small bandgap (2.4 eV). It possesses the necessary band levels for water reduction and oxidation in the presence of sacrificial agents [105]. ZnS, which has an energy gap of 3.6 eV, can be doped with different metal cations to exhibit strong photocatalytic activity for H_2 evolution from aqueous solutions that contain S^{2-} and/or SO_3^{2-} as electron donors when exposed to visible light [106]. Under visible light, H_2 evolution activity has been demonstrated in solid solutions including $AgGa_{0.9}In_{0.1}S_2$, CdS–ZnS, $AgInS_2$–ZnS, $CuInS_2$–ZnS, and $CuInS^{2-}AgInS_2$–ZnS [49, 104].

Despite this, in the presence of visible light, sulfur photocatalysts show excellent band levels for water splitting. But rather than O_2 evolution, they encounter difficulties with instability and deactivation, which are mainly caused by photo-corrosion or self-oxidation [107]. The bandgap of sulfide materials is narrowed by a VB that results from N2p and S3p orbitals in addition to O2p orbitals. These materials have a bandgap energies in the range of 1.7–2.5 eV, which corresponds to an absorption band of 500–750 nm [38]. Unlike (oxy)nitrides and (oxy)sulfides, materials doped with sulfur or nitrogen are referred to as VB-controlled photocatalysts. Instead of distinct impurity levels, the constituent anion component creates a VB. In strong acids such as aqua regia and highly concentrated H_2SO_4, valence-band-controlled photocatalysts, such as those doped with nitrogen or sulfur, are unstable [84]. (Oxy)nitrides may be partially degraded by photogenerated holes during the initial phases of the photocatalytic process, which results in reduced N_2 evolution. Nevertheless, as reaction progresses, N_2 generation becomes fully controlled. While (oxy)nitrides exhibit stable H_2O reduction and oxidation processes, achieving the required total water splitting performance is hindered by low levels of H_2 evolution. Several (oxy)nitrides that have d0 metal cations, such as Ti^{4+}, Nb^{5+}, and Ta^{5+}, are ineffective in H_2 and O_2 evolution when sacrificial agents are absent [108, 109]. In conclusion, problems arise from the instability of sulfide photocatalysts despite their potential band levels for water splitting. Valence-band-controlled photocatalysts, particularly those doped with nitrogen or sulfur, face stability issues in strong acids, and (oxy)nitrides have limitations in achieving total water splitting performance, especially when sacrificial agents are absent. Ongoing research aims to address these challenges and optimize the effectiveness of these materials in photocatalytic applications.

In 2009, Wang *et al* [51] introduced g-C_3N_4 as a metal-free photocatalyst for water splitting that is active under visible light. g-C_3N_4 is characterized by a medium bandgap of 2.7 eV, which allows it to absorb visible light. It exhibits strong physiochemical stability and has a changeable band structure. g-C_3N_4 can be synthesized in a single step by polymerizing inexpensive feedstocks such as cyanamide, urea, thiourea, melamine, and dicyandiamide [110]. The graphitic planes of g-C_3N_4 consist of tri-s-triazine units linked together by planar amino groups.

The response of g-C_3N_4 under visible light is attributed to electron transfer from the VB, filled by N2p orbitals, to the CB, populated by C2p orbitals. This electronic structure allows g-C_3N_4 to harness visible light for photocatalytic processes. The copolymerization of dicyandiamide with barbituric acid is mentioned as a way of extending the absorption band of g-C_3N_4 to 750 nm. Even without a cocatalyst, g-C_3N_4 has demonstrated photocatalytic H_2 evolution from water, particularly at wavelengths greater than 420 nm [108]. This suggests that g-C_3N_4 has inherent photocatalytic activity for water splitting in the presence of visible light. In summary, g-C_3N_4, with its medium bandgap, stable physiochemical properties, and adjustable band structure, has emerged as a notable visible-light-activated photocatalyst for water splitting. Its synthesis from readily available feedstocks and the ability to extend its absorption band further contribute to its potential in photocatalytic applications.

A new class of photocatalysts active under visible light has the potential to be formed from certain elemental semiconductors, such as Si, P, S, and Se, in addition to compound photocatalysts [111]. Among these, crystalline Si (1.12 eV) has attracted substantial research. Silicon may not be considered to be a material of interest for specific applications in which stability in aqueous environments and suitable band-edge positions are critical. Researchers and engineers often explore alternative materials or modify the properties of silicon to overcome these limitations and enhance its performance in various applications [112]. The photocatalytic activity of red P was initially shown by Wang et al [113]. They discovered that crystalline monoclinic red P may generate H_2 from water. Recent research suggests that amorphous sulfur may hold promise as a visible-light-activated photocatalyst. Ongoing studies and advancements in materials science may further elucidate the potential applications and optimization of this materials for photocatalysis [114]. However, it has low hydrophilicity, which reduces its photocatalytic activity. This issue might be solved by modifying the sulfur crystal structure. A material is heated to a certain temperature during the annealing process and subsequently cooled, usually to eliminate imperfections, promote crystallization, or improve the material's properties. In the case of amorphous selenium, the transition to the crystalline form at higher annealing temperatures may impact its photocatalytic properties or stability. This information underscores a challenge associated with using amorphous selenium as a photocatalyst and highlights the importance of understanding and controlling the annealing process to maintain the desired amorphous structure for optimal performance [115].

The development of primary photocatalysts is still in its early stages and suffers from numerous challenges. The more intriguing features of elemental semiconductors might be studied if unique methodologies could be developed. Creating heterojunctions of elemental photocatalysts or forming heterostructures with corresponding sulfides, phosphides, and selenides can lead to unexpected photocatalytic activity. This approach involves combining different materials to take advantage of their complementary properties, potentially synergizing their photo-catalytic performance. Investigating specific cocatalysts that may work well with elemental photocatalysts has been mentioned as another option for enhancing performance. Cocatalysts can facilitate charge separation and improve the overall efficiency of photocatalytic processes. A recent addition to metal-free photocatalysts involves organic compounds with C–C/C–H bonds in their π-conjugated organic frameworks. These organic compounds, when utilized as photocatalysts, contribute to the expanding range of materials for light-driven reactions [116–119]. Figure 5.9 depicts the chronology of important advancements in photocatalyst research.

5.9 Overall water splitting using photocatalysts

While designing a photocatalytic water splitting system, factors such as light absorption, photogenerated charge separation, and the use of suitable cocatalysts for catalytic processes should be considered. It is easy to categorize photocatalysts capable of overall water splitting into two main methods, as shown in Figure 5.10. One method involves employing a single photocatalyst to split water into H_2 and O_2 without the need for redox mediators. It is true that the two-step excitation process described is similar to the Z-scheme charge separation approach found in natural photosynthesis systems. It involves separate HEPs and OEPs in addition to an electronic redox mediator.

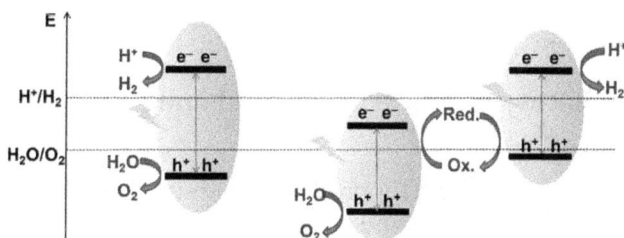

Figure 5.10. Schemes for single-photocatalyst photocatalytic water splitting and two-step photoexcitation photocatalyst systems using Z-schemes.

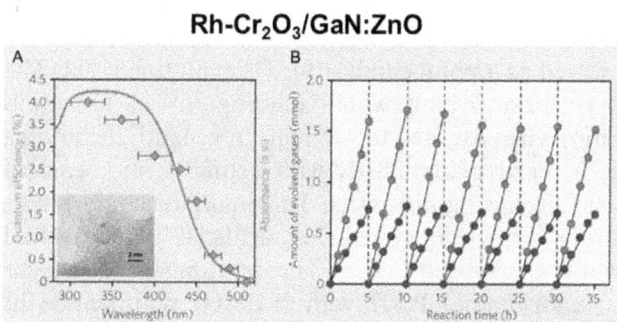

Figure 5.11. Rh–Cr_2O_3/GaN:ZnO—an example of a single photocatalyst for photocatalytic overall water splitting. Adapted from [13], Copyright (2006), with permission from Nature.

5.9.1 Photocatalyst systems with a single photocatalyst

It has been found that there are several single-photocatalyst systems for general water splitting. For instance, a La-doped $NiO-NaTaO_3$ photocatalyst system demonstrated excellent photocatalytic activity for water splitting, with a notable quantum efficiency of 56% at 270 nm [120]. It had a nano-step surface structure for the spatial separation of electrons and holes. Using diluted $CaCl_2$ solution, a Zn-ion-doped Ga_2O_3 photocatalyst was created which exhibited a substantially improved AQE of 71% when illuminated at 254 nm [121]. An Al-doped strontium titanate ($SrTiO_3$) treated with $SrCl_2$ flux has been reported; the AQE of this material was 30% at a wavelength of 360 nm when used for the overall water splitting process [122]. Moreover, gallium nitride (GaN) was loaded with zinc oxide (ZnO) and featured a cocatalyst combination of rhodium (Rh) and chromium oxide (Cr_2O_3). This material was described as responsive to visible light (figure 5.11), suggesting its ability to utilize visible light for the water splitting process. The reported AQE for this GaN:ZnO photocatalyst loaded with Rh/Cr_2O_3 cocatalyst was as high as 5.9%, and the measurement was obtained at a wavelength of 420 nm [13]. Other materials reported include $In_{1x}Ni_xTaO_4$ [123], $LaMg_xTa_{1x}O_{1+3x}N_{23x}$ [124], nitrogen-doped graphene oxide quantum dots [125], and InGaN/GaN nanowire [126]. While progress has been made in total water splitting using photocatalysts, the efficiency of particulate photocatalysts in achieving solar-to-hydrogen conversion remains a significant hurdle. Overcoming this challenge is crucial for realizing the full potential of photocatalytic water splitting as a sustainable and efficient method for hydrogen production.

5.10 Hybrid photosynthesis system: combining natural and artificial approaches

Natural systems offer a paradigm for processing and storing solar energy in the form of chemical fuels such as cellulose and glucose [127]. In recent years, it has also been possible to create hybrid systems by combining some natural photosystem components with synthetic photocatalysts. A two-step photoexcitation natural–artificial hybrid photosystem was built, wherein the $[Fe(CN_6)]^{3-} = [Fe(CN_6)]^{4-}$ donor/acceptor pair shown in figure 5.12 acted as an electronic mediator, a photosystem II (PS-II) membrane made from fresh spinach served as a catalyst for O_2 evolution, and $SrTiO_3$:Rh served as a photocatalyst for H_2 evolution [128]. The hybrid system may effectively perform overall water splitting (at H_2/O_2 levels close to the stoichiometric ratio) when exposed to sunlight. In general, the integration of natural biophoton systems, particularly PS-II, with artificial photocatalysts for efficient charge transfer has promising prospects. This interface could hold significance for the development of biohybrid systems or artificial photosynthesis technologies aimed at sustainable energy conversion. The development of a hybrid photoanode that combines cyanobacterial PS-II with a Fe_2O_3 photoanode has also greatly increased the efficiency of photoelectrochemical water splitting [129]. This is because effective electron transfer from PS-II to Fe_2O_3.

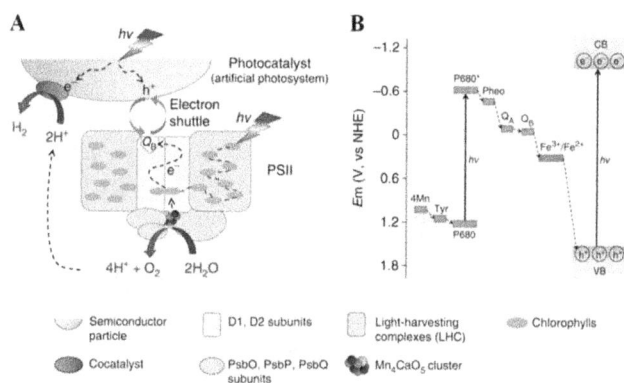

Figure 5.12. PS-II and artificial photocatalysts integrated into two-step photoexcitation natural–artificial hybrid photosystems. Adapted from [128], Copyright (2014), with permission from Nature.

5.11 Conclusions and future perspectives

The most recent developments in metal-free photocatalysts for water-to-hydrogen synthesis were reviewed in this chapter. To make things easier for the reader, we separated the proposed photocatalytic systems into four groups: elemental, binary, ternary, and organic. Among metal-free photocatalysts, carbon nitride is the most well-known and has AQE values of more than 60%, according to recent studies. As a photocatalyst of the future, carbon nitride exhibits excellent potential in terms of scalability, stability, and selectivity. On the other hand, the development of organic and elemental photocatalysts is still in its infancy. For improved quantum efficiency and long-term stable operation, a thorough investigation is needed to enhance their bulk and interfacial electron–hole separation, followed by their transport, and the mediators' redox behavior. Improved photocatalytic performance may be achieved by fine-tuning crystal morphologies and structural features at the nanoscale level, resulting in nanostructured photocatalysts with the desired physicochemical properties. Because of their highly specialized surface areas, easily accessible and interconnected porous networks, efficient charge transfer kinetics, oxidant–reductant interactions, and the active sites of their guest species, nanostructured photocatalysts are an efficient means of collecting light. The combined effects of these mechanisms greatly increase photocatalytic activity. While a lot of work has gone into studying different carbon nitride nanostructures, further investigation is needed to determine the relationship between structure and property and how it affects catalytic performance in other kinds of metal-free photocatalysts. Moreover, the interest in understanding the photophysics and surface processes of organic photocatalysts in aqueous solutions reflects a pursuit of knowledge critical for advancing the field of photocatalysis, especially in the context of sustainable water-related applications. Moreover, for polymeric photocatalysts, a comprehensive understanding of exciton dissociation and its impact on material characteristics is still lacking. The use of a donor–acceptor (D-A) system might potentially enhance exciton dissociation. The complete dissociation of excitons is facilitated by the additional energy produced during an excited donor electron's transfer to the acceptor. Interfacial dipoles and exciton delocalization may also encourage exciton dissociation.

A complicated set of photophysical and electrocatalytic processes are involved in the synthesis of solar fuel by photocatalysis. The six components of photocatalytic reactions are as follows: mass transfer of reactants and products, photon absorption, exciton separation, carrier diffusion, carrier transport, and catalytic efficiency. The fundamental efficiencies are impacted by these six characteristics in different ways. To obtain a high AQE, all of these characteristics need be optimized, even if they occur at various timescales and spatial resolutions. Addressing this topic seems to be necessary since different linear, planarized, non-crosslinked, and nonporous conjugated polymers have been reported to have photocatalytic activity that equals or exceeds that of microporous polymer photocatalysts. The processes involved in producing solar fuels through photocatalysis are both complex and sophisticated. The integration of photophysical and electrocatalytic reactions is crucial to developing efficient and sustainable technologies for solar-driven fuel generation. The impacts of the six steps of photocatalytic reactions on fundamental efficiency are diverse. Although they happen at different timescales and spatial resolutions, each of these features has to be adjusted to achieve high AQE values.

There are no standards by which to evaluate and contrast the different claimed photocatalyst efficiencies. Several parameters, including the sample mass, reactor setup, light source, and filters utilized, affect the overall amount of hydrogen created per unit of time. As such, when examined in different laboratories, the same material may yield different findings. The authors contend that stating the mass catalyst $(mol\ h^{-1}\ g^{-1})$ and the rate of hydrogen generation in units per time exacerbates the issue, since the mass of the catalyst and hydrogen production do not follow a linear relationship. Reporting AQE values presents a similar problem, as they are often not measured according to a strict protocol. The production of solar fuels provides opportunities to stop catastrophic climate change. It is essential to comprehend what functions and why in order to evaluate solar fuel initiatives. In order to arrive at a definitive answer, we must first apply our combined expertise to conduct a thorough global analysis of energy-innovation initiatives in order to identify optimal approaches.

References and further reading

[1] Lewis N S and Nocera D G 2006 Powering the planet: chemical challenges in solar energy utilization *Proc. Natl Acad. Sci. USA* **103** 15729–35
[2] Trancik J E 2014 Back the renewables boom *Nature* **507** 300–2
[3] Zou X and Zhang Y 2015 Noble metal-free hydrogen evolution catalysts for water splitting *Chem. Soc. Rev.* **44** 5148–80
[4] Cook T R, Dogutan D K, Reece S Y, Surendranath Y, Teets T S and Nocera D G 2010 Solar Energy supply and storage for the legacy and nonlegacy worlds *Chem. Rev.* **110** 6474–502

[5] Subbaraman R, Tripkovic D, Strmcnik D, Chang K C, Uchimura M, Paulikas A P, Stamenkovic V and Markovic N M 2011 Enhancing hydrogen evolution activity in water splitting by tailoring Li^+–$Ni(OH)_2$–Pt interfaces *Science (1979)* **334** 1256–60

[6] Xiao Y, Hu T, Zhao X, Hu F X, Yang H B and Li C M 2020 Thermo-selenizing to rationally tune surface composition and evolve structure of stainless steel to electrocatalytically boost oxygen evolution reaction *Nano Energy* **75** 104949

[7] Yao Q, DIng Y and Lu Z H 2020 Noble-metal-free nanocatalysts for hydrogen generation from boron- and nitrogen-based hydrides *Inorg. Chem. Front.* **7** 3837–74

[8] Chen S, Huang D, Xu P, Xue W, Lei L, Cheng M, Wang R, Liu X and Deng R 2020 Semiconductor-based photocatalysts for photocatalytic and photoelectrochemical water splitting: will we stop with photocorrosion? *J. Mater. Chem. A Mater.* **8** 2286–322

[9] Bonaccorso F, Colombo L, Yu G, Stoller M, Tozzini V, Ferrari A C, Ruoff R S and Pellegrini V 2015 Graphene, related two-dimensional crystals, and hybrid systems for energy conversion and storage *Science (1979)* **347** 1246501

[10] Jensen L (ed.) 2021 *The Sustainable Development Goals Report 2021* United Nations Statistics Division Development Data and Outreach Branch, New York https://unstats.un.org/sdgs/report/2021/

[11] Bonde J, Moses P G, Jaramillo T F, Nørskov J K and Chorkendorff I 2008 Hydrogen evolution on nano-particulate transition metal sulfides *Faraday Discuss.* **140** 219–31

[12] Mamiyev Z Q and Balayeva N O 2015 Preparation and optical studies of PbS nanoparticles *Opt. Mater. (Amst)* **46** 522–5

[13] Maeda K, Teramura K, Lu D, Takata T, Saito N, Inoue Y and Domen K 2006 Photocatalyst releasing hydrogen from water *Nature* **440** 295 295

[14] Balayeva N O and Mamiyev Z Q 2016 Synthesis and characterization of Ag_2S/PVA–fullerene (C_{60}) nanocomposites *Mater. Lett.* **175** 231–5

[15] Baran T, Wojtyła S, Dibenedetto A, Aresta M and Macyk W 2015 Zinc sulfide functionalized with ruthenium nanoparticles for photocatalytic reduction of CO_2 *Appl. Catal. B* **178** 170–6

[16] Ganapathy M, Chang C T and Alagan V 2022 Facile preparation of amorphous $SrTiO_3$-crystalline PbS heterojunction for efficient photocatalytic hydrogen production *Int. J. Hydrogen Energy* **47** 27555–65

[17] Reber J F and Meier K 1984 Photochemical production of hydrogen with zinc sulfide suspensions *J. Phys. Chem.* **88** 5903–13

[18] Keimer B and Moore J E 2017 The physics of quantum materials *Nat. Phys.* **13** 1045–55

[19] Sivula K and Van De Krol R 2016 Semiconducting materials for photoelectrochemical energy conversion *Nat. Rev. Mater.* **1** 15010

[20] Amirav L and Alivisatos A P 2010 Photocatalytic hydrogen production with tunable nanorod heterostructures *J. Phys. Chem. Lett.* **1** 1051–4

[21] Shiga Y, Umezawa N, Srinivasan N, Koyasu S, Sakai E and Miyauchi M 2016 A metal sulfide photocatalyst composed of ubiquitous elements for solar hydrogen production *Chem. Commun.* **52** 7470–3

[22] Hou H, Yuan Y, Cao S, Yang Y, Ye X and Yang W 2020 $CuInS_2$ nanoparticles embedded in mesoporous TiO_2 nanofibers for boosted photocatalytic hydrogen production *J. Mater. Chem. C Mater.* **8** 11001–7

[23] Caudillo-Flores U, Kubacka A, Berestok T, Zhang T, Llorca J, Arbiol J, Cabot A and Fernández-García M 2020 Hydrogen photogeneration using ternary $CuGaS_2$–TiO_2–Pt nanocomposites *Int. J. Hydrogen Energy* **45** 1510–20

[24] Singh J and Soni R K 2021 Enhanced sunlight driven photocatalytic activity of In_2S_3 nanosheets functionalized MoS_2 nanoflowers heterostructures *Sci. Rep.* **11** 14

[25] Armaroli N and Balzani V 2016 Solar electricity and solar fuels: status and perspectives in the context of the energy transition *Chem.—A Eur. J.* **22** 32–57

[26] Wang W, Yu J C, Xia D, Wong P K and Li Y 2013 Graphene and g-C 3 N 4 nanosheets cowrapped elemental α-sulfur as a novel metal-free heterojunction photocatalyst for bacterial inactivation under visible-light *Environ. Sci. Technol.* **47** 8724–32

[27] Markard J 2018 The next phase of the energy transition and its implications for research and policy *Nat. Energy* **3** 628–33

[28] Maeda K 2011 Photocatalytic water splitting using semiconductor particles: history and recent developments *J. Photochem. Photobiol., C* **12** 237–68

[29] Rahman M Z, Kibria M G and Mullins C B 2020 Metal-free photocatalysts for hydrogen evolution *Chem. Soc. Rev.* **49** 1887–931

[30] Brousse T *et al* 2017 Materials for electrochemical capacitors *Springer Handbook of Electrochemical Energy* (Berlin: Springer) 495–561

[31] Turner J A 2004 Sustainable hydrogen production *Science (1979)* **305** 972–4

[32] Schlapbach L and Züttel A 2001 Hydrogen-storage materials for mobile applications *Nature* **414** 353–8

[33] Navarro R M, Peña M A and Fierro J L G 2007 Hydrogen production reactions from carbon feedstocks: fossil fuels and biomass *Chem. Rev.* **107** 3952–91

[34] Kumaravel V and Kang M 2020 Photocatalytic hydrogen evolution *Catalysts* **10** 6–7

[35] Fujishima A and Honda K 1972 Electrochemical photolysis of water at a semiconductor electrode *Nature* **238** 37–8

[36] Kumaravel V, Imam M D, Badreldin A, Chava R K, Do J Y, Kang M and Abdel-Wahab A 2019 Photocatalytic hydrogen production: role of sacrificial reagents on the activity of oxide, carbon, and sulfide catalysts *Catalysts* **9** 276

[37] Chen X, Shen S, Guo L and Mao S S 2010 Semiconductor-based photocatalytic hydrogen generation *Chem. Rev.* **110** 6503–70

[38] Kudo A and Miseki Y 2009 Heterogeneous photocatalyst materials for water splitting *Chem. Soc. Rev.* **38** 253–78

[39] Osterloh F E 2008 Inorganic materials as catalysts for photochemical splitting of water *Chem. Mater.* **20** 35–54

[40] Moriya Y, Takata T and Domen K 2013 Recent progress in the development of (oxy) nitride photocatalysts for water splitting under visible-light irradiation *Coord. Chem. Rev.* **257** 1957–69

[41] Hisatomi T, Kubota J and Domen K 2014 Recent advances in semiconductors for photocatalytic and photoelectrochemical water splitting *Chem. Soc. Rev.* **43** 7520–35

[42] Li X, Low J and Yu J 2016 Photocatalytic Hydrogen Generation *Photocatalysis: Applications* ed D D Dionysiou *et al* (London: The Royal Society of Chemistry) 10

[43] Chun W J, Ishikawa A, Fujisawa H, Takata T, Kondo J N, Hara M, Kawai M, Matsumoto Y and Domen K 2003 Conduction and valence band positions of Ta_2O_5, TaOn, and Ta_3N_5 by UPS and electrochemical methods *J. Phys. Chem.* B **107** 1798–803

[44] Hitoki G, Ishikawa A, Takata T, Kondo J N, Hara M and Domen K 2002 Ta_3N_5 as a novel visible light-driven photocatalyst ($\lambda < 600nm$) *Chem. Lett.* **31** 736–7

[45] Hitoki G, Takata T, Kondo J N, Hara M, Kobayashi H and Domen K 2002 An oxynitride, TaON, as an efficient water oxidation photocatalyst under visible light irradiation ($\lambda \leqslant 500$ nm) *Chem. Commun.* **2** 1698–9

[46] Maeda K, Takata T, Hara M, Saito N, Inoue Y, Kobayashi H and Domen K 2005 GaN: ZnO solid solution as a photocatalyst for visible-light-driven overall water splitting *J. Am. Chem. Soc.* **127** 8286–7

[47] Liu M, Jing D, Zhou Z and Guo L 2013 Twin-induced one-dimensional homojunctions yield high quantum efficiency for solar hydrogen generation *Nat. Commun.* **4** 2278

[48] Zhang K and Guo L 2013 Metal sulphide semiconductors for photocatalytic hydrogen production *Catal. Sci. Technol.* **3** 1672–90

[49] Tsuji I, Kato H, Kobayashi H and Kudo A 2004 Photocatalytic H_2 evolution reaction from aqueous solutions over band structure-controlled $(AgIn)_xZn_{2(1-x)}S_2$ solid solution photocatalysts with visible-light response and their surface nanostructures *J. Am. Chem. Soc.* **126** 13406–13

[50] Liao L *et al* 2014 Efficient solar water-splitting using a nanocrystalline CoO photocatalyst *Nat. Nanotechnol.* **9** 69–73

[51] Wang X, Maeda K, Thomas A, Takanabe K, Xin G, Carlsson J M, Domen K and Antonietti M 2009 A metal-free polymeric photocatalyst for hydrogen production from water under visible light *Nat. Mater.* **8** 76–80

[52] Cao S, Low J, Yu J and Jaroniec M 2015 Polymeric photocatalysts based on graphitic carbon nitride *Adv. Mater.* **27** 2150–76

[53] Wang X, Blechert S and Antonietti M 2012 Polymeric graphitic carbon nitride for heterogeneous photocatalysis *ACS Catal.* **2** 1596–606

[54] Rahman M Z and Edvinsson T 2019 What is limiting pyrite solar cell performance? *Joule* **3** 2290–3

[55] Rahman M Z and Edvinsson T 2019 How to make a most stable perovskite solar cell *Matter* **1** 562–4

[56] Pender J P, Guerrera J V, Wygant B R, Weeks J A, Ciufo R A, Burrow J N, Walk M F, Rahman M Z, Heller A and Mullins C B 2019 Carbon nitride transforms into a high lithium storage capacity nitrogen-rich carbon *ACS Nano* **13** 9279–91

[57] Linsebigler A L, Lu G and Yates J T 1995 Photocatalysis on TiO_2 surfaces: principles, mechanisms, and selected results *Chem. Rev.* **95** 735–58

[58] Chen X and Mao S S 2007 Titanium dioxide nanomaterials: synthesis, properties, modifications and applications *Chem. Rev.* **107** 2891–959

[59] Li R and Li C 2017 Photocatalytic Water Splitting on Semiconductor-Based Photocatalysts *Advances in Catalysis* ed C Song (Amsterdam: Elsevier) **60** 1–57

[60] Serpone N 1997 Relative photonic efficiencies and quantum yields in heterogeneous photocatalysis *J. Photochem. Photobiol. A Chem.* **104** 1–12

[61] Sayama K and Arakawa H 1997 Effect of carbonate salt addition on the photocatalytic decomposition of liquid water over $Pt–TiO_2$ catalyst *J. Chem. Soc., Faraday Trans.* **93** 1647–54

[62] Ciamician G 1912 The photochemistry of the future *Science* **36** 385–94

[63] Kisch H 2013 Semiconductor photocatalysis—mechanistic and synthetic aspects *Angew. Chem. Int. Ed.* **52** 812–47

[64] Ravelli D, Dondi D, Fagnoni M and Albini A 2009 Photocatalysis. A multi-faceted concept for green chemistry *Chem. Soc. Rev.* **38** 1999–2011

[65] Fagnoni M, Dondi D, Ravelli D and Albini A 2007 Photocatalysis for the formation of the C–C bond *Chem. Rev.* **107** 2725–56

[66] Ohtani B 2010 Photocatalysis A to Z: what we know and what we do not know in a scientific sense *J. Photochem. Photobiol.*, C **11** 157–78

[67] Rahman M Z 2014 Advances in surface passivation and emitter optimization techniques of c-Si solar cells *Renew. Sustain. Energy Rev.* **30** 734–42

[68] Yue M, Lambert H, Pahon E, Roche R, Jemei S and Hissel D 2021 Hydrogen energy systems: a critical review of technologies, applications, trends and challenges *Renew. Sustain. Energy Rev.* **146** 111180

[69] Rahman M Z, Tang Y and Kwong P 2018 Reduced recombination and low-resistive transport of electrons for photo-redox reactions in metal-free hybrid photocatalyst *Appl. Phys. Lett.* **112** 253902

[70] Hisatomi T, Takanabe K and Domen K 2015 Photocatalytic water-splitting reaction from catalytic and kinetic perspectives *Catal. Lett.* **145** 95–108

[71] Rahman M Z and Mullins C B 2019 Understanding charge transport in carbon nitride for enhanced photocatalytic solar fuel production *Acc. Chem. Res.* **52** 248–57

[72] Kroeze J E, Savenije T J and Warman J M 2004 Electrodeless determination of the trap density, decay kinetics, and charge separation efficiency of dye-sensitized nanocrystalline TiO_2 *J. Am. Chem. Soc.* **126** 7608–18

[73] Enright B and Fitzmaurice D 1996 Spectroscopic determination of electron and hole effective masses in a nanocrystalline semiconductor film *J. Phys. Chem.* **100** 1027–35

[74] Maeda K 2013 Z-scheme water splitting using two different semiconductor photocatalysts *ACS Catal.* **3** 1486–503

[75] Chen S, Takata T and Domen K 2017 Particulate photocatalysts for overall water splitting *Nat. Rev. Mater.* **2** 17050

[76] Bard A J 1980 Photoelectrochemistry *Science (1979)* **207** 139–44

[77] Serpone N and Emeline A V 2012 Semiconductor photocatalysis—past, present, and future outlook *J. Phys. Chem. Lett.* **3** 673–7

[78] Fujishima A, Zhang X and Tryk D A 2008 TiO_2 photocatalysis and related surface phenomena *Surf. Sci. Rep.* **63** 515–82

[79] Sato J, Saito N, Yamada Y, Maeda K, Takata T, Kondo J N, Hara M, Kobayashi H, Domen K and Inoue Y 2005 RuO_2-loaded β-Ge_3N_4 as a non-oxide photocatalyst for overall water splitting *J. Am. Chem. Soc.* **127** 4150–1

[80] Arai N, Saito N, Nishiyama H, Inoue Y, Domen K and Sato K 2006 Overall water splitting by RuO_2-dispersed divalent-ion-doped GaN photocatalysts with d^{10} electronic configuration *Chem. Lett.* **35** 796

[81] Gueymard C A 2004 The sun's total and spectral irradiance for solar energy applications and solar radiation models *Sol. Energy* **76** 423–53

[82] Niishiro R, Kato H and Kudo A 2005 Nickel and either tantalum or niobium-codoped TiO_2 and $SrTiO_3$ photocatalysts with visible-light response for H_2 or O_2 evolution from aqueous solutions *Phys. Chem. Chem. Phys.* **7** 2241

[83] Sakata Y, Matsuda Y, Yanagida T, Hirata K, Imamura H and Teramura K 2008 Effect of metal ion addition in a Ni supported Ga_2O_3 photocatalyst on the photocatalytic overall splitting of H_2O *Catal. Lett.* **125** 22–6

[84] Asahi R, Morikawa T, Ohwaki T, Aoki K and Taga Y 2001 Visible-light photocatalysis in nitrogen-doped titanium oxides *Science (1979)* **293** 269–71

[85] Yashima M, Lee Y and Domen K 2007 Crystal structure and electron density of tantalum oxynitride, a visible light responsive photocatalyst *Chem. Mater.* **19** 588–93

[86] Li X, Kikugawa N and Ye J 2008 Nitrogen-doped lamellar niobic acid with visible light-responsive photocatalytic activity *Adv. Mater.* **20** 3816–9

[87] Fang J, Cao S W, Wang Z, Shahjamali M M, Loo S C J, Barber J and Xue C 2012 Mesoporous plasmonic Au–TiO$_2$ nanocomposites for efficient visible-light-driven photocatalytic water reduction *Int. J. Hydrogen Energy* **37** 17853–61

[88] Kowalska E, Mahaney O O P, Abe R and Ohtani B 2010 Visible-light-induced photocatalysis through surface plasmon excitation of gold on titania surfaces *Phys. Chem. Chem. Phys.* **12** 2344–55

[89] Méndez-Medrano M G, Kowalska E, Lehoux A, Herissan A, Ohtani B, Rau S, Colbeau-Justin C, Rodríguez-López J L and Remita H 2016 Surface modification of TiO$_2$ with Au nanoclusters for efficient water treatment and hydrogen generation under visible light *J. Phys. Chem. C* **120** 25010–22

[90] Bae E, Choi W, Park J, Shin H S, Kim S B and Lee J S 2004 Effects of surface anchoring groups (carboxylate vs phosphonate) in ruthenium-complex-sensitized TiO$_2$ on visible light reactivity in aqueous suspensions *J. Phys. Chem. B* **108** 14093–101

[91] Darwent J R and Mills A 1982 Photo-oxidation of water sensitized by WO$_3$ powder *J. Chem. Soc., Faraday Trans. 2* **78** 359–67

[92] Erbs W, Desilvestro J, Borgarello E and Grätzel M 1984 Visible-light-induced O$_2$ generation from aqueous dispersions of WO$_3$ *J. Phys. Chem.* **88** 4001–6

[93] Scaife D E 1980 Oxide semiconductors in photoelectrochemical conversion of solar energy *Sol. Energy* **25** 41–54

[94] Kato H and Kudo A 2002 Visible-light-response and photocatalytic activities of TiO$_2$ and SrTiO$_3$ photocatalysts codoped with antimony and chromium *J. Phys. Chem. B* **106** 5029–34

[95] Kim H G, Hwang D W and Lee J S 2004 An undoped, single-phase oxide photocatalyst working under visible light *J. Am. Chem. Soc.* **126** 8912–3

[96] Hosogi Y, Shimodaira Y, Kato H, Kobayashi H and Kudo A 2008 Role of Sn^{2+} in the band structure of SnM$_2$O$_6$ and Sn$_2$M$_2$O$_7$ (M = Nb and Ta) and their photocatalytic properties *Chem. Mater.* **20** 1299–1307

[97] Abe R, Sayama K and Arakawa H 2003 Significant influence of solvent on hydrogen production from aqueous I$_3^-$/I$^-$ redox solution using dye-sensitized Pt/TiO$_2$ photocatalyst under visible light irradiation *Chem. Phys. Lett.* **379** 230–5

[98] Cargnello M, Gasparotto A, Gombac V, Montini T, Barreca D and Fornasiero P 2011 Photocatalytic H$_2$ and added-value by-products-the role of metal oxide systems in their synthesis from oxygenates *Eur. J. Inorg. Chem.* **2011** 4309–23

[99] Simon Q *et al* 2011 Plasma-assisted synthesis of Ag/ZnO nanocomposites: first example of photo-induced H$_2$ production and sensing *Int. J. Hydrogen Energy* **36** 15527–37

[100] Gasparotto A *et al* 2011 F-doped Co$_3$O$_4$ photocatalysts for sustainable H$_2$ generation from water/ethanol *J. Am. Chem. Soc.* **133** 19362–5

[101] Simon Q, Barreca D, Gasparotto A, MacCato C, Montini T, Gombac V, Fornasiero P, Lebedev O I, Turner S and Van Tendeloo G 2012 Vertically oriented CuO/ZnO nanorod

arrays: from plasma-assisted synthesis to photocatalytic H_2 production *J. Mater. Chem.* **22** 11739–47

[102] Zheng N, Bu X, Vu H and Feng P 2005 Open-framework chalcogenides as visible-light photocatalysts for hydrogen generation from water *Angew. Chem. Int. Ed.* **44** 5299–303

[103] Zheng N, Bu X and Feng P 2005 $Na_5(In_4S)(InS_4)_3 \cdot 6H_2O$, a zeolite-like structure with unusual SIn_4 tetrahedra *J. Am. Chem. Soc.* **127** 5286–7

[104] Tsuji I, Kato H and Kudo A 2005 Visible-light-induced H_2 evolution from an aqueous solution containing sulfide and sulfite over a $ZnS–CuInS_2–AgInS_2$ solid-solution photocatalyst *Angew. Chem. Int. Ed.* **44** 3565–8

[105] Acar C, Dincer I and Zamfirescu C 2014 A review on selected heterogeneous photocatalysts for hydrogen production *Int. J. Energy Res.* **38** 1903–20

[106] Kakuta N, Park K H, Finlayson M F, Ueno A, Bard A J, Campion A, Fox M A, Webber S E and White J M 1985 Photoassisted hydrogen production using visible light and coprecipitated $ZnS \cdot CdS$ without a noble metal *J. Phys. Chem.* **89** 732–4

[107] Shangguan W and Yoshida A 2002 Photocatalytic hydrogen evolution from water on nanocomposites incorporating cadmium sulfide into the interlayer *J. Phys. Chem.* B **106** 12227–30

[108] Zhang J, Chen X, Takanabe K, Maeda K, Domen K, Epping J D, Fu X, Antonieta M and Wang X 2010 Synthesis of a carbon nitride structure for visible-light catalysis by copolymerization *Angew. Chem.—Int. Ed.* **49** 441–4

[109] Hara M, Takata T, Kondo J N and Domen K 2004 Photocatalytic reduction of water by TaON under visible light irradiation *Catal. Today* **90** 313–7

[110] Kessler F K, Zheng Y, Schwarz D, Merschjann C, Schnick W, Wang X and Bojdys M J 2017 Functional carbon nitride materials-design strategies for electrochemical devices *Nat. Rev. Mater.* **2** 17030

[111] Liu G, Niu P and Cheng H M 2013 Visible-light-active elemental photocatalysts *ChemPhysChem.* **14** 885–92

[112] Peng C, Gao J, Wang S, Zhang X, Zhang X and Sun X 2011 Stability of hydrogen-terminated surfaces of silicon nanowires in aqueous solutions *J. Phys. Chem.* C **115** 3866–71

[113] Wang F, Ng W K H, Yu J C, Zhu H, Li C, Zhang L, Liu Z and Li Q 2012 Red phosphorus: an elemental photocatalyst for hydrogen formation from water *Appl. Catal.* B **111–2** 409–14

[114] Liu G, Niu P, Yin L and Cheng H M 2012 α-sulfur crystals as a visible-light-active photocatalyst *J. Am. Chem. Soc.* **134** 9070–3

[115] Chiou Y D and Hsu Y J 2011 Room-temperature synthesis of single-crystalline Se nanorods with remarkable photocatalytic properties *Appl. Catal.* B **105** 211–9

[116] Wang Y *et al* 2019 Current understanding and challenges of solar-driven hydrogen generation using polymeric photocatalysts *Nat. Energy* **4** 746–60

[117] Ghosh S, Kouamé N A, Ramos L, Remita S, Dazzi A, Deniset-Besseau A, Beaunier P, Goubard F, Aubert P H and Remita H 2015 Conducting polymer nanostructures for photocatalysis under visible light *Nat. Mater.* **14** 505–11

[118] Ghosh S, Kouame N A, Remita S, Ramos L, Goubard F, Aubert P H, Dazzi A, Deniset-Besseau A and Remita H 2015 Visible-light active conducting polymer nanostructures with superior photocatalytic activity *Sci. Rep.* **5** 18002

[119] Yuan X, Dragoe D, Beaunier P, Uribe D B, Ramos L, Méndez-Medrano M G and Remita H 2020 Polypyrrole nanostructures modified with mono- and bimetallic nanoparticles for photocatalytic H$_2$ generation *J. Mater. Chem.* A **8** 268–77

[120] Kato H, Asakura K and Kudo A 2003 Highly efficient water splitting into H$_2$ and O$_2$ over lanthanum-doped NaTaO$_3$ photocatalysts with high crystallinity and surface nanostructure *J. Am. Chem. Soc.* **125** 3082–9

[121] Sakata Y, Hayashi T, Yasunaga R, Yanaga N and Imamura H 2015 Remarkably high apparent quantum yield of the overall photocatalytic H$_2$O splitting achieved by utilizing Zn ion added Ga$_2$O$_3$ prepared using dilute CaCl$_2$ solution *Chem. Commun.* **51** 12935–8

[122] Ham Y, Hisatomi T, Goto Y, Moriya Y, Sakata Y, Yamakata A, Kubota J and Domen K 2016 Flux-mediated doping of SrTiO$_3$ photocatalysts for efficient overall water splitting *J. Mater. Chem.* A **4** 3027–33

[123] Zou Z, Ye J, Sayama K and Arakawa H 2001 Direct splitting of water under visible light irradiation with an oxide semiconductor photocatalyst *Nature* **414** 625–7

[124] Pan C, Takata T, Nakabayashi M, Matsumoto T, Shibata N, Ikuhara Y and Domen K 2015 A complex perovskite-type oxynitride: the first photocatalyst for water splitting operable at up to 600 nm *Angew. Chem.—Int. Ed.* **54** 2955–9

[125] Yeh T F, Teng C Y, Chen S J and Teng H 2014 Nitrogen-doped graphene oxide quantum dots as photocatalysts for overall water-splitting under visible light illumination *Adv. Mater.* **26** 3297–303

[126] Kibria M G, Nguyen H P T, Cui K, Zhao S, Liu D, Guo H, Trudeau M L, Paradis S, Hakima A R and Mi Z 2013 One-step overall water splitting under visible light using multiband InGaN/GaN nanowire heterostructures *ACS Nano* **7** 7886 93

[127] Zhang C, Chen C, Dong H, Shen J R, Dau H and Zhao J 2015 A synthetic Mn$_4$Ca-cluster mimicking the oxygen-evolving center of photosynthesis *Science (1979)* **348** 690–3

[128] Wang W, Chen J, Li C and Tian W 2014 Achieving solar overall water splitting with hybrid photosystems of photosystem II and artificial photocatalysts *Nat. Commun.* **5** 4647

[129] Liu G, Ye S, Yan P, Xiong F, Fu P, Wang Z, Chen Z, Shi J and Li C 2016 Enabling an integrated tantalum nitride photoanode to approach the theoretical photocurrent limit for solar water splitting *Energy Environ. Sci.* **9** 1327–34

IOP Publishing

Photocatalysts for Energy and Environmental Sustainability

Vijay B Pawade and Bharat A Bhanvase

Chapter 6

Photocatalysis for organic degradation using perovskite materials

Phuong Hoang Nguyen, Thi Minh Cao and Viet Van Pham

6.1 Introduction

In recent years, the issue of organic compound pollution, especially that caused by persistent organic pollutants (POPs), has become increasingly severe [1, 2]. These compounds are often used in producing various industrial products, leading to their widespread release into the environment. Organic compounds can be found in multiple water sources, including rivers, lakes, and groundwater. They can also enter the food chain through the consumption of contaminated water, causing harm to human and animal health [3]. In addition, long-term exposure to organic compounds can lead to a variety of health problems, including cancer, reproductive disorders, and immune system damage [4]. POPs and industrial chemicals that were manufactured in great quantities during the 20th century directly affect human and animal health. When examining the effects of pollution on morbidity and mortality in relation to other risk factors, the results indicate that pollution remains one of the most significant contributors to disease and premature death worldwide. According to figure 6.1, the number of deaths attributed to pollution is comparable to the number of deaths caused by smoking. This highlights the substantial and ongoing impact of pollution on global health outcomes. Exposure to high levels of persistent organic pollutants (POPs) can have detrimental health effects on both humans and animals. These effects are well documented and may include various adverse outcomes, such as cancer, nervous system damage, reproductive disorders, and immune system disruption. It is important to recognize that POPs have the potential to cause significant harm to living organisms, highlighting the need for effective measures to reduce exposure and mitigate their impact on health [5].

To address this issue, it is necessary to implement effective wastewater treatment processes in production facilities, enhance the monitoring and assessment of water quality to detect pollutants promptly, and implement other measures such as the use

doi:10.1088/978-0-7503-5697-8ch6

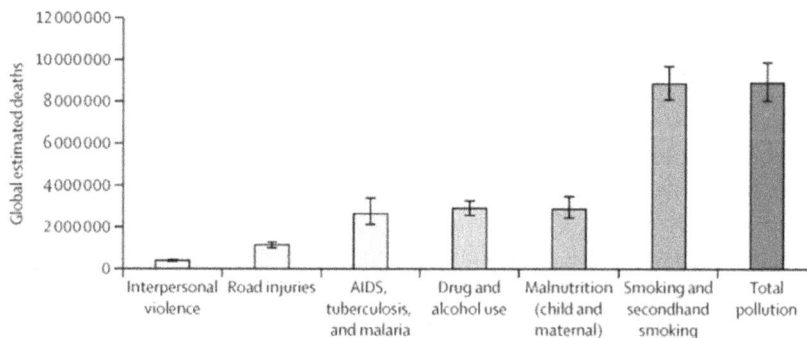

Figure 6.1. Global estimated deaths by major risk factor or cause. Reprinted from The Lancet [6], Copyright (2022), with permission from Elsevier.

of wastewater treatment technologies, a reduction in the chemicals used in production, and the use of recycled water sources. Furthermore, increasing education and awareness about environmental pollution is also critical in reducing pollution. Governments and nongovernmental organizations need to improve the information, education, and training offered to communities to raise awareness about environmental issues and enhance cooperation and consultation with businesses and organizations in implementing effective pollution reduction measures.

In terms of technical approaches, several methods can be used to reduce organic pollutants, which can be broadly categorized into prevention, treatment, and remediation approaches. For instance, biological treatment based on the use of microorganisms to degrade organic pollutants; physical and chemical treatment; phytoremediation; and adsorption. Therein, advanced oxidation processes (AOPs) can be employed to remove or degrade organic pollutants. These include activated carbon adsorption, advanced oxidation processes (e.g. ozonation, UV irradiation, photocatalysis, and chemical oxidation), air stripping, and membrane filtration. Among advance oxidation methods, photocatalysis and other processes based on photocatalysis, i.e. the photo-Fenton method, the use of a peroxymonosulfate (PMS) activator, etc. are promising technologies for the treatment of water contaminated with organic compounds. These processes use light to activate a photocatalyst, typically TiO_2, to generate highly reactive oxygen species that can degrade organic pollutants. One of the significant advantages of photocatalysis is that it is a green technology that does not require chemicals or produce harmful byproducts. In addition, photocatalysis can treat a wide range of organic contaminants, including pesticides, pharmaceuticals, and dyes. Furthermore, photocatalysis is a relatively simple and low-cost technology that can easily be scaled up for industrial applications. Overall, the use of photocatalysis in treating water contaminated with organic compounds offers several advantages over traditional treatment methods and shows great potential for widespread adoption in the future [7].

Perovskite materials have recently attracted significant attention in photocatalysis due to their unique optical and electronic properties [8]. These materials have a perovskite crystal structure that consists of a metal cation, an organic or inorganic cation, and an anion. Perovskite materials have a high absorption coefficient, excellent charge carrier mobility, and long carrier lifetimes, which make them ideal for photocatalytic applications. Several studies have investigated the application of perovskite materials in photocatalysis to degrade organic pollutants, split water, and reduce carbon dioxide. For example, researchers have reported the successful use of perovskite materials in the photocatalytic degradation of organic dyes and the production of hydrogen gas from water [7]. These promising results suggest that perovskite materials have the potential to become a new class of photocatalysts for a wide range of environmental and energy applications.

This chapter provides a complete overview of the concept, properties, and application of perovskites as photocatalysts for organic pollutant removal. The primary methods used to synthesize perovskites are summarized in this chapter. In addition, this chapter describes some of the most important applications of perovskites in the photocatalysis field, including photocatalysis, the photo-Fenton reaction, and PMS activation for the degradation of organic pollutants in wastewater. Some conclusions and prospects are also briefly introduced, which could lead to breakthrough developments in perovskite materials design.

6.2 An overview of perovskite materials

6.2.1 An introduction to perovskites

Perovskites, named after Russian mineralogist Lev Perovski, are compounds with a crystal structure that follows the formula ABX_3, where A and B are cations and X is an anion (figure 6.2). The perovskite structure is a crystal structure that was first discovered in the mineral calcium titanate ($CaTiO_3$). It is characterized by a three-dimensional network of corner-sharing BX_6 octahedra that form a cubic lattice. The A cation occupies the center of the cube and is surrounded by octahedra.

The arrangement of the ideal perovskite structure can be described as follows:

$$(r_A + r_O) = \frac{\sqrt{2}}{2}A = \sqrt{2}(r_A + r_O), \tag{6.1}$$

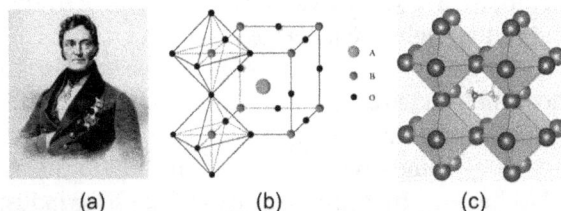

| (a) | (b) | (c) |

Figure 6.2. (a) Russian mineralogist Lev Perovski (1792–1856). This Count Lev Alekseevich Perovski image has been obtained by the author(s) from the Wikimedia website, where it is stated to have been released into the public domain. It is included within this book on that basis. (b) The structure of ABO_3 perovskite-type oxides. Reprinted from [10], Copyright (2021), with permission from Elsevier. (c) A methylammonium cation ($CH_3NH_3^+$) occupies the central A site, surrounded by 12 nearest-neighbor iodide ions in corner-sharing PbI_6 octahedra. Reproduced from [11]. CC BY 4.0.

where r_A, r_B, and r_O represent the ionic radii of the cations A and B and the anion O, respectively. Although the idealized structure is primitive cubic, differences in the sizes of the cations can cause distortions. This typically involves tilting of the octahedral units (known as octahedral tilting). The stability of the perovskite structure is described by a tolerance factor t, which represents the range of relative ionic sizes [3]:

$$t = \frac{(r_A + r_O)}{\sqrt{2}\,(r_B + r_O)}.$$

(6.2)

Empirically, if the tolerance factor is between 0.95 and 1.0, the structure is cubic. If $0.8 < t < 0.9$, the compound transforms into a tetragonal, rhombohedral, or orthorhombic crystal [9]. In photocatalytic studies, lattice distortion significantly affects the crystal field, altering dipoles and electronic band structures, which in turn impact the behaviors of photogenerated charge carriers, including excitation, transfer, and redox reactions, during the photocatalytic process.

Figure 6.3 shows that energy-dispersive x-ray (EDX) mapping of the elements in perovskite oxides ($BaTiO_3$ and $SrTiO_3$) exhibits visible particles on the particle clusters, and it also shows a uniform arrangement of Mo and S atoms on those clusters. $SrTiO_3$ and $BaTiO_3$ perovskite-type oxides offer a viable alternative to the predominantly used TiO_2 due to their favorable band positions, optical

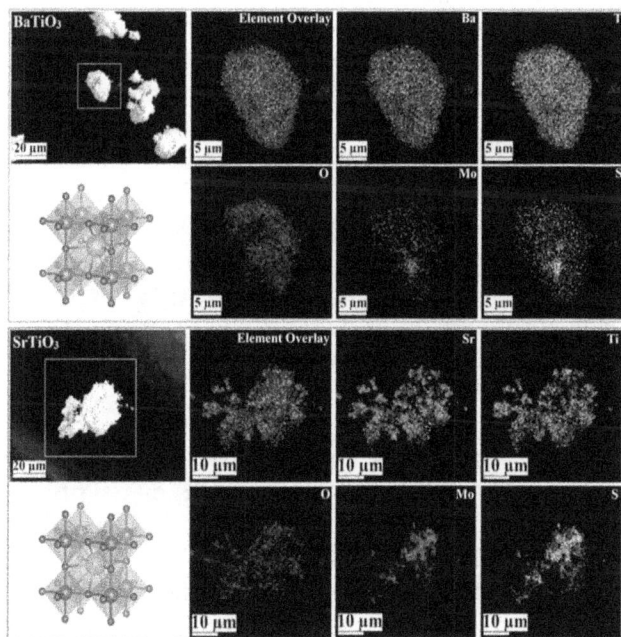

Figure 6.3. Elemental mapping images of photodeposited $BaTiO_3$ and $SrTiO_3$ nanocrystal clusters. Reprinted from [12], Copyright (2023), with permission from Elsevier.

characteristics, and crystallographic properties. These qualities make them suitable for effective photochemical energy conversion reactions.

In addition to the oxide-based perovskites, another type called 'perovskite halides' is a group of materials with the formula ABX_3, where A represents an organic and/or inorganic cation with a single oxidation state, B represents a group 4A metal with a valence of two (e.g. Pb^{2+}, Ge^{2+}, Sn^{2+}, etc.), and X corresponds to a halide anion, e.g. I^-, Cl^-, Br^-, etc [8]. These materials have gained tremendous attention due to their excellent optoelectronic properties, such as high light absorption coefficients, long carrier diffusion lengths, and high charge carrier mobilities.

In general, perovskites have shown great promise as low-cost and high-performance materials for solar cells. They have achieved remarkable power conversion efficiencies (PCEs) that exceed 25% and are comparable to the efficiencies of traditional silicon solar cells. In addition to their use in photovoltaics, metal halide perovskites have also been investigated for use in other optronics applications (light-emitting diodes (LEDs), photodetectors, photodiodes, and lasers). Overall, metal halide perovskites represent a promising class of materials with significant potential for advancing the field of solar energy and optoelectronics. Ongoing research and development efforts aim to address their stability issues and further enhance their performance for practical applications.

The bandgap energy of halide perovskites is one of the important factors that affect their optical properties. The bandgap is the energy required to push an electron from the valence band (VB) to the conduction band (CB), creating an electron–hole pair. The smaller the bandgap, the greater the semiconductor's ability to absorb light. Figure 6.4 shows the bandgaps of oxide-type perovskites (a) and halide-type perovskites (b) with the redox levels for the formation of reactive oxygen species (ROS) and $OH^\bullet$ radicals. For organic–inorganic perovskites ($MAPbI_3$), research studies have observed a small bandgap (about 1.5 eV) and a high light absorption coefficient of 105 cm^{-1} [13]. This means that $MAPbI_3$ one μm thick can absorb all light with a wavelength shorter than 800 nm in the solar spectrum. Moreover, by modifying factors such as size, composition substitution, phase segregation, phase transition, and deformation, it is possible to extensively adjust the range of light absorption in perovskite halides from UV to near-infrared wavelengths. The halogen anions that appear in the chemical formula $APbX_3$ for halide perovskites affect their CB energy levels. For instance, the bandgap of $MAPbI_3$ can be extensively tuned simply by changing the ratio of the different halogen ions. Similarly, continuous tuning of the spectrum can be achieved in inorganic halogen perovskites, as shown, for example, by $CsPbBr_3$, whose bandgap can be tuned from 1.88 to 3.03 eV by partially replacing Br with Cl or I. Furthermore, perovskite halides are a group of materials with unique photocatalytic properties. Their CBM and VBM positions are suitable for promoting many photocatalytic reactions [13].

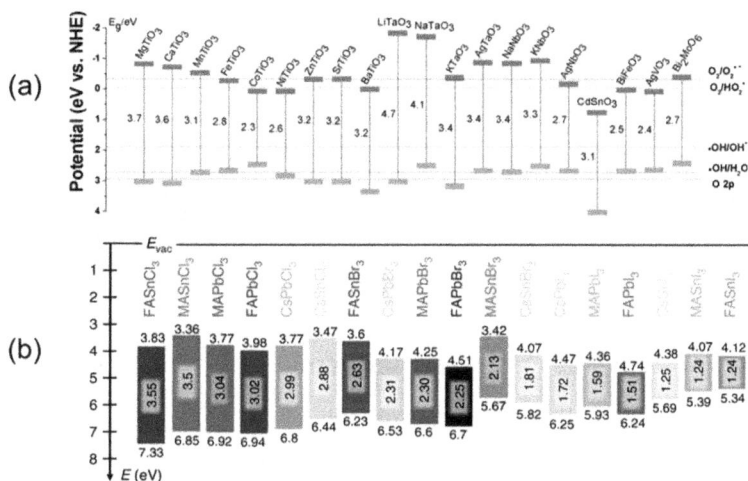

Figure 6.4. (a) E_g values (in eV) and the positions of the CB (upper) and the VB (lower) for various perovskite semiconductors at a pH of 7 vs. normal hydrogen electrode (NHE). Reprinted from [14], Copyright (2019), with permission from Elsevier. (b) A schematic energy level diagram of the 18 metal halide perovskites. Reproduced from [15]. CC BY 4.0.

6.2.2 The properties of perovskites

Perovskites have many advantageous properties for photovoltaics and photocatalysis due to their outstanding properties (figure 6.5).

6.2.2.1 Optoelectronic properties

Perovskite materials exhibit exceptional optoelectronic properties, making them suitable for various applications. Deschler *et al* [16] investigated the optoelectronic properties of perovskite materials and demonstrated that perovskite materials possess a high absorption coefficient, enabling efficient light harvesting across a wide spectrum, including visible and near-infrared wavelengths. In addition, they observed efficient charge transport due to long carrier diffusion lengths in these materials, making them suitable for application in solar cells and photocatalysis [17].

6.2.2.2 Tunable bandgap

Perovskite materials have quite diverse bandgap values that range from 2.1 to 4.0 eV (table 6.1) and a specific crystal structure that corresponds to a three-dimensional arrangement of metal cations, such as lead (Pb), tin (Sn), or other elements, surrounded by an octahedral arrangement of oxygen anions. Perovskite materials possessing the ABO_3 structure offer numerous benefits that make them a highly promising option for photocatalysis. These advantages include straightforward synthesis methods, adjustable physicochemical properties, well-suited frameworks, diverse levels of lattice distortion, and localized sites for surface oxidation and reduction reactions. These characteristics contribute to their potential to be excellent choices for photocatalytic processes.

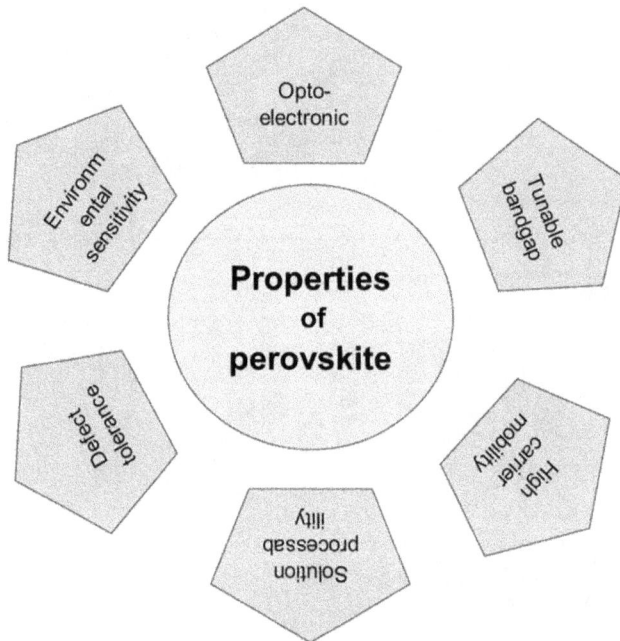

Figure 6.5. Some properties of perovskites.

Table 6.1. The bandgaps of some perovskite materials. Reproduced from [19]. CC BY 4.0.

Material	Bandgap (eV)
$SrTiO_3$	3.20
$NaTaO_3$	4.00
$CaTiO_3$	3.62
$BiFeO_3$	2.40
$LaFeO_3$	2.00
$NaNbO_3$	3.48
$LaCoO_3$	2.10
Bi_2WO_6	2.70
$La_2Ti_2O_7$	3.28

One of the remarkable features of perovskite materials is their tunable bandgap, which can be adjusted to absorb light across a wide range of energies. This tunability is advantageous for applications such as solar cells and light-emitting devices. Prasanna *et al* synthesized and investigated [18] tin and lead iodide perovskite semiconductors with the composition AMX_3, where M is a metal and X is a halide. The introduction of a smaller A-site cation into a perovskite structure can induce two distinct distortions: tilting of the MX_6 octahedra or isotropic

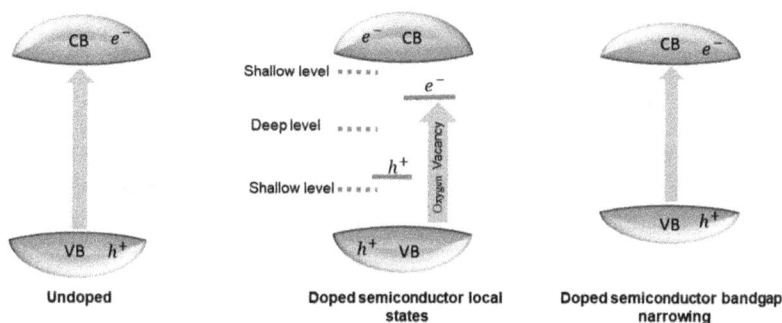

contraction of the lattice. The former effect generally increases the bandgap, while the latter decreases it. In the case of lead iodide perovskites, replacing the larger formamidinium cation with the smaller cesium cation resulted in an expanded bandgap, which was attributed to octahedral tilting. Conversely, tin-based perovskites, being slightly smaller than lead, displayed no octahedral tilting upon cesium substitution but rather lattice contraction, leading to a progressive reduction in the bandgap. This study demonstrated a systematic approach to fine-tuning the bandgap and band positions of metal halide perovskites by controlling the cation composition. This research also showcased the development of solar cells that have bandgaps optimized for harvesting infrared light at up to 1040 nm, achieving a stabilized power conversion efficiency of 17.8%. The findings provide insights into cation-based bandgap tuning in 3D metal halide perovskites and have implications for the development of perovskite semiconductors for optoelectronic applications [18].

In addition, the control of the band positions and bandgaps of perovskites is also of interest in the visible-light-driven photocatalysis field. Figure 6.6 shows the effects on the bandgap of perovskites of doping to form new states, i.e. shallow and deep levels or oxygen vacancy states. Besides, the narrow bandgap of perovskites helps them become more effective in photocatalytic reactions under visible light conditions.

6.2.2.3 High carrier mobility
Perovskite materials exhibit high charge carrier mobility, allowing efficient movement of electrons and holes through the material. This high mobility is crucial to obtain high-performance electronic and optronics devices. Liu *et al* [20] synthesized InP and InAs nanocrystals (NCs) and used various inorganic ligands (molecular metal chalcogenide complexes (MCCs) and chalcogenide ions to cap the NCs). These inorganic ligands effectively replaced the organic ligands on the NC surfaces and stabilized the colloidal solutions of InP and InAs NCs in polar solvents. Moreover, they facilitated efficient charge transport between individual NCs. The films of InP and InAs NCs capped with MCCs exhibited high electron mobility, with $Cu_7S_4^-$ MCC ligands achieving electron mobility exceeding 15 cm^2 V^{-1} s^{-1}. In addition, Lim *et al* [21] observed similar long-range mobilities of 0.3 to 6.7 cm^2 V^{-1} s^{-1} in perovskites such as $FA_{0.83}Cs_{0.17}Pb(I_{0.9}Br_{0.1})_3$, $(FA_{0.83}MA_{0.17})_{0.95}Cs_{0.05}Pb(I_{0.9}Br_{0.1})_3$, and

$CH_3NH_3PbI_{3-x}Cl_x$ polycrystalline films, highlighting their potential for use in various optoelectronic applications in which efficient charge transport is crucial.

6.2.2.4 Processability in solution

Perovskite materials can be processed in solution, which offers advantages for large-scale manufacturing processes. They can be deposited using techniques such as spin-coating, inkjet printing, or spray coating, enabling the fabrication of thin films with good uniformity and low-cost manufacturing potential. A study by Hu focused on scaling up perovskite photovoltaic technology through the exploration of film formation, device architectures, deposition methods, and scalable manufacturing approaches. This study emphasized the importance of precise control over nucleation and crystal growth during the film formation process to achieve high-performance perovskite solar cells [22].

6.2.2.5 Defect tolerance

Perovskite materials show a certain degree of tolerance toward defects, which can be beneficial for photocatalytic performance. Defects in the crystal structure can trap charge carriers, reducing their mobility. However, in perovskites, these trapped carriers can be released and continue to contribute to the device's functionality. Steirer et al [23] explored the defect tolerance characteristic of perovskite materials, with a specific focus on methylammonium lead triiodide ($MAPbI_3$) perovskite, and indicated that defects in the crystal structure influenced the performance and stability of energy-related devices utilizing this material. Furthermore, MAPbI3 perovskite exhibited a certain level of tolerance towards specific defects in its crystal structure. These defects have the potential to trap charge carriers, resulting in a decrease in their mobility. However, perovskite materials possess the capability to release these trapped charge carriers, enabling them to continue contributing to the device's overall performance [24].

Defect engineering provides a convenient and effective approach for improving the photocatalytic efficiency of perovskite materials. The substitution of different materials can introduce defects into the perovskite structure. These defects serve as sites that can trap charge carriers, leading to enhanced charge separation and increased photocatalytic efficiency. In addition, the existence of the intrinsic factors, e.g. Sn^{4+} that are easy formation of Sn vacancies, impacting to the defect chemistry and subsequent recombination of charge carriers, prolonged exposure of these materials to extrinsic factors (e.g., air, O_2) can severely degrade their key optoelectronic properties [17].

6.2.2.6 Environmental sensitivity

Perovskite materials can be sensitive to moisture, heat, and light exposure. They tend to degrade in the presence of moisture and are susceptible to thermal stress. However, ongoing research aims to improve their stability and durability for practical applications. Ahangharnejhad et al presented a method for protecting metal halide perovskite solar cells from degradation in high-humidity environments using a sputtered SiO_2 barrier coating. The SiO_2 protective layer significantly enhanced the devices' resistance to extreme humidity conditions, increasing their lifetimes by approximately 60 times for $CH_3NH_3PbI_3$ and around 600 times for triple-cation perovskite material. Real-time laser-beam-induced current (LBIC) measurements validated the effectiveness of this approach in preventing degradation at the edges of scribed lines, making it suitable for the production of monolithically integrated modules [25].

6.3 Methods used to synthesize perovskite materials

There are several methods for making perovskite materials, i.e. wet chemical methods (sol–gel process hydrothermal route, coprecipitation method, electrochemical synthesis, and emulsion or microemulsion synthesis, etc.), solid-state reaction methods, template-assisted methods, etc. These methods can be adjusted to control the properties of the perovskite material, such as its crystal structure, morphology, and surface area, which can affect its photocatalytic activity [26].

6.3.1 Wet chemical methods

Wet chemical synthesis, also known as solution-based synthesis, is a broad term that refers to a wide range of chemical processes used to create or modify materials in a liquid or dissolved form. It involves the reaction of chemical precursors dissolved in a solvent to form a desired product. The reaction can be carried out at various temperatures, pressures, and reaction times, depending on the specific synthesis requirements. The reactants may undergo various chemical transformations, such as precipitation, hydrolysis, redox reactions, or complexation, to form the desired product. Wet chemical synthesis is widely used in various fields, including chemistry, materials science, nanotechnology, and pharmaceuticals. It offers several advantages, such as simplicity, scalability, and the ability to control the composition, size, shape, and structure of the resulting materials.

6.3.1.1 The sol–gel method

The sol–gel technique is commonly used to prepare nanosized perovskites with a high surface area. It involves the mixture of precursor compounds in specific solvents to form a stable 'sol,' which is then transformed into a gel-like system (both liquid and solid phases). The gel is then dried to evaporate the solvent and undergoes heating to obtain the desired perovskite materials [27]. Parida *et al* [28] synthesized $LaFeO_3$ perovskite nanomaterials using the sol–gel method at 130 °C. This heating process caused changes in the color and viscosity of the sol. When further heated, the brown, porous dry gel automatically ignited due to the thermal redox reaction. Finally, a solid was created by a self-burning process and then activated at different temperatures (500 to 900 °C) for 2 h to obtain a visible-light photocatalytic material. Wang *et al* [29] prepared perovskite materials including $La(NO_3)_3$ and $B(NO_3)_2$ (B = Mn, Fe, Co, Ni, Cu, Zn) by a sol–gel method using an ethylene glycol and alcohol mixture as a complex. In general, this synthesis method is quite simple: drying at 100 °C and annealing and sintering processes (500 °C–700 °C) were performed to obtain the final product.

6.3.1.2 *The hydrothermal method*

The hydrothermal method is one of the standard techniques used to synthesize perovskite particles with controlled shapes. This method subjects a reaction mixture to high-temperature and -pressure conditions in a stainless steel autoclave, and the final product is collected after cooling. Zhao *et al* [30] synthesized $CaTiO_3$ particles with a controlled morphology. Their study pointed out that rectangular $CaTiO_3$ particles exhibited the highest photocatalytic activity for the degradation of methylene blue under UV irradiation due to their high specific surface area compared to the surface areas of cubic and spherical particles.

6.3.1.3 *The coprecipitation method*

The coprecipitation method is another important technique used to synthesize perovskite materials. It is conducted using a suitable precipitating agent to coprecipitate two or more precursor components homogeneously [31]. Moshtaghi *et al* synthesized $CaSnO_3$ nano-cubes using a simple and green coprecipitation method [32]. To obtain the final product, $CaSnO_3$, the researchers utilized a complex precursor called $Ca(sal)_2$ as a calcium source and $SnCl_2$ as a source of tin. The synthesis procedure employed water as an environmentally friendly solvent, and various alkaline agents were used as precipitating agents. After the reaction, the resulting precipitate was subjected to centrifugation, followed by washing and drying. Subsequently, the obtained material was annealed at a temperature of 900 °C for 5 h. Using this process, the researchers successfully obtained the desired final product, $CaSnO_3$. Junwu *et al* [33] synthesized $LaCoO_3$ nanocrystals using the coprecipitation method, in which NaOH was used as the precipitating agent. After precipitation, the resulting precipitates were centrifuged, washed, and dried at 100 °C. The dried residue was then annealed at 600 °C to obtain the product.

Figure 6.7 illustrates the general method of precipitation synthesis. The precursors A and B are dissolved in solvents and then precipitated using a NaOH solution

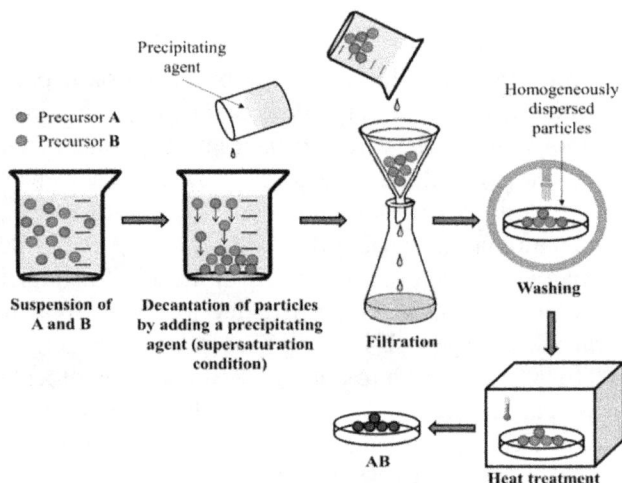

Figure 6.7. A general schematic of the coprecipitation method of perovskite preparation. Reprinted from [31], Copyright (2016), with permission from Elsevier.

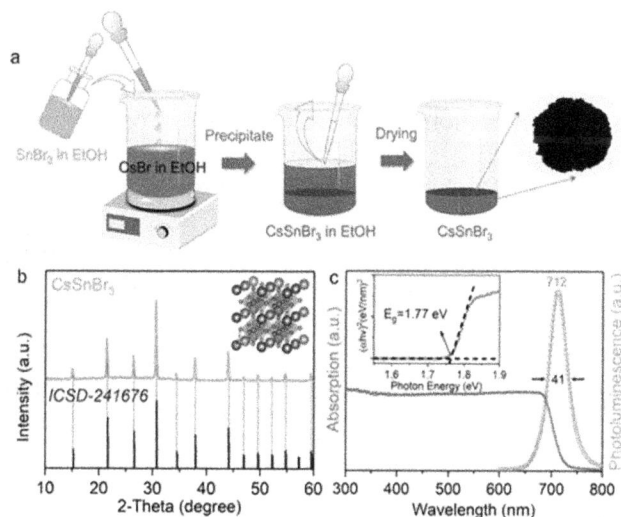

Figure 6.8. (a) A schematic of the coprecipitation process used to prepare CsSnBr₃ perovskite. (b) The x-ray diffraction pattern of CsSnBr₃. (c) The UV–vis absorption and PL spectra of CsSnBr₃. Reprinted with permission from [34]. Copyright (2023) American Chemical Society.

in water. The resulting precipitates are formed, filtered, and washed. Finally, the particles are annealed at a high temperature to form the final product.

Figure 6.8 represents a stepwise coprecipitation strategy for the synthesis of a stable $CsSnBr_3$ perovskite by a facile and low-cost strategy at room temperature [34]. Herein, ethanol as the solvent and salicylic acid as the additive were proposed for the synthesis of a $CsSnBr_3$ perovskite with a highly stable cubic phase. The results showed that the use of the ethanol solvent and the salicylic acid additive can not only effectively prevent the oxidation of Sn^{2+} during the synthesis processes but also stabilize the synthesized $CsSnBr_3$ perovskite. The absorption properties of the perovskite film remained consistent, and its photoluminescent (PL) intensity was significantly preserved at around 69% even after being stored for ten days. This performance was superior to that of a bulk $CsSnBr_3$ perovskite film synthesized using the spin-coating method, which experienced a decrease in PL intensity to 43% after only 12 h of storage. Clearly, this method is an efficient route for the synthesis of perovskite materials at room temperature.

6.3.2 Biotemplate-supported synthesis

The family of microorganisms provides various biological templates that can be used as an inspiration for designing and synthesizing various perovskite materials with different micro- and nanostructures and enhanced specific surface area. Different biological samples, such as algae, bacteria, fungi, viruses, etc. can be used, which offer the benefits of fast growth with control over the material's shape, size, and structure. In addition, biological materials, such as plant leaves, microorganisms, etc. provide good templates for the development of materials, and the unique 3D designs of these samples can be mimicked in the desired materials [26].

Figure 6.9 depicts the introduction of a small quantity of the precious metal Pt into $LaNiO_3$, leading to the formation of $LaNi_{1-x}Pt_xO_3$ with a perovskite structure. In this process, Pt is dispersed atom by atom within the crystal lattice of the perovskite and is supported on SiO_2 which has a high specific surface area. After reduction, the resulting product is $Pt/LaNiO_3/SiO_2$. This approach enables Pt to be incorporated into the perovskite lattice, resulting in a uniform dispersion of Pt atoms at the atomic level. In addition, there is a strong interaction between the Pt atoms and the perovskite material, which effectively prevents the sintering of the Pt atoms [35].

The advantages and disadvantages of the methods used to synthesize perovskite materials, including physical and chemical methods, are listed in table 6.2.

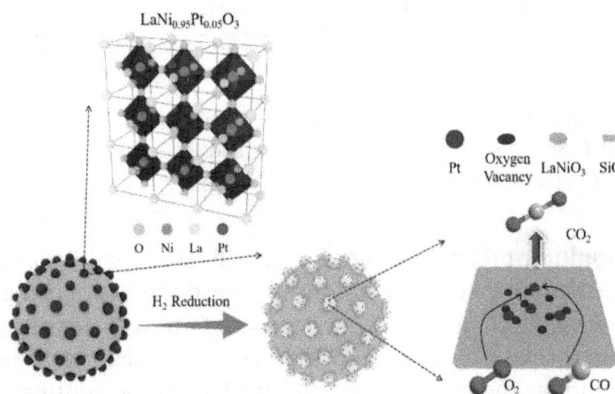

Figure 6.9. A schematic illustration of a process used to load highly dispersed Pt on $LaNiO_3$ which is on the surface of SiO_2 by reducing $LaNi_{1-x}Pt_xO_3/SiO_2$. Reprinted from [35], Copyright (2020), with permission from Elsevier.

Table 6.2. A summary of the advantages and disadvantages of common perovskite synthesis methods. Reprinted from [36], Copyright (2022), with permission from Elsevier.

Synthesis methods	Synthesis sub methods	Advantages	Disadvantages
Physical	High-temperature treatment	• Highly scalable • Simple	• Irregular and nonuniform morphology • Difficult to obtain single-phase and complex perovskites • Generally low specific surface area • High energy consumption
	• Mechanochemical • Reaction	• Highly scalable • Simple	• Difficult to control the morphology • Difficult to obtain single-phase and complex perovskites • High energy consumption

(Continued)

Table 6.2. (*Continued*)

Synthesis methods	Synthesis sub methods	Advantages	Disadvantages
	Radiation	• Short crystallization time • Produces uniform perovskite	• High energy consumption • Requires specialized instrument
Chemical	Coprecipitation	• Highly scalable • Simple	• Generally low crystallinity without additional post-treatment
	Sol–gel	• Highly scalable • Homogeneous and tunable morphology • Mild conditions	• Complex synthesis route • May involve toxic chemicals
	Hydrothermal/ solvothermal	• Homogeneous and tunable morphology	• Requires high temperature and pressure • May involve toxic chemicals

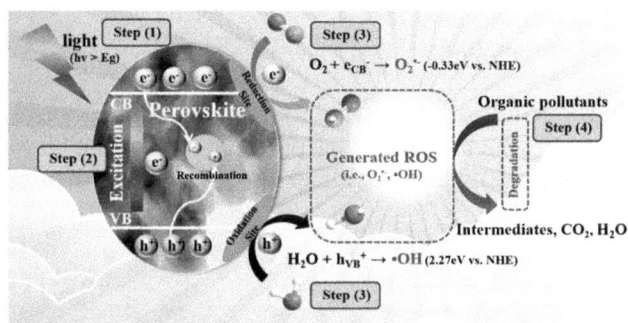

Figure 6.10. The mechanism of ROS production and pollutant degradation performed by perovskite photocatalysts. Reprinted from [10], Copyright (2021), with permission from Elsevier.

6.4 Perovskite photocatalysts for organic degradation

Perovskite-based photocatalytic agents are considered to be good candidates for the effective removal of organic pollutants from wastewater. Recently, many studies have focused on improving their strengths, enhancing the stability of perovskites, improving quantum efficiency, etc [37–40]. Figure 6.10 displays the photocatalytic processes of perovskites, in four steps: (i) light energy interacts with the perovskite material with the requirement that the energy of irradiation must be greater than E_g of the semiconductor. (ii) This activates electrons in the VB region of the semiconductor, causing them to jump to the CB. This process leaves behind positive charges in the VB, which are called holes (h^+). These charges are key factors in carrying out the oxidation (h^+) and reduction reactions (e^-) in photocatalysis. (iii) Complex chemical reactions occur, creating components with strong reducing and oxidizing properties, such as $^\bullet OH$ and $^\bullet O_2^-$. (iv) Reactive radicals degrade the organic pollutants to form the intermediates CO_2 and H_2O (step 4).

The photocatalytic activities of selected perovskite-based and P25 photocatalysts for the degradation of organic dyes are listed in table 6.3.

6.4.1 Photocatalysis

As mentioned above, materials made from perovskites can absorb solar energy to create oxidizing sites (electron acceptors) and reducing areas (electron donors) for the purpose of oxidating organic pollutants into CO_2, H_2O, and other oxoanions. In photocatalytic degradation, photogenerated holes (h^+) can directly oxidize organic compounds into various intermediates or CO_2 and H_2O. Generally, OH^- derived from adsorbed H_2O is oxidized to hydroxyl radicals ($OH^\bullet$) by transferring h^+ from organic compounds to CO_2 and H_2O. Equations (6.3)–(6.5) describe the photocatalytic degradation reactions of organic compounds:

$$\text{Organic} + h^+ \rightarrow \text{intermediates}/CO_2 + H_2O \tag{6.3}$$

$$h^+ + OH^- \rightarrow OH^\bullet \tag{6.4}$$

$$\text{Organic} + OH^\bullet \rightarrow CO_2 + H_2O \tag{6.5}$$

However, the introduction of O_2 can inhibit the recombination of h^+ and e^- by forming $^\bullet O_2^-$ radicals, which then further protonate to form hydroperoxyl radicals ($^\bullet HO_2$) and subsequently hydroperoxide (H_2O_2), as depicted in equations (6.6)–(6.9):

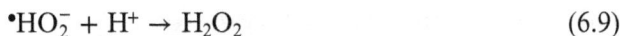

$$(O_2)_{ads} + e^- \rightarrow {}^\bullet O_2^- \tag{6.6}$$

$$O_2 + OH \rightarrow HOO \tag{6.7}$$

$$HOO + e^- \rightarrow {}^\bullet HO_2^- \tag{6.8}$$

$${}^\bullet HO_2^- + H^+ \rightarrow H_2O_2 \tag{6.9}$$

Figure 6.11 illustrates the use of a hybrid material based on a lead-free halide perovskite which is activated by light. This hybrid material exhibits enhanced photoelectrical effects and utilizes synergistic surface activation mechanisms. As a result, it supports an efficient process for converting aromatic C(sp3)–H bonds under visible light and in the presence of air.

Table 6.3. The photocatalytic activity of selected perovskite-based and P25 photocatalysts for the degradation of organic dyes [41].

Photocatalyst	BG (eV)	Dye type and/or concentration	Degradation rate $(1 - C/C_0, \%)$	Catalyst loading	Light type	Preparation method
P25	3.2	RB, 20 μmol l^{-1}	3	5 g l^{-1}	$\lambda > 400$ nm	Commercial
LaFeO$_3$(500)			24			Sol–gel
LaFeO$_3$(600)			16			
LaFeO$_3$(700)			13			
LaFeO$_3$(800)			9			
LaFeO$_3$ nanocubes	2.01	RB	76.81 at 180 min	1 g l^{-1}	$\lambda > 400$ nm	Hydrothermal
LaFeO$_3$ nanorods	2.05		88.36 at 180 min			
LaFeO$_3$ nanospheres	2.1		90.8 at 180 min			
P25	3.2		5.1 at 180 min			Commercial
LaNiO$_3$	2.26	MO, 10 mg l^{-1}	74.9 at 5h	2 g l^{-1}	$\lambda > 400$ nm	Sol–gel
CeCo$_{0.05}$Ti$_{0.95}$O$_{3.97}$	1.57	Nile blue, 30 ppm	91 at 3h	3 g l^{-1}	254–310 and 410–500 nm	Sol–gel
NaTaO$_3$	4.13	MB, 2×10^{-4} mol l^{-1}	60 at 200 min	50 mg	$\lambda > 420$ nm	Hydrothermal
2.47 mol% Cr-doped NaTaO$_3$	3.95		70 at 200 min			
6.35 mol% Cr-doped NaTaO$_3$	3.28		~35 at 200 min			
P25	3.2	RB, 5×10^{-6} mol l^{-1}	75 at 150 min	2 g l^{-1}	$\lambda > 400$ nm	Commercial
K$_2$La$_2$Ti$_3$O$_{10}$	3.63		10 at 150 min			Steady-state reaction (SSR)
N-K$_2$La$_2$Ti$_3$O$_{10}$	3.59		20 at 150 min			SSR + NH$_3$ gas
CN-K$_2$La$_2$Ti$_3$O$_{10}$	2.92		90 at 150 min			SSR + urea
P25	3.2	MO, 0.02 g l^{-1}	~6 at 30 min	0.8 g l^{-1}	$\lambda > 400$ nm	Commercial
K$_2$La$_2$Ti$_3$O$_{10}$	3.69		~2 at 30 min			SSR
N-K$_2$La$_2$Ti$_3$O$_{10}$	3.44		~30 at 30 min			SSR + NH$_3$ gas
LaFeO$_3$		MB, 10 mg l^{-1}	~50 at 90 min	0.5 g l^{-1}	$\lambda > 400$ nm	
La$_{0.9}$Ca$_{0.1}$FeO$_3$			~80 at 90 min			Reverse microemulsion

Catalyst	Band gap (eV)	Dye/pollutant	Degradation	Loading	Light source	Synthesis method
$La_2Ti_2O_7$		MO, 1×10^{-5} mol l^{-1}	~55 at 90 min	1 g l^{-1}	UV, $\lambda = 254$ nm	Polymeric complex
$La_{1.5}Pr_{0.5}Ti_2O_7$			~25 at 90 min			
$La_{1.5}Gd_{0.5}Ti_2O_7$			90 at 90 min			
$La_{1.5}Er_{0.5}Ti_2O_7$			~50 at 90 min			
$LaFeO_3$	2.36	MB, 10 mg l^{-1}	~100 at 90 min	2 g l^{-1}	$\lambda > 400$ nm	Microwave-assisted route
P25	3.2		20 at 90 min			Commercial
$LaFeO_3$	2.1	RB	490 at 180 min	1 g l^{-1}	$\lambda > 400$ nm	Hydrothermal
P25	3.2	MO, 10 mg l^{-1}	~85 at 100 min	0.2 g l^{-1}	UV	Commercial
$LaCoO_3$	2.07		~85 at 100 min			Surface-ion adsorption method
P25	3.2	MB, 10 mg l^{-1}	~85 at 100 min			Commercial
$LaCoO_3$	2.07		~85 at 100 min			Surface-ion adsorption method
KNb_3O_8	3.06	Acid red G, 50 mg l^{-1}	63.03 at 60 min	1 g l^{-1}	UV, $\lambda = 253.7$ nm	SSR
0.3 wt% Cu-doped KNb_3O_8			93.23 at 60 min			
2 wt% Cu-doped KNb_3O_8			83.92 at 60 min			

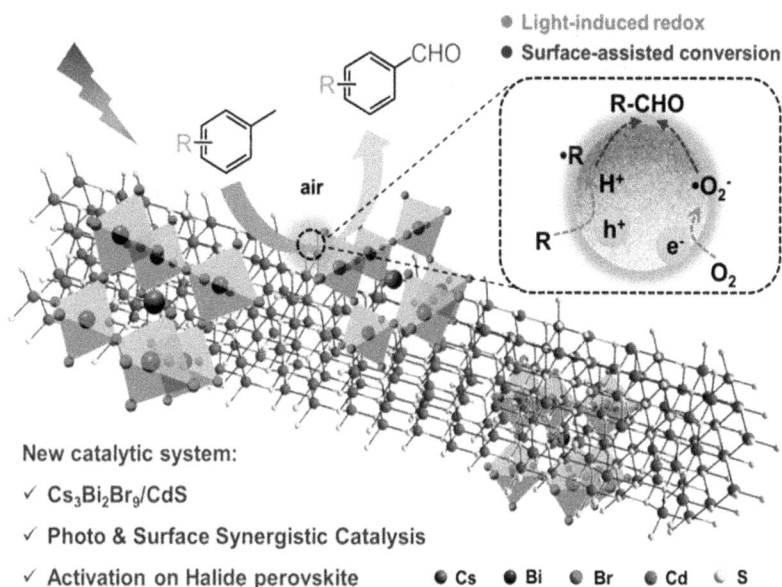

Figure 6.11. A schematic illustration of the photocatalytic conversion of aromatic C(sp3)–H bonds by a $Cs_3Bi_2Br_9/CdS$ hybrid photocatalyst under air and visible light. The conversion involves the synergistic activation of photo and surface catalysis. Reprinted from [42], Copyright (2023), with permission from Elsevier.

6.4.2 Fenton and photo-Fenton catalysis

The Fenton and photo-Fenton processes are both AOPs that involve the generation of radicals for oxidation. The distinction between them lies in the origin of these radicals, as shown in figure 6.12. Therein, hydroxyl radicals ($^{\bullet}OH$) can be generated through the catalytic decomposition of hydrogen peroxide (H_2O_2) by ferric ions in the Fenton reaction. Meanwhile, the photo-Fenton reactions occur in the presence of UV or visible light, and the oxidative radicals derived from Fe^{3+}/H_2O_2 are produced through the action of the light source. In photocatalytic oxidation processes, electron–hole pairs are generated within the catalyst upon exposure to UV light, and the oxidative radicals are formed at the interface between the catalyst and water [43].

The Fenton method has attracted great interest due to its convenience and effectiveness. Notably, the Fenton method can produce many hydroxyl radicals ($^{\bullet}OH$) by introducing a divalent iron solution and hydrogen peroxide [44], as shown in equation (6.10).

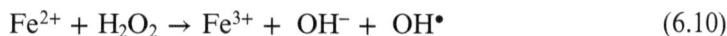

$$Fe^{2+} + H_2O_2 \rightarrow Fe^{3+} + OH^- + OH^{\bullet} \qquad (6.10)$$

The main result of the Fenton reaction, which is primarily governed by equation (6.10), is the formation of hydroxyl radicals ($HO^{\bullet}$). However, in the photo-Fenton process, two additional reactions occur simultaneously [45, 46]. First, there is the photolysis of hydrogen peroxide (H_2O_2) through the absorption of light energy ($h\nu$),

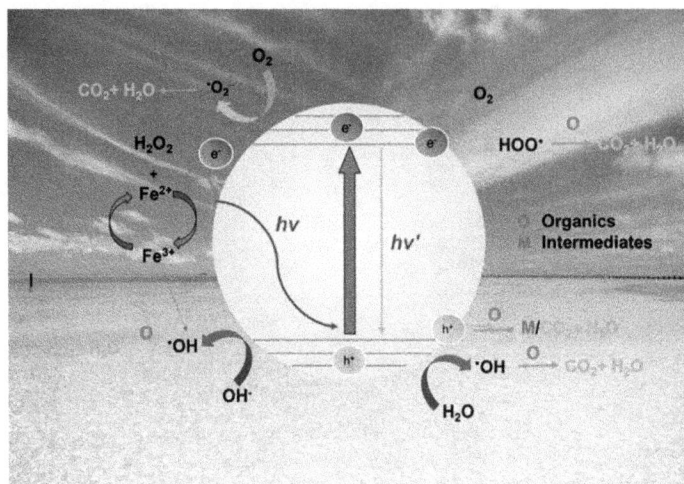

Figure 6.12. A schematic illustration of the photo-Fenton reactions of perovskites.

leading to the formation of two hydroxyl radicals ($HO^\bullet$) as shown in equation (6.11). This photolytic reaction occurs when H_2O_2 absorbs photons and undergoes a cleavage reaction, resulting in the generation of highly reactive hydroxyl radicals.

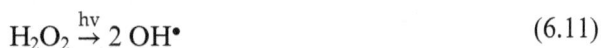

$$H_2O_2 \xrightarrow{h\nu} 2\,OH^\bullet \tag{6.11}$$

Second, there is the photoreduction of ferric ions (Fe^{3+}) in the presence of water (H_2O) and light energy ($h\nu$). This reaction, described by equation (6.11), involves the absorption of photons by Fe^{3+}, leading to the reduction of Fe^{3+} to ferrous ions (Fe^{2+}), the generation of hydroxyl radicals ($HO^\bullet$), and the release of a hydrogen ion (H^+). This photoreduction process contributes to the overall generation of hydroxyl radicals in the photo-Fenton process.

$$Fe^{3+} + H_2O \quad h\nu \quad Fe^{2+} + OH^\bullet + H^+ \tag{6.12}$$

Together, the Fenton reaction (equation (6.10)), the photolysis of H_2O_2 (equation (6.18)), and the photoreduction of Fe^{3+} (equation (6.11)) result in the production of hydroxyl radicals ($HO^\bullet$) in the photo-Fenton process. These highly reactive radicals play a crucial role in the degradation of organic pollutants by oxidizing and breaking down their chemical bonds.

6.4.3 PMS activation

PMS is frequently employed as an oxidizing agent in AOPs to generate sulfate radicals. These radicals can be activated through various methods. Instead of relying on energy-based activation methods such as solar/UV radiation, ultrasound, electricity, or heat, catalytic activation is more commonly employed. This is because it offers advantages such as lower cost and higher efficiency in decomposing PMS using transition-metal ions and heterogeneous catalysts. In AOP systems, the

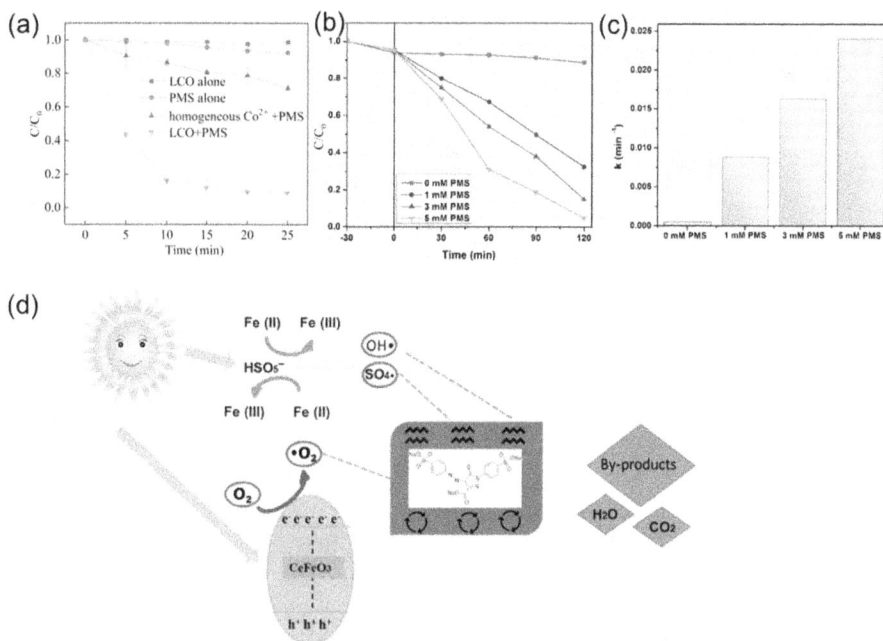

Figure 6.13. (a) BPA degradation in different reaction systems. Reaction conditions: [BPA] = 0.05 mM, [PMS] = 1.0 mM, [LCO] = 0.5 g l^{-1}, initial pH = 6.8. Reproduced from [47] with permission from the Royal Society of Chemistry. (b–c) Photocatalytic degradation curves of CeFeO$_3$ samples in the PMS/Vis system. (d) The proposed mechanism of the photoactivation of PMS over CeFeO$_3$ samples. Reprinted from [48], Copyright (2023), with permission from Elsevier.

presence of heterogeneous perovskite catalysts allows the generation of numerous ROS, including HO$^\bullet$ radicals, singlet oxygen, and $^\bullet$O$_2^-$. The activation of PMS by heterogeneous catalysts has gained significant attention because it enables easier separation of the catalysts, which helps to prevent secondary pollution issues that may arise in homogeneous systems.

Zhong *et al* synthesized La$_2$CoO$_4$+δ (LCO) perovskite and active PMS to degrade bisphenol A (BPA) [47]. As shown in figure 6.13a, LCO+PMS had a highly efficient photocatalytic performance, which was superior to those of LCO alone, PMS, or homogeneous Co^{2+}+PMS systems. Based on figures 6.13(b) and (c), it can be observed that the activity of bare PMS was relatively low, achieving only 12.6% degradation. This suggests that the activation of PMS on the surface of CeFeO$_3$ plays a more prominent role compared to light irradiation, which is illustrated in figure 6.13(c). Two primary interactions involving persulfate ions can be credited for the heightened activity observed in CeFeO$_3$/PMS/Vis systems. First, visible light can activate PMS, leading to its activation and the generation of reactive species. Second, the presence of ferrite sites within CeFeO$_3$ aids in the activation of PMS, facilitating the desired reactions. These interactions lead to the generation of free radicals, specifically sulfate radicals ($^\bullet$SO$_4^-$) and hydroxyl radicals

(HO$^\bullet$), which are responsible for the decomposition of dye molecules. The proposed overall activation reactions are summarized in equations (6.13)–(6.15):

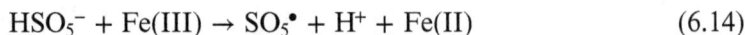

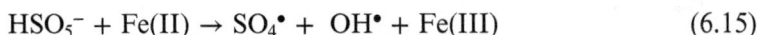

$$HSO_5^- + hv \rightarrow SO_4^\bullet + OH^\bullet \tag{6.13}$$

$$HSO_5^- + Fe(III) \rightarrow SO_5^\bullet + H^+ + Fe(II) \tag{6.14}$$

$$HSO_5^- + Fe(II) \rightarrow SO_4^\bullet + OH^\bullet + Fe(III) \tag{6.15}$$

Figure 6.13(c) shows that the visible-light photons are primarily absorbed by the surface of $CeFeO_3$ in the $CeFeO_3$/PMS/vis system. This absorption leads to the generation of photoexcited electrons in the conduction band and holes in the valence band of $CeFeO_3$. This $CeFeO_3$ can react with adsorbed oxygen molecules, leading to the formation of $^\bullet O_2^-$ radicals through a process known as electron transfer. These $^\bullet O_2^-$ radicals are highly reactive and can participate in subsequent oxidation reactions.

Simultaneously, PMS is also irradiated with light, resulting in the formation of sulfate radicals ($^\bullet SO_4^-$) and hydroxyl radicals ($^\bullet OH$). These radicals are generated through the activation of PMS by the absorbed light energy. The presence of PMS contributes to the overall radical generation in the system. In addition, the Fe^{3+} state within the $CeFeO_3$ perovskite structure can undergo oxidation, leading to the formation of additional radicals. This further enhances the photocatalytic oxidation processes occurring in the system.

Clearly, the combined action of $^\bullet O_2^-$ radicals, sulfate radicals, hydroxyl radicals, and other reactive species generated within the $CeFeO_3$/PMS/vis system contributes to the overall positive effect of the radicals. These radicals participate in various oxidation reactions, effectively degrading organic pollutants, such as dye molecules. Overall, the interplay between photoexcited electrons in $CeFeO_3$, $^\bullet O_2^-$ radicals and PMS activation through light irradiation results in the generation of a variety of radicals that drive the photocatalytic oxidation processes within the system, leading to the enhanced degradation of pollutants.

6.5 Conclusions

Perovskite materials possess exceptional crystallinity and unique features, making them highly promising as semiconductor photocatalysts for the degradation of organic pollutants. The ability to fine-tune their physicochemical properties by adjusting their chemical composition offers ample opportunities for developing novel nanocomposites. However, pristine perovskites suffer from limitations, i.e. low photocatalytic efficiency, poor stability, inadequate utilization of solar energy, rapid electron–hole recombination, and low redox potential. To overcome these challenges, researchers have capitalized on the flexibility of regulating perovskite properties to develop efficient photocatalysts for water remediation. Numerous perovskite-based nanocomposites have been synthesized and studied using various methods and strategies, including cationic substitution with dopants, downsizing and morphological modifications, and coupling with AOPs. While designing and preparing novel perovskite-based nanocomposites, particularly through partial and full cationic substitution, is an elegant approach, it remains a challenging task.

The photocatalytic activity of perovskites is improved by controlling the bandgap, defects, morphology, etc. In addition, the photo-Fenton method and PMS activation by perovskite-based nanomaterials were discussed in this chapter. Perovskite-based nanomaterials have demonstrated encouraging outcomes as activators for PMS in applications. They have exhibited promising results in the removal of diverse organic pollutants, including endocrine disruptors, antibiotics, and dyes However, it is worth noting that metal halide perovskites face challenges related to their stability, especially in the presence of moisture and heat. To achieve broader success in practical applications utilizing perovskite-based nanomaterials, the following research directions have been identified:

(a) developing new perovskites and coupling perovskites with other semiconductors or metal nanoparticles;
(b) investigating the photocatalysis of realistic wastewater samples;
(c) upscaling perovskite catalysts;
(d) improving their stability and long-term performance to make them commercially viable;
(e) developing environmentally friendly synthesis methods;
(f) decorating 3D-printed materials with perovskite photocatalysts for use in industrial production, and
(g) improving catalyst separation (membrane filtration is commonly used for catalyst separation but using membranes for perovskite-based catalyst recovery can be costly and complex due to membrane fouling).

References

[1] Lomonaco T et al 2020 Release of harmful volatile organic compounds (VOCs) from photo-degraded plastic debris: a neglected source of environmental pollution J. Hazard. Mater. **394** 122596
[2] Chaudhry G R and Chapalamadugu S J 1991 Biodegradation of halogenated organic compounds Microbiol Rev. **55** 59–79
[3] Singh S, Yadav R, Sharma S and Singh A N 2023 Arsenic contamination in the food chain: a threat to food security and human health J. Appl. Biol. Biotechnol. **11** 24–33
[4] Carpenter D O 2013 Effects of Persistent and Bioactive Organic Pollutants on Human Health (New York, NY: Wiley) DOI:10.1002/9781118679654
[5] Balk S J, Carpenter D O and Corra L et al 2010 Persistent Organic Pollutants: Impact on Child Health 9789241501101 World Health Organization Technical document
[6] Fuller R et al 2022 Pollution and health: a progress update Lancet Planet. Health **6** e535–47 (Corrected 18 May 2022 at https://doi.org/10.1016/S2542-5196(22)00145-0)
[7] Ahmed S et al 2011 Advances in heterogeneous photocatalytic degradation of phenols and dyes in wastewater: a review Water, Air, Soil Pollut. **215** 3–29
[8] Huang H, Pradhan B, Hofkens J, Roeffaers M B and Steele J A 2020 Solar-driven metal halide perovskite photocatalysis: design, stability, and performance ACS Energy Lett. **5** 1107–23

[9] Chen Z-Y, Huang N-Y and Xu Q 2023 Metal halide perovskite materials in photocatalysis: design strategies and applications *Coord. Chem. Rev.* **481** 215031

[10] Wei K, Faraj Y, Yao G, Xie R and Lai B 2021 Strategies for improving perovskite photocatalysts reactivity for organic pollutants degradation: a review on recent progress *Chem. Eng. J.* **414** 128783

[11] Eames C *et al* 2015 Ionic transport in hybrid lead iodide perovskite solar cells *Nat. Commun.* **6** 7497

[12] Kuru T *et al* 2023 Photodeposition of molybdenum sulfide on $MTiO_3$ (M: Ba, Sr) perovskites for photocatalytic hydrogen evolution *J. Photochem. Photobiol. A: Chem.* **436** 114375

[13] Luo J *et al* 2021 Halide perovskite composites for photocatalysis: a mini review *EcoMat* **3** e12079

[14] Kong J, Yang T, Rui Z and Ji H 2019 Perovskite-based photocatalysts for organic contaminants removal: current status and future perspectives *Catal. Today* **327** 47–63

[15] Tao S *et al* 2019 Absolute energy level positions in tin- and lead-based halide perovskites *Nat. Commun.* **10** 2560

[16] Deschler F *et al* 2014 High photoluminescence efficiency and optically pumped lasing in solution-processed mixed halide perovskite semiconductors *J. Phys. Chem. Lett.* **5** 1421–6

[17] Dey K, Roose B and Stranks S D 2021 Optoelectronic properties of low-bandgap halide perovskites for solar cell applications *Adv. Mater.* **33** 2102300

[18] Prasanna R *et al* 2017 Band gap tuning via lattice contraction and octahedral tilting in perovskite materials for photovoltaics *J. Am. Chem. Soc.* **139** 11117–24

[19] Irshad M *et al* 2022 Photocatalysis and perovskite oxide-based materials: a remedy for a clean and sustainable future *RSC Adv.* **12** 7009–39

[20] Liu W, Lee J-S and Talapin D V 2013 III–V nanocrystals capped with molecular metal chalcogenide ligands: high electron mobility and ambipolar photoresponse *J. Am. Chem. Soc.* **135** 1349–57

[21] Lim J *et al* 2022 Long-range charge carrier mobility in metal halide perovskite thin-films and single crystals via transient photo-conductivity *Nat. Commun.* **13** 4201

[22] Hu H *et al* 2020 Nucleation and crystal growth control for scalable solution-processed organic–inorganic hybrid perovskite solar cells *J. Mater. Chem.* A*8 1578–603*

[23] Steirer K X *et al* 2016 Defect tolerance in methylammonium lead triiodide perovskite *ACS Energy Lett.* **1** 360–6

[24] Farhad F T *et al* 2018 $MAPbI_3$ and $FAPbI_3$ perovskites as solar cells: Case study on structural, electrical and optical properties *Results Phys.* **10** 616–27

[25] Hosseinian Ahangharnejhad R *et al* 2021 Protecting perovskite solar cells against moisture-induced degradation with sputtered inorganic barrier layers *ACS Appl. Energy Mater.* **4** 7571–8

[26] Kumar A, Kumar A and Krishnan V J A c 2020 Perovskite oxide based materials for energy and environment-oriented photocatalysis *ACS Catal.* **10** 10253–315

[27] Salavati-Niasari M *et al* 2016 Synthesis, characterization, and morphological control of $ZnTiO_3$ nanoparticles through sol–gel processes and its photocatalyst application *Adv. Powder Technol.* **27** 2066–75

[28] Parida K, Reddy K, Martha S, Das D and Biswal N 2010 Fabrication of nanocrystalline $LaFeO_3$: an efficient sol–gel auto-combustion assisted visible light responsive photocatalyst for water decomposition *Int. J. Hydrog. Energy* **35** 12161–8

[29] Wang S *et al* 2018 Sol-gel preparation of perovskite oxides using ethylene glycol and alcohol mixture as complexant and its catalytic performances for CO oxidation *ChemistrySelect* **3** 12250–7

[30] Zhao H, Duan Y and Sun X 2013 Synthesis and characterization of $CaTiO_3$ particles with controlled shape and size *New J. Chem.* **37** 986–991,

[31] Athayde D D *et al* 2016 Review of perovskite ceramic synthesis and membrane preparation methods *Ceram. Int.* **42** 6555–71

[32] Moshtaghi S, Gholamrezaei S and Niasari M S 2017 Nano cube of $CaSnO_3$: facile and green co-precipitation synthesis, characterization and photocatalytic degradation of dye *J. Mol. Struct.* **1134** 511–9

[33] Junwu Z *et al* 2007 Solution-phase synthesis and characterization of perovskite LaCoO3 nanocrystals via a co-precipitation route *J. Rare Earths* **25** 601–4

[34] Cao L *et al* 2023 Highly ambient stable $CsSnBr_3$ perovskite via a new facile room-temperature 'Coprecipitation' strategy *ACS Appl. Mater. Interfaces* **15** 30409–16

[35] Zhang S *et al* 2020 SiO2 supported highly dispersed Pt atoms on $LaNiO_3$ by reducing a perovskite-type oxide as the precursor and used for CO oxidation *Catal. Today* **355** 222–30

[36] Koo P-L, Jaafar N F, Yap P-S and Oh W-D 2022 A review on the application of perovskite as peroxymonosulfate activator for organic pollutants removal *J. Environ. Chem. Eng.* **10** 107093

[37] Peng Y, Albero J and Garcia H 2019 Surface silylation of hybrid benzidinium lead perovskite and its influence on the photocatalytic activity *ChemCatChem* **11** 6384–90

[38] Schanze K S, Kamat P V, Yang P and Bisquert J 2020 Progress in perovskite photocatalysis *ACS Energy Lett.* **5** 2602–4

[39] Li Q and Lian T 2019 Ultrafast charge separation in two-dimensional CsPbBr3 perovskite nanoplatelets *J. Phys. Chem. Lett.* **10** 566–73

[40] DuBose J T and Kamat P V 2022 Efficacy of perovskite photocatalysis: challenges to overcome *ACS Energy Lett.* **7** 1994–2011

[41] Wang W, Tadé M O and Shao Z 2015 Research progress of perovskite materials in photocatalysis- and photovoltaics-related energy conversion and environmental treatment *Chem. Soc. Rev.* **44** 5371–408

[42] Yang Y *et al* 2023 Synergistic surface activation during photocatalysis on perovskite derivative sites in heterojunction *Appl. Catal. B* **323** 122146

[43] Orak C, Atalay S, Ersöz G J S S and Technology 2017 Photocatalytic and photo-Fenton-like degradation of methylparaben on monolith-supported perovskite-type catalysts *Sep. Sci. Technol.* **52** 1310–20

[44] Huang C-W *et al* 2022 Solar-light-driven $LaFe_xNi_{1-x}O_3$ perovskite oxides for photocatalytic Fenton-like reaction to degrade organic pollutants *Beilstein J. Nanotechnol.* **13** 882–95

[45] Rojas-Cervantes M L and Castillejos E 2019 Perovskites as catalysts in advanced oxidation processes for wastewater treatment *Catalysts* **9** 230

[46] Shinichi E *et al* 2014 Fenton chemistry at aqueous interfaces *PNAS* **111** 623628

[47] Zhong X, Wu W, Jie H and Jiang F 2023 $La_2CoO_{4+\delta}$ perovskite-mediated peroxymono-sulfate activation for the efficient degradation of bisphenol A *RSC Adv.* **13** 3193–203

[48] Tuna Ö and Bilgin Simsek E 2023 Promoted peroxymonosulfate activation into ferrite sites over perovskite for sunset yellow degradation: optimization parameters by response surface methodology *Opt. Mater.* **142** 114122

Chapter 7

Emerging trends and future prospects in photocatalysis-based environmental remediation and hydrogen production

İbrahim Hakkı Karakaş and Zeynep Karcıoğlu Karakaş

Photocatalysis-based environmental remediation and water splitting technologies are critical to address urgent global challenges such as pollution, environmental degradation, and the need for sustainable energy sources. These technologies have the potential to provide sustainable solutions in an era of increasing emphasis on cleaner production technologies and a growing preference for more environmentally responsible approaches to environmental remediation and energy production processes. In this chapter, we investigate recent advances in photocatalysis-based environmental remediation and hydrogen production technologies and provide insights into the challenges faced in applying the processes or innovations that will likely emerge. Moreover, this chapter sheds light on issues such as enhancing catalytic efficiency, recent advances in the design of new photocatalytic materials, and expanding their application areas. Furthermore, a comprehensive overview of current emerging trends and challenges in photocatalysis and future prospects is presented, and the multifaceted role of photocatalysis technology in environmental remediation and sustainable hydrogen production is explained.

7.1 Introduction

7.1.1 Emerging trends and future prospects in photocatalysis

Photocatalysis-based environmental remediation technologies are a promising and rapidly developing research field with many exciting developments and potential future directions. Photocatalysis is a process based on the use of light-activated catalysts to accelerate chemical reactions. Photocatalytic processes often remove pollutants from the air or water or produce H_2 via water splitting. It is expected that photocatalysis-based environmental remediation applications will gain more attention in the future, as some of the existing limitations will be overcome [1].

Nowadays, many researchers are investigating the development of new photocatalytic materials with improved efficiency, stability, and high selectivity. These studies focus on the photocatalytic use of new materials such as graphene-based catalysts, metal–organic frameworks (MOFs), MXene-based materials, and perovskites. These materials can be used directly as photocatalysts or in combination with other materials [2].

Nanotechnology plays a vital role in photocatalytic processes. When nanoparticles and nanocomposites are used as photocatalysts, the high surface areas of these materials have the potential to significantly increase catalytic activity. Photocatalytic processes can be applied for various environmental purposes, such as air purification, water treatment, water splitting, and the remediation of contaminated soil. Using these processes, it is possible to degrade a wide range of pollutants, such as organic dyes, pharmaceutics, pesticides, and volatile organic compounds (VOCs). New studies of the usability of photocatalytic processes in removing different contaminants continue to be carried out daily. It is expected that the types of pollutants that photocatalytic processes can remove will soon increase, and thus their application areas will become much more diversified. Today, photocatalytic processes can be used for water treatments, the removal of organic and inorganic pollutants, and photocatalytic disinfection [3].

Despite current challenges, research and development efforts are ongoing to address potential issues and make large-scale photocatalytic water splitting a reality. Innovative reactor designs, advanced catalyst materials, improved mass transfer techniques, and a better understanding of system dynamics are some focal areas that are being used to overcome these challenges. As technology advances, the scalability and efficiency of photocatalytic water splitting processes are expected to increase, and these processes are expected to become increasingly important in the renewable energy and hydrogen production sectors.

Photocatalytic processes have also been used in air quality improvement studies to improve indoor and outdoor air quality by removing some organic or inorganic pollutants from the air, including VOCs and nitrogen oxides (NOx). Some of these studies have also developed into commercial products such as self-cleaning coatings. Many new commercial products with similar properties are imminently expected to emerge. As the technology in this field matures and becomes more cost-effective, its adoption for commercial purposes is expected to increase further.

On the other hand, there is also growing interest in integrating photocatalytic processes with renewable energy sources such as solar and wind energy to improve energy efficiency and sustainability. Integrating photocatalytic processes with renewable energy requires careful planning and coordination [4].

Increasing the efficiency of photocatalysis under visible light is an important focus. Since visible light is more abundant than UV light, this would allow for more efficient solar energy utilization. The use of natural daylight instead of artificial light sources in photocatalytic processes has both advantages and some limitations [5]. Natural daylight is a sustainable and easily accessible energy source.

In photocatalytic processes, it helps to lower operating costs and environmental impact by reducing the need for electricity or artificial light sources. The use of renewable energy sources makes photocatalytic processes compatible with clean energy and sustainability goals. This provides a significant advantage for the usability of photocatalytic processes, especially in large-scale applications. Moreover, the use of natural daylight in photocatalysis can reduce the carbon footprint by reducing energy consumption and thus minimizing the greenhouse gas emissions caused by energy production.

Furthermore, natural daylight has the potential to provide higher energy efficiency, as it contains a broader spectrum of light than the UV–vis light sources currently used as light sources for photocatalytic reactions. In addition to the advantages of using daylight in photocatalytic processes, there are also some limiting challenges. Natural daylight varies in intensity and availability due to the weather conditions, time of day, and geographical location. This dictates that daylight-driven photocatalytic processes can only be actively applied when daylight is available. Therefore, these variations can significantly affect the consistency and predictability of photocatalytic reactions. Moreover, the spectrum and intensity of natural sunlight vary with the geographical location, time of year, and time of day. These differences can negatively affect the performance of photocatalytic materials optimized for specific wavelengths of light. Energy storage systems such as batteries may be required to sustain continuous photocatalytic reactions in the absence of daylight. These systems not only make construction more complex but also significantly increase costs [6–8].

Moreover, some photocatalytic reactions may require light sources of exceptionally high intensity. It may not be possible to provide this irradiance using natural daylight alone. Thus, for such applications, it may be necessary to use additional artificial light sources. The choice between natural daylight and artificial light sources in photocatalysis depends on the application and requirements. For example, natural sunlight can be used for water purification or improving air quality in remote or off-grid areas. In contrast, artificial light sources should be used for precise and consistent reactions in controlled environments.

In practice, many photocatalytic applications aim for an optimal mix of natural daylight and artificial lighting. In these studies, the system utilizes sunlight for photocatalysis and is supplemented with artificial light sources when needed. This approach combines the advantages of sustainability and cost savings with the reliability and control offered by artificial light [9].

Photocatalytic water splitting is a technology that has the potential to play an important role in renewable energy production. Photocatalytic water splitting uses direct UV or sunlight to split water molecules into gaseous hydrogen and oxygen. Hydrogen produced by this method can be used as a clean and renewable energy source [10–12].

Researchers worldwide are currently working to improve the efficiency of photocatalytic water splitting systems. These studies often focus on developing new materials for photocatalysts and designing new reactors to capture more of the solar spectrum and convert it into chemical energy [13, 14].

Today, metal oxides such as titanium dioxide (TiO_2), zinc oxide (ZnO), and many different semiconductor materials are used as photocatalysts in photocatalytic water splitting processes. New and more efficient photocatalysts are expected to be developed as a result of advances in materials science in the near future. Moreover, it may be possible to produce composites of some of the currently used photocatalysts with other materials and thus improve the catalytic activity of these materials. In addition, some recent studies have shown that new materials, such as perovskite compounds and quantum dots, can be very effective in photocatalytic water splitting processes. Further research and development efforts in this field will lead to even more significant progress [15–18].

It is also essential to develop new technologies that will make it possible to provide the light required for photocatalytic water splitting processes directly from the Sun. In today's photocatalytic water splitting processes, UV lamps that emit continuous radiation at a specific wavelength are generally preferred as light sources. However, these lamps' energy consumption increases these processes' environmental footprint. Improving the integration of photocatalytic water splitting with solar technologies such as photovoltaic cells and concentrated solar power can make large-scale hydrogen production more efficient, environmentally friendly, and practical [11, 19–21].

7.2 Scalability challenges for photocatalytic applications

Researchers and engineers have been working to design scalable and cost-effective photocatalysis processes for large-scale applications. However, scaling up photocatalytic water splitting processes for industrial applications presents significant challenges [22]. Scaling up photocatalytic water splitting processes can present challenges in transitioning from primarily laboratory-scale experiments to industrial or commercial applications. As the scale increases, ensuring efficient and homogeneous light absorption or distribution across a larger reactor becomes more difficult. Maximizing the utilization of sunlight or artificial light sources requires innovative engineering solutions. Moreover, mass transfer limitations can also be an issue in scaling up processes. Mass transfer limitations become more pronounced in larger reactors. Especially in large-scale heterogeneous catalysis systems, it can be much more challenging to deliver reactants to the catalyst surface and efficiently remove the products (hydrogen and oxygen) from the reactor than in laboratory-scale systems. Another challenge associated with scaling up processes is to ensure that operating parameters such as temperature, pH, and flow rates, which have the potential to directly affect process efficiency, are homogeneous in all regions of the reactor. In addition, in large-scale systems, some parts of the reactor must be sufficiently exposed to light [23–25].

For high-efficiency photocatalytic processes, larger photocatalytic reactors need to be designed and manufactured to ensure that the ideal values of parameters such as optimized temperature, pH, and flow rate remain constant and, at the same time, ensure the homogeneous distribution of light throughout the reactor. For this purpose, new reactors should be designed by considering efficient mixing and flow models [26]. In addition, problems with uniform and homogeneous catalyst distribution throughout the reactor may arise in large reactors. A homogeneous catalyst distribution is fundamental in large-scale systems to achieve stable and highly efficient photocatalytic reactions. If homogeneous catalyst distribution is not achieved, this can lead to variations in reaction rates and a decrease in overall efficiency. To avoid these problems, designing and manufacturing new reactors is essential [27].

Moreover, large quantities of catalysts will be required for large-scale systems. Given that the materials currently used as photocatalysts are nanoscale advanced materials, it can be challenging to produce photocatalytic materials in the quantities needed for large-scale applications. The methods used to synthesize and prepare photocatalysts at the laboratory scale may need to be more easily scalable to meet commercial demands. In this case, scaled up processes for these catalysts may also need to be meticulously designed. In addition, factors such as the stability and durability of the photocatalysts produced may also become an issue during scale-up [28, 29]. Commercial applications often require longer catalyst lifetimes, and problems due to photocatalyst degradation or fouling can become more pronounced in larger systems. Significant progress is expected in the development of reusable photocatalyst materials that have improved chemical and mechanical stability.

Another aspect that needs to be considered concerning photocatalytic water splitting processes is economic viability. The scale-up of photocatalytic water splitting processes must be economically viable. Factors such as the reactor's design, manufacture, and maintenance and the cost of production, stability, and reusability of catalyst materials are directly related to process economics [30]. To assess the economic viability of the process, all the costs incurred need to be evaluated together with the amount of hydrogen produced by this process and any other possible benefits. Moreover, meeting regulatory standards for environmental and safety considerations can require more work in large-scale systems. This is because it is crucial to avoid negative environmental impacts when scaling up photocatalytic processes.

Despite these challenges, much research and development work is underway to overcome these issues and make large-scale photocatalytic water splitting a reality. Scientific studies in this field are focused on designing innovative reactors, developing advanced catalyst materials, and improving mass transfer processes. In the near future, due to increasing environmental concerns and constantly evolving technology, it is predicted that the interest in hydrogen, considered a green energy source, will increase even more. Therefore, it is expected to become increasingly essential to scale up and increase the efficiency of hydrogen production processes by photocatalytic water splitting.

The possible negative environmental impacts of photocatalytic water splitting processes must be addressed further. Currently, researchers are focused on developing methods that use sustainable materials for hydrogen production and minimize the release of pollutants during the hydrogen production process. In light of the data obtained from these studies, it is predicted that more concrete advances will be made, and much cleaner production processes will be designed. In hydrogen technologies, safe hydrogen storage is as necessary as its production. New safe and efficient hydrogen storage technologies should also be developed to store the hydrogen gas produced by photocatalytic water splitting methods in hydrogen fuel cells. Photocatalytic processes are likely to be used in many different commercial-scale applications in the near term. Photocatalytic processes have the potential to overcome various environmental and industrial challenges. Currently, the main factor limiting the availability of photocatalytic methods on a commercial scale is the high cost of these processes [31]. However, shortly, due to advances in materials science, reactor design, and manufacturing techniques, the cost of photocatalytic systems is expected to decrease significantly, making them economically viable for large-scale applications. Some applications in which photocatalytic processes are expected to be applied commercially are summarized below [32, 33]. Photocatalytic processes are likely to play an essential role in water treatment on a commercial scale. They have the potential to simultaneously and efficiently remove many types of contaminants, including organic pollutants, pharmaceuticals, and micro-organisms that are likely to be present in water. Photocatalytic processes can simultaneously disinfect water as well as remove contaminants. Therefore, photocatalysis-based treatment technologies can be used in large-scale water treatment systems to improve water quality [33].

Air purification is another area in which photocatalysis-based technologies have been used extensively and successfully. Photocatalysis is well suited for large-scale applications such as indoor air purification in commercial buildings, factories, and public spaces. Coatings and materials with photocatalytic properties have the potential to help reduce the concentration of VOCs and pollutants in both indoor and outdoor air [34–36]. Photocatalysis can also be applied in green hydrogen production through water splitting, which has commercial applications in clean energy and fuel cell technologies. Large-scale hydrogen production plants are predicted to adopt photocatalytic processes as a sustainable and environmentally friendly method of production. Due to photocatalytic technology, it is possible to produce products with self-cleaning surfaces. Self-cleaning surfaces coated with photocatalytic materials, such as glass or building exteriors, are expected to be more widely used in commercial buildings. These materials can significantly reduce the maintenance costs of facilities and improve their esthetics [37].

Thanks to photocatalysis-based technologies, it is possible to remove contaminants from soil areas and groundwater contaminated with various pollutants. Moreover, photocatalytic processes can be applied in agriculture to improve soil quality, remove pesticides, and enhance the decomposition of organic matter. Thus, developing more sustainable and efficient agricultural practices on a commercial scale may be possible. Moreover, photocatalysis-based processes will likely be used for pollutant control and process optimization in industrial processes in various fields such as chemistry, textiles, and food.

In conclusion, the applicability of photocatalysis-based processes in commercial-scale applications in the future will be shaped by the data to be obtained from ongoing research and development efforts. Accordingly, the future of photocatalytic processes depends on the adoption of possible new technologies that will emerge as a result of scientific studies by the relevant industries and the reduction of the costs associated with these processes. The integration of photocatalytic processes into various commercial sectors is expected to expand considerably due to increasing environmental concerns and energy-related challenges.

7.3 Scalability challenges for photocatalytic H_2 production applications

More and more fossil fuels are being consumed every day to meet the ever-increasing energy demand on a global scale. This situation rapidly increases the environmental risks arising from fossil fuels and also leads to the rapid depletion of these resources. Therefore, interest in green or renewable energy sources is growing daily. Hydrogen is seen as a green energy source. However, many current methods for hydrogen production need to be more economically and technically scalable to meet the requirements of large-scale production. Moreover, the methods currently applied for H_2 storage have some significant limitations regarding safety and usefulness. The future of large-scale hydrogen production by photocatalytic oxidation-based methods has great potential. It represents a sustainable and environmentally friendly method for producing hydrogen, which is a clean energy carrier. However, there are several factors to consider and some challenges to overcome when assessing the potential and applicability of these methods for large-scale production [1, 38].

As with other technologies, the scale-up for the large-scale production of H_2 by photocatalytic methods depends on the economic viability of the process. The process must be made cost-effective for photocatalytic hydrogen production to be preferable to other methods currently being applied for hydrogen production.

On the other hand, while hydrogen is considered a clean energy source, the potential environmental impact of the production methods is also significant. Photocatalytic oxidation is considered to be more environmentally friendly, as it does not produce carbon emissions. However, the environmental concerns associated with the production and use of materials must be carefully considered and potential risks minimized.

Scaling up hydrogen production through photocatalytic oxidation for large-scale production is seen as a promising method for creating a sustainable environment. Scaling up H_2 production by photocatalytic methods for large-scale production can only be achieved by overcoming some of the current economic and environmental challenges. Accordingly, advances in materials science, reactor design, and renewable energy technologies are expected to make this method more efficient and cost-effective.

'Scale-up' refers to increasing the capacity or size of a system or production process. In the context of chemical manufacturing, scale-up refers to the transition from a smaller-scale laboratory setup to much larger commercial-scale production. Scale-up is crucial for moving new technologies from the research and development stage into practical applications. There are some key considerations to bear in mind when scaling up for the large-scale production of H_2 production by photocatalytic methods.

Photocatalytic oxidation processes typically use light from a light source to drive the reaction. This light can come from an artificial lighting system or the Sun. However, artificial light sources are not sustainable due to their energy consumption. Accordingly, using sunlight as a light source is an economical and environmentally preferable approach. However, the utilization of sunlight for large-scale H_2 production in photocatalysis-based systems can only be made possible by developing new, more efficient, and cost-effective methods for capturing and converting sunlight. Therefore, with the advances in photovoltaics and the development of better photocatalysts, these processes are soon expected to become more effective and preferable for large-scale H_2 production.

For large-scale H_2 production in photocatalysis-based systems, one of the most critical factors affecting the efficiency of the process is the material used as the photocatalyst. Researchers are continuously working to develop new catalysts that are not only efficient but also cost-effective and readily available and can maintain their stability over long periods. Another important parameter in increasing the yields of reactions in H_2 production by photocatalytic methods is the reactor design. Most laboratory-scale systems are relatively controllable under laboratory conditions and in small volumes. However, when these systems are scaled up for large-scale production, some significant problems arise. Therefore, photocatalysis-based methods must develop highly efficient and easily controllable reactors for large-scale H_2 production applications.

In photocatalytic processes, the materials used in reactor manufacture and the photocatalysts used in photocatalytic processes are expected to have properties such as chemical stability, long-term preservation of catalytic activity, and reusability over a wide range of pH values. They should be durable in harsh photocatalytic reactions and resistant to degradation over time. Developing materials that can withstand long-term use is essential for long-term scalability. Heterogeneous photocatalytic systems have a relatively simple structure and are more economical than panel photocatalysts or photoelectrochemical systems [31]. Nevertheless, heterogeneous photocatalytic systems based on particulate photocatalysts have limited use in large-scale applications [39–42]. These limitations mainly stem from three different issues:

(1) Heterogeneous systems require a lot of water. This makes the operation of these processes difficult and uneconomical.
(2) The settling of particles to the bottom of the vessel is a major problem. If this occurs, it decreases the photocatalyst's interaction with light, thereby decreasing the process efficiency. Therefore, continuous and effective mixing is required.
(3) Difficulties in separating and recovering the photocatalyst from the suspension at the end of the process [1].

When designing reactors to be used for heterogeneous photocatalytic systems, operational parameters such as the water flow rate, catalyst dosage, and pollutant concentration, which potentially have direct effects on the efficiency of the photocatalytic process, must be taken into account [31, 39].

7.4 Advances in photocatalytic materials and innovative synthesis approaches

'Photocatalytic materials' is a general term for substances that have the potential to initiate or accelerate chemical reactions when exposed to light. They are widely used in applications such as water splitting for environmental remediation and hydrogen production. The development of efficient and cost-effective photocatalysts is an exciting area of research.

Titanium dioxide (TiO_2) and zinc oxide (ZnO) are the best-known photocatalysts; these materials are also called conventional photocatalysts. Numerous scientific studies have been conducted on the photocatalytic properties of these materials, and their stability and photocatalytic efficiency in the degradation of many pollutants and water splitting processes have been repeatedly proven. To increase the efficiency of these photocatalysts, composite or hybrid forms with other materials have also been produced. However, researchers continue working with these materials to improve efficiency and find new applications. In this context, hybrid catalysts produced by doping conventional photocatalysts such as titanium dioxide (TiO_2) with metals or nonmetals are known to improve charge separation and extend light absorption into the visible range, which can lead to increased photocatalytic efficiency. Therefore, it is expected that many scientific studies will be carried out in the future to develop new TiO_2-based hybrid or composite photocatalysts [43].

MOFs are a group of materials that have recently been extensively used for different purposes in many other applications. MOFs are composed of metal ions coordinated with organic ligands. These materials offer great potential for photocatalytic applications due to their high surface area and tunable properties. Studies are being carried out to improve the light absorption properties of MOFs. This is expected to further enhance the photocatalytic properties of these materials. MOFs can also be combined with other materials used as catalysts, significantly increasing their efficiency [44–47].

Perovskites are another group of materials whose photocatalytic properties have been studied in detail. Perovskite materials, which stand out due to their high potential for use in solar cell technology, have also been extensively investigated for their efficiency in photocatalytic processes. Perovskites also have great potential as photocatalysts due to their high photosensitivity. The high photosensitivity of perovskite materials allows them to absorb light emitted over an extensive wavelength range. Therefore, they are potentially efficient photocatalysts [48–50].

Quantum dots are another group of materials used as photocatalysts today. Quantum dots are semiconducting nanocrystals with a tunable bandgap structure. They can be incorporated into the structure of various photocatalysts to improve light efficiency and charge separation. Thanks to their flexible and tunable bandgap structure, these materials enable precise control of the light absorption properties of the materials they are incorporated into. Thus, they can make it possible to prepare highly efficient photocatalysts [51, 52].

Two-dimensional (2D) materials such as graphene, graphene oxide (GO), and transition-metal dichalcogenides (MXenes) can exhibit photocatalytic effects themselves or can be used to enhance the activity of other photocatalysts. These materials are widely used in photocatalysis-based applications due to their high surface area and advanced electronic and optical properties.

Graphene, GO, and reduced GO are carbon-based materials with different structures. These materials are widely used in photocatalytic processes for various purposes. Pure graphene has limited photocatalytic activity due to its zero bandwidth. Its zero bandgap means that graphene cannot absorb visible light efficiently. However, it can exhibit a little photocatalytic activity under UV light. Graphene is often used as a support material in photocatalytic systems to improve the efficiency of other photocatalysts [53, 54]. Its high surface area and excellent electron transport properties enhance the interaction between pollutants and the catalyst material. GO has a broader bandgap range than that of pure graphene. As a result, it can absorb visible light. GO can be used directly in photocatalytic processes or as a support material or cocatalyst for some photocatalysts. It can exhibit photocatalytic activity, especially for the degradation of some organic pollutants. GO is used in environmental remediation applications such as water treatment, pollutant degradation, and organic removal. When combined with other photocatalysts, it has the potential to be used in more advanced oxidation processes. Reduced graphene oxide (rGO) is obtained by the chemical reduction of GO. It has much improved photocatalytic activity compared to those of graphene and GO. The reduction process reverts GO to a graphene-like structure, reducing its bandgap. The reduction in bandgap allows rGO to absorb and utilize visible light more efficiently. rGO finds use in environmental remediation applications such as the degradation of organic pollutants, hydrogen production, and CO_2 reduction. Its enhanced photocatalytic properties make it a valuable material for various photocatalytic systems [55, 56].

In summary, graphene has limited photocatalytic activity but has the potential to be used as a support or cocatalyst. GO has a relatively wider bandgap and can thus exhibit photocatalytic activity under visible light. rGO, obtained by chemical reduction of GO, has the potential to exhibit even better photocatalytic activity. These materials have significant potential in various photocatalytic processes, especially in environmental remediation and sustainable energy production. It is expected that more scientific studies of the use of these materials in photocatalytic processes will be carried out to increase or optimize their efficiency in photocatalytic processes.

Carbon nitride (C_3N_4) is another carbon-based material extensively used in photocatalytic processes. Carbon nitride is a polymeric material with a graphitic structure that exhibits photocatalytic activity in visible light. Like other carbon-derived materials, carbon nitrides are seen as promising, especially for applications such as photocatalytic water splitting and the degradation of organic pollutants [57–59].

MXenes are another material group that has recently attracted great interest due to their high activity in photocatalytic processes. MXenes are a class of 2D materials that have recently attracted much attention because of their unique structural, morphological, and electrical properties. MXenes include transition-metal carbides, nitrides, and carbonitrides. As a result of their tunable surfactant groups and high surface areas, MXenes are widely used in many environmental remediation applications. It is known that MXenes can be used as adsorbents, photocatalysts, sensors, and membranes in these environmental remediation processes [60, 61].

In many studies, MXene-based materials have been used as photocatalysts in some photocatalytic processes, exhibiting excellent photocatalytic activity. Most MXenes have a wide bandgap, one of the most fundamental properties for photocatalytic activity. As such, MXenes are materials with high potential for use in photocatalysis. MXenes and MXene-based composites show promise in environmental applications such as removing pollutants from water and air. Moreover, some MXenes and their composites have been extensively investigated for their potential use in hydrogen production via photocatalytic water splitting. In studies conducted for this purpose, it has been reported that they can split water into hydrogen and oxygen using sunlight and thus have the potential to constitute a sustainable approach to producing hydrogen as a clean energy source. Moreover, another critical feature of MXene-class materials is their ease of functionalization. It is estimated that the photocatalytic properties of MXenes can be further improved through various surface modifications or by combining them with other materials such as semiconductors or noble metal nanoparticles. Although MXenes are considered a promising class of materials for the future, there are still challenges in increasing their photocatalytic efficiency and optimizing their stability. Researchers continue to work actively to improve the performance of MXenes and to understand the mechanism of MXene-based photocatalytic processes [61, 62].

In summary, MXenes have exhibited promising photocatalytic properties in photocatalytic processes such as environmental remediation and hydrogen production by water splitting. In the coming years, MXenes or MXene-based materials will likely be used more intensively in photocatalytic processes, so the photocatalytic properties of MXenes will be better understood and utilized more efficiently.

Plasmonic photocatalysis is another basic approach recently applied to increase the efficiency of photocatalysts. It is known that plasmonic nanoparticles, such as gold and silver nanoparticles, can enhance photocatalytic activity through localized surface plasmon resonance [63]. When combined with a photocatalyst, plasmonic nanoparticles are known to significantly improve light absorption and charge separation in photocatalytic systems and thus notably increase the catalytic activity of photocatalysts [64–66].

Hybrid, composite, or heterojunction materials that combine different photocatalyst materials can also exhibit effective photocatalytic activities. A common approach to obtaining hybrid photocatalysts is to incorporate other materials by self-assembly. For example, it is known that when metallic nanoparticles are combined with semiconductors, the resulting photocatalyst can facilitate charge transfer and improve photocatalytic performance. Therefore, many different photocatalysts are now obtained by combining various materials.

Photocatalytic processes are undergoing rapid change and development. Researchers are constantly investigating different photocatalytic materials and synthesis methods to develop highly efficient and more functional photocatalysts.

To synthesize the materials used as photocatalysts, bottom-up approaches such as the sol–gel method, hydrothermal synthesis, and chemical vapor deposition are generally preferred, allowing precise control over the structures and properties of the product materials. Through these methods, the morphological properties of the produced catalysts, such as shape and size (which have the potential to directly affect their catalytic activity), can also be precisely controlled. Bottom-up synthesis methods are expected to be extensively used in the synthesis of new photocatalysts that are likely to emerge in the future.

Another synthesis approach that is expected to be widely used in the production of photocatalysts is bioinspired synthesis. Bioinspired synthesis is a synthesis approach based on the creation of new materials inspired by natural photosynthetic systems. It is known that many photochemical processes occur spontaneously in nature [67]. Researchers aim to design new materials and processes inspired by these natural processes and to produce new photocatalysts. Moreover, the development of artificial systems that simulate natural photosynthesis is expected to become an essential application of photocatalysis in the coming years. These systems will convert solar energy into chemical fuels such as hydrogen or directly into valuable products [68].

The synthesis and characterization of nanostructured materials with high surface area and well-defined structures is an important research area. Nanomaterials can be produced by controlling their size, shape, and chemical composition. It is known that the photocatalytic properties of nanomaterials, like many other properties, are significantly affected by morphological properties such as their shape and size. Therefore, it may be possible to tailor nanomaterials for specific reactions. Like all nanotechnology-based products and processes, nanomaterials' production and synthesis methods are a constantly evolving research field. As a result of new technologies, it is possible to design and produce nanostructures with special photocatalytic properties is possible. In the future, it is expected that advanced synthesis methods such as atomic-layer deposition and chemical vapor deposition will be used more intensively in synthesizing photocatalyst materials. Thus, more effective new photocatalysts will be developed.

Machine learning and computational chemistry approaches are other synthesis approaches currently attracting interest for use in the synthesis of new materials. As a result of these methods, it is possible to predict the properties and optimize the structures of new photocatalysts that are intended to be synthesized before proceeding to the experimental stage. Machine learning and computational chemistry approaches are expected to be used more intensively in the synthesis of photocatalysts. Thus, the objective is to design more advanced photocatalysts and, at the same time, to save time and resources in the laboratory [69–72].

Another synthesis approach used extensively for designing more advanced photocatalysts is cocrystallization. This method is based on developing composite materials by cocrystallizing different chemical compounds. Photocatalysts produced by this method are expected to exhibit synergistic effects that enhance photocatalytic performance in addition to the individual photocatalytic activity of the materials [73, 74].

Another synthesis approach that has attracted interest in nanomaterials is *in situ* synthesis. Today, most of the materials used as photocatalysts are nanomaterials. All methods for producing nanomaterials can be categorized into two main groups: top-down and bottom-up synthesis. Top-down synthesis methods are generally more straightforward and better suited to being scaled up. However, the nanomaterials produced by this method are relatively large. They are also inhomogeneous in shape and size. Moreover, the yield is low, and the synthesis process offers limited control [28, 29, 75, 76]. However, bottom-up methods involving highly complex chemical processes make it possible to synthesize small and homogeneous nanoparticles [77, 78]. Bottom-up synthesis methods can be carried out in a single step (one-pot) or multiple stages. Among the bottom-up approaches, one-pot [79]and *in situ* synthesis methods are the most suitable for the large-scale, low-cost production of nanomaterials [80]. In particular, one-pot synthesis methods are considered economical and environmentally friendly, as they involve few reaction steps [81, 82]. The *in situ* synthesis method is a synthesis approach that aims to synthesize nanomaterials directly in the application environment, thereby providing photocatalysts with enhanced integration capability and efficiency [83, 84]. Moreover, the use of *in situ* synthesis methods makes it possible to precisely control the morphological properties of the product material, such as its shape and size [85, 86]. In the coming years, *in situ* synthesis approaches are expected to be applied more intensively and effectively to the synthesis of new photocatalysts.

It is possible to obtain photocatalysts under laboratory conditions, even if the synthesis processes are complex. However, large-scale synthesis processes require large reactors, and the methods of ensuring homogeneous and high-temperature heating and precise control of mass transfer in these systems can be much more complex [87]. Moreover, large-scale nanomaterial production may require several methods to be combined [88]. Furthermore, possible by-products or wastes may be generated during these processes. The potential adverse environmental impacts of production processes need to be carefully considered [85, 89]

The future of photocatalyst materials and their synthesis methods will likely be shaped by combining these approaches, driven by the need for more efficient and sustainable catalysts for various applications, including renewable energy production and environmental remediation. It is crucial to design advanced photocatalysts to address the global challenges of clean energy and pollution control.

7.5 The future evolution of light sources for photocatalytic processes

The choice of a light source for a photocatalytic process is a crucial parameter, as it directly affects the efficiency and applicability of the process. The chosen light sources for future photocatalysis processes are likely to be significantly influenced by several factors, such as energy efficiency, spectral range, and cost-effectiveness. Some potential trends related to the light sources used in future photocatalytic processes are summarized below.

LEDs are light sources that have high energy efficiency. Catalysts with different chemical structures are known to be activated when exposed to different wavelengths of radiation. LED light sources have the advantage that they can be tuned to emit specific wavelengths to optimize the photocatalytic activity of other materials [90–92]. Therefore, they are likely to be widely used as light sources, especially in applications requiring variable or precise light spectrum control.

Xenon and mercury lamps emit light over a wider spectral range, including the UV and visible-light regions. These lamps, which emit continuous and stable radiation, have become the traditional light sources used for photocatalytic processes. Although they are less energy efficient than LEDs in terms of the light intensity they emit, they are widely preferred when a wide spectral range is required. Therefore, xenon and mercury lamps are expected to be extensively used as light sources for photocatalytic processes in the future.

One of the approaches that has recently attracted interest in the field of photocatalysis is the use of plasmonic nanoparticles as light sources. It is expected that much more extensive research will be carried out on photocatalytic technology based on plasmonic particles, which is still under development. Plasmonic nanoparticles have the potential to significantly increase the absorption and utilization efficiency of light in photocatalytic reactions. Therefore, gold and silver nanoparticles, known for their strong plasmonic properties, have the potential to play an essential role in increasing the efficiency of photocatalytic processes.

The light source is a factor that directly impacts the sustainability of photocatalytic processes. Obtaining the light required for the process from natural sunlight is a sustainable and cost-effective approach. Therefore, photocatalytic processes that efficiently utilize the sunlight spectrum will likely become more prominent in the near term. On the other hand, technologies that allow more concentrated sunlight to be used are developing rapidly. Concentrated solar energy systems have the potential to provide intense, stable, and continuous light for specific applications. Therefore, much work is expected to be done on integrating concentrated solar energy systems into photocatalytic processes. Pulsed light sources such as pulsed xenon lamps and lasers can be used in specific photocatalytic applications in which intense, short bursts of light are required. They can potentially be used for fast, high-energy reactions [91, 93].

Future photocatalytic processes may also require the use of hybrid light sources based on combining natural sunlight with artificial light to optimize energy efficiency. In this way, the adverse effects of variations in sunlight, which can vary continuously throughout the day, on the ongoing photocatalytic process can be significantly minimized.

As in almost all fields, the general trend when choosing light sources for photocatalytic processes is to use environmentally friendly and nonhazardous light sources. The cost of light sources is pivotal to the commercial viability of photocatalytic processes. Their price is likely to be reduced over time due to advances in manufacturing and technology. The choice of light source for a photocatalytic process varies depending on the application's requirements, including the photocatalyst's spectral absorption characteristics and the desired reaction kinetics. Thus, some applications, especially in sectors such as water treatment and air quality improvement, may require the development of specialized light sources tailored to specific catalysts and reactions. Like many other fields, light technology is constantly evolving and being renewed by advancing technology. Therefore, in the future, in parallel with advances in technology and materials science, it is expected that new, more efficient, cost-effective, and sustainable light sources will be developed and integrated into photocatalytic systems. Moreover, if optical systems based on the guidance or management of light, such as lenses, mirrors, and light guides, are integrated into photocatalytic systems, the efficiency of these processes will increase significantly. Thus, more sustainable strategies can be designed.

Studies of hydrogen production by the photocatalytic splitting of water or photocatalysis-based environmental remediation processes are usually carried out on an experimental scale in small-volume glass or quart reactors. In these systems, also known as heterogeneous photocatalysis processes, the entire active surface of the photocatalyst can be used. Thus, the photocatalysts used are highly effective, and these experimental-scale studies observe excellent yields and high reaction rates. However, due to the nature of heterogeneous mixtures for large-scale applications, it is also possible that the particles used as photocatalysts may settle to the bottom of the reactor, and thus, their effectiveness may decrease. To avoid this, it is essential to ensure that adequate mixing is performed from the beginning to the end of the photocatalytic process. This mixing can lead to extra energy consumption and additional costs in large-volume systems. As the efficiency of the photocatalyst decreases over time, the photocatalyst needs to be regenerated or replaced. However, it is a challenging process to separate these photocatalysts, which are primarily nanomaterials, from the water environment and collect them for reuse. Although methods such as magnetic separation, centrifugation, etc. can be used in laboratory-scale studies, they would need to be more effective for production-scale applications. Therefore, it is anticipated that photocatalysts immobilized on a fixed substrate may be more helpful and significantly increase the usability of photocatalytic processes in large-scale applications. The main expectations here are to facilitate the separation of spent photocatalysts from the aqueous solution and to provide continuous system operability. However, these systems have limiting factors, such as their inability to utilize the entire surface of the catalyst and the reduction of the irradiated catalyst surface, which reduce their efficiency [94–96].

Plasmonic photocatalysis can also be used in the photocatalytic degradation of organic pollutants in water and air. Plasmonic materials can improve the performance of solar cells and other energy conversion technologies by enhancing light absorption and charge separation. Moreover, plasmonic nanoparticles are used to develop sensitive sensors that can detect small concentrations of various substances.

Plasmonic photocatalysis has excellent potential to improve the efficiency of photocatalytic reactions. However, plasmonic photocatalysis is still an emerging research field and poses several challenges, including optimizing plasmonic nano-materials, controlling plasmon resonance, and developing practical and scalable applications. Many scientific studies are expected to overcome these challenges in the future. Thus, it is predicted that this process will find more widespread applications.

7.5.1 Plasmonic photocatalysis

Plasmonic photocatalysis is a field of research based on combining particles with plasmonic properties and photocatalytic materials to improve the efficiency of photocatalytic reactions. The plasmonic effect occurs when electromagnetic radiation interacts with free electrons in the structure of a noble metal, such as gold or silver, resulting in the formation of surface plasmons on the metal surface. These surface plasmons have the potential to significantly enhance light absorption and energy transfer in photocatalytic processes. Plasmonic photocatalytic technology is based on the use of plasmonic particles, such as gold or silver nanoparticles, in combination with photocatalysts. Plasmonic nanoparticles exhibit a unique property known as surface plasmon resonance (SPR), which allows materials to absorb and re-scatter specific wavelengths of light incident on them. Plasmonic photocatalysis technology often combines plasmonic nanoparticles with conventional photocatalysts such as titanium dioxide or zinc oxide to form hybrid materials. These hybrid photocatalysts can utilize absorbed light to initiate photochemical reactions [66].

The mechanism of plasmonic photocatalysis takes place in several basic steps. In the first step, plasmonic nanoparticles absorb incident light due to SPR, forming localized surface plasmons. The energy of these plasmons can be tuned to match the desired wavelength, which is usually in the visible or near-infrared range. In the second stage, light interacts with the plasmonic nanoparticles, generating intense and localized electric fields around the plasmonic metal nanoparticles. These localized electric fields concentrate the light energy and initiate the photochemical process, resulting in the formation of electron (e^-) and hole (h+) pairs in the internal structure of the photocatalyst. In the final stage, the enhanced charge transfer and high energy concentration resulting from the plasmonic effect increase reaction rates and make the photocatalytic process more efficient [63, 65].

One of the potential application areas of plasmonic photocatalysis is photocatalytic water splitting processes. Plasmonic photocatalysts significantly increase the efficiency of reactions that produce hydrogen and oxygen gases by splitting water [66]. Plasmonic photocatalysis can also be used in the photocatalytic degradation of organic pollutants in water and air. Plasmonic materials can improve the performance of solar cells and other energy conversion technologies by enhancing light absorption and charge separation. Moreover, plasmonic nanoparticles are used to develop sensitive sensors that can detect small concentrations of various substances.

Plasmonic photocatalysis has excellent potential to improve the efficiency of photocatalytic reactions. However, plasmonic photocatalysis is still an emerging research field and poses several challenges, including optimizing plasmonic nanomaterials, controlling plasmon resonance, and developing practical and scalable applications. Many scientific studies are expected to overcome these challenges in the future. Thus, it is predicted that this process will find more widespread applications.

7.5.2 A comparison of visible-light photocatalysis and plasmonic photocatalysis

Visible-light photocatalysis and plasmonic photocatalysis differ significantly in terms of the light source used and the mechanism involved. In visible-light photocatalysis, visible light is the energy source. This makes it compatible with natural sunlight and typical indoor lighting. Visible-light photocatalysis uses photocatalytic materials to absorb visible light to form electron–hole pairs. These photoexcited charge carriers then reach the surface of the photocatalyst, where they participate in various chemical reactions. The active chemical species resulting from these reactions lead to the degradation of pollutants. A schematic representation of the visible-light photocatalytic mechanism is shown in figure 7.1. Visible-light photocatalysis is used in multiple applications, including water purification, air quality improvement, and organic synthesis.

In plasmonic photocatalysis processes, plasmonic nanoparticles such as gold or silver nanoparticles generate localized surface plasmons. These nanoparticles can absorb light at specific resonant wavelengths, usually in the visible or near-infrared range. Plasmonic photocatalysis requires the use of plasmonic nanoparticles in combination with a conventional photocatalyst. Plasmonic nanoparticles absorb light and form localized surface plasmons, generating intense and highly localized electric fields. These electric fields concentrate light energy near the active sites of the photocatalyst, promoting electron–hole pair formation and facilitating charge transfer. Thus, they increase reaction rates. Plasmonic photocatalysis has applications such as water splitting, pollutant degradation, and energy conversion. It is beneficial when visible or near-infrared light needs to be used efficiently or when visible light is insufficient for the photocatalytic reaction.

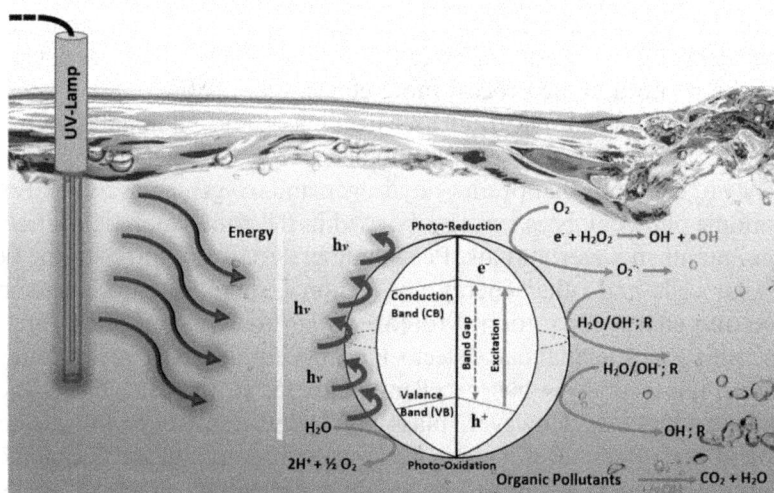

Figure 7.1. A schematic representation of the visible-light photocatalysis mechanism. Adapted from [97], Copyright (2022), with permission from Elsevier.

The main difference between visible-light photocatalysis and plasmonic photo-catalysis processes is the light source used. Visible-light photocatalysis utilizes the visible light used in natural and indoor lighting. Plasmonic photocatalysis, on the other hand, uses plasmonic nanoparticles to concentrate light at specific wavelengths across a wide range of the electromagnetic wave spectrum, including the visible or near-infrared regions. Another critical difference between these two processes is that they differ in their mechanisms. Plasmonic photocatalysis relies on generating localized surface plasmons and enhancing local electric fields to improve reaction rates. In contrast, visible-light photocatalysis uses visible light directly to excite charge carriers in the photocatalyst. On the other hand, both methods rely on the use of photocatalysts. However, plasmonic photocatalytic processes require the use of plasmonic nanoparticles in addition to photocatalysts. When a choice between these methods has to be made, an assessment should be performed depending on the specific requirements of the photocatalytic reaction and the available light source.

7.6 Conclusions

Currently, only a few photocatalysts can be used in large-scale photocatalytic applications involving environmental applications or water splitting. The main reason for the limited availability of commercially available photocatalysts is that many new photocatalysts developed need to meet essential criteria, such as low cost, high efficiency, chemical stability, stability, and environmental friendliness. Therefore, photocatalyst production processes must be carefully considered, and the operating parameters must be optimized for commercial-scale production.

The techniques proposed for the synthesis of photocatalysts in recent years have different advantages and weaknesses in terms of catalyst stability, efficiency, cost, production, structure, and catalytic performance. Many catalyst materials and production methods for these materials have been proposed in the studies carried out. However, most of the proposed synthesis techniques have various difficulties, such as uncontrollable catalyst morphology, a high risk of contamination, complex purification processes, labor-intensive methods, multistep processes, expensive starting chemicals, waste generation, environmental unfriendliness, low yield, and high energy consumption. Moreover, for a photocatalyst, critical factors such as the bandgap, the types of charge carriers, the crystallinity of the materials, the specific surface area, and the stability of the photocatalyst are all closely related to the synthesis technique used to produce the catalyst. All these factors potentially affect the electron–hole recombination process, the synergy between the pollutant and the photocatalyst, the availability of active sites, and the reusability of the photo-catalyst. Considering all these aspects, it is predicted that the obstacles, limitations, and challenges encountered in developing new photocatalysts will continue to be an attractive research area for researchers for some time.

In summary, photocatalytic processes have great potential in many application areas, such as environmental sustainability, energy production, water treatment, and chemical synthesis. However, some critical challenges must be overcome to secure the wider adoption and development of this technology.

There is a need to improve the efficiency of materials used in photocatalytic processes and to design and manufacture more advanced materials capable of absorbing light over a wider electromagnetic wavelength range. Similarly, more work needs to be done on the design and efficiency of photocatalytic reactors. Parameters such as scalability, energy efficiency, and operating costs should also be considered when designing reactors. On the other hand, although many scientific studies have been conducted to identify the mechanisms of photocatalytic processes, these processes still need to be fully explained. Artificial intelligence may be an essential tool in understanding the mechanisms of photocatalysis and designing new materials. Therefore, more qualified studies on these topics will be needed. If the challenges of photocatalytic processes can be overcome, they have great potential for use in many applications, such as wastewater treatment, converting sunlight into electrical energy, controlling air pollution, and sustainable energy production by water splitting.

In conclusion, photocatalysis offers significant benefits in various application areas, but more research and development work needs to be done before it can be adapted to large-scale systems. Although the future of this technology is bright, a sustained effort is required to overcome the scientific and technical challenges. The wider adoption of photocatalysis would be an excellent contribution to a sustainable future.

References

[1] Kuspanov Z, Bakbolat B, Baimenov A, Issadykov A, Yeleuov M and Daulbayev C 2023 Photocatalysts for a sustainable future: innovations in large-scale environmental and energy applications *Sci. Total Environ.* **885** 163914

[2] Wang Q, Gao Q, Al-Enizi A M, Nafady A and Ma S 2020 Recent advances in MOF-based photocatalysis: environmental remediation under visible light *Inorg. Chem. Front.* **7** 300–39

[3] Koe W S, Lee J W, Chong W C, Pang Y L and Sim L C 2020 An overview of photocatalytic degradation: photocatalysts, mechanisms, and development of photocatalytic membrane *Environ. Sci. Pollut. Res. Int.* **27** 2522–65

[4] Hoang S and Gao P X 2016 Nanowire array structures for photocatalytic energy conversion and utilization: a review of design concepts, assembly and integration, and function enabling *Adv. Energy Mater.* **6** 1600683

[5] Borges M E, Sierra M, Cuevas E, García R and Esparza P 2016 Photocatalysis with solar energy: sunlight-responsive photocatalyst based on TiO_2 loaded on a natural material for wastewater treatment *Sol. Energy* **135** 527–35

[6] Sun J-H, Dong S-Y, Feng J-L, Yin X-J and Zhao X-C 2011 Enhanced sunlight photocatalytic performance of Sn-doped ZnO for methylene blue degradation *J. Mol. Catal. A: Chem.* **335** 145–50

[7] Cai T, Liu Y, Wang L, Zhang S, Zeng Y, Yuan J, Ma J, Dong W, Liu C and Luo S 2017 Silver phosphate-based Z-Scheme photocatalytic system with superior sunlight photocatalytic activities and anti-photocorrosion performance *Appl. Catal.* B **208** 1–13

[8] Chen P, Blaney L, Cagnetta G, Huang J, Wang B, Wang Y, Deng S and Yu G 2019 Degradation of ofloxacin by perylene diimide supramolecular nanofiber sunlight-driven photocatalysis *Environ. Sci. Technol.* **53** 1564–75

[9] Enesca A and Isac L 2020 The influence of light irradiation on the photocatalytic degradation of organic pollutants *Materials (Basel)* **13** 2494

[10] Maeda K 2011 Photocatalytic water splitting using semiconductor particles: history and recent developments *J. Photochem. Photobiol., C* **12** 237–68

[11] Jafari T, Moharreri E, Amin A S, Miao R, Song W and Suib S L 2016 Photocatalytic water splitting—the untamed dream: a review of recent advances *Molecules* **21** 900

[12] Maeda K and Domen K 2010 Photocatalytic water splitting: recent progress and future challenges *J. Phys. Chem. Lett.* **1** 2655–61

[13] Gupta A, Likozar B, Jana R, Chanu W C and Singh M K 2022 A review of hydrogen production processes by photocatalytic water splitting—from atomistic catalysis design to optimal reactor engineering *Int. J. Hydrogen Energy* **47** 33282–307

[14] Goto Y, Hisatomi T, Wang Q, Higashi T, Ishikiriyama K, Maeda T, Sakata Y, Okunaka S, Tokudome H and Katayama M 2018 A particulate photocatalyst water-splitting panel for large-scale solar hydrogen generation *Joule* **2** 509–20

[15] Schneider J, Matsuoka M, Takeuchi M, Zhang J, Horiuchi Y, Anpo M and Bahnemann D W 2014 Understanding TiO_2 photocatalysis: mechanisms and materials *Chem. Rev.* **114** 9919–86

[16] Xu C, Anusuyadevi P R, Aymonier C, Luque R and Marre S 2019 Nanostructured materials for photocatalysis *Chem. Soc. Rev.* **48** 3868–902

[17] Yang X and Wang D 2018 Photocatalysis: from fundamental principles to materials and applications *ACS Appl. Energy Mater.* **1** 6657–93

[18] Luo B, Liu G and Wang L 2016 Recent advances in 2D materials for photocatalysis *Nanoscale* **8** 6904–20

[19] Gholipour M R, Dinh C-T, Béland F and Do T-O 2015 Nanocomposite heterojunctions as sunlight-driven photocatalysts for hydrogen production from water splitting *Nanoscale* **7** 8187–208

[20] Idriss H 2020 The elusive photocatalytic water splitting reaction using sunlight on suspended nanoparticles: is there a way forward? *Catal. Sci. Technol.* **10** 304–10

[21] Kumar P, Boukherroub R and Shankar K 2018 Sunlight-driven water-splitting using two-dimensional carbon based semiconductors *J. Mater. Chem.* A **6** 12876–931

[22] Tentu R D and Basu S 2017 Photocatalytic water splitting for hydrogen production *Curr. Opin. Electrochem.* **5** 56–62

[23] Braham R J and Harris A T 2009 Review of major design and scale-up considerations for solar photocatalytic reactors *Ind. Eng. Chem. Res.* **48** 8890–905

[24] Ray A K 1999 Design, modelling and experimentation of a new large-scale photocatalytic reactor for water treatment *Chem. Eng. Sci.* **54** 3113–25

[25] Alalm M G, Djellabi R, Meroni D, Pirola C, Bianchi C L and Boffito D C 2021 Toward scaling-up photocatalytic process for multiphase environmental applications *Catalysts* **11** 562

[26] Reilly K, Wilkinson D P and Taghipour F 2018 Photocatalytic water splitting in a fluidized bed system: computational modeling and experimental studies *Appl. Energy* **222** 423–36

[27] Abdel-Maksoud Y, Imam E and Ramadan A 2016 TiO_2 solar photocatalytic reactor systems: selection of reactor design for scale-up and commercialization—analytical review *Catalysts* **6** 138

[28] Burke D W, Sun C, Castano I, Flanders N C, Evans A M, Vitaku E, McLeod D C, Lambeth R H, Chen L X and Gianneschi N C 2020 Acid exfoliation of imine-linked covalent organic frameworks enables solution processing into crystalline thin films *Angew. Chem.* **132** 5203–9

[29] Chen X, Li Y, Wang L, Xu Y, Nie A, Li Q, Wu F, Sun W, Zhang X and Vajtai R 2019 High-lithium-affinity chemically exfoliated 2D covalent organic frameworks *Adv. Mater.* **31** 1901640

[30] Frowijn L S and van Sark W G 2021 Analysis of photon-driven solar-to-hydrogen production methods in the Netherlands *Sustain. Energy Technol. Assess.* **48** 101631

[31] Hisatomi T and Domen K 2019 Reaction systems for solar hydrogen production via water splitting with particulate semiconductor photocatalysts *Nat. Catal.* **2** 387–99

[32] Cao F, Wei Q, Liu H, Lu N, Zhao L and Guo L 2018 Development of the direct solar photocatalytic water splitting system for hydrogen production in Northwest China: design and evaluation of photoreactor *Renew. Energy* **121** 153–63

[33] Ruiz-Aguirre A, Villachica-Llamosas J, Polo-López M, Cabrera-Reina A, Colón G, Peral J and Malato S 2022 Assessment of pilot-plant scale solar photocatalytic hydrogen generation with multiple approaches: valorization, water decontamination and disinfection *Energy* **260** 125199

[34] Ren H, Koshy P, Chen W-F, Qi S and Sorrell C C 2017 Photocatalytic materials and technologies for air purification *J. Hazard. Mater.* **325** 340–66

[35] Zhao J and Yang X 2003 Photocatalytic oxidation for indoor air purification: a literature review *Build. Environ.* **38** 645–54

[36] Hay S O, Obee T, Luo Z, Jiang T, Meng Y, He J, Murphy S C and Suib S 2015 The viability of photocatalysis for air purification *Molecules* **20** 1319–56

[37] He F, Jeon W and Choi W 2021 Photocatalytic air purification mimicking the self-cleaning process of the atmosphere *Nat. Commun.* **12** 2528

[38] Qian H, Hou Q, Zhang W, Nie Y, Lai R, Ren H, Yu G, Bai X, Wang H and Ju M 2022 Construction of electron transport channels and oxygen adsorption sites to modulate reactive oxygen species for photocatalytic selective oxidation of 5-hydroxymethylfurfural to 2, 5-diformylfuran *Appl. Catal.* B **319** 121907

[39] Jing D, Guo L, Zhao L, Zhang X, Liu H, Li M, Shen S, Liu G, Hu X and Zhang X 2010 Efficient solar hydrogen production by photocatalytic water splitting: from fundamental study to pilot demonstration *Int. J. Hydrogen Energy* **35** 7087–97

[40] Chen J, Xu W, Zuo H, Wu X, Jiaqiang E, Wang T, Zhang F and Lu N 2019 System development and environmental performance analysis of a solar-driven supercritical water gasification pilot plant for hydrogen production using life cycle assessment approach *Energy Convers. Manage.* **184** 60–73

[41] Chen S, Ma G, Wang Q, Sun S, Hisatomi T, Higashi T, Wang Z, Nakabayashi M, Shibata N and Pan Z 2019 Metal selenide photocatalysts for visible-light-driven Z-scheme pure water splitting *J. Mater. Chem.* A **7** 7415–22

[42] Chen X, Cai S, Yu E, Li J, Chen J and Jia H 2019 Photothermocatalytic performance of ACo_2O_4 type spinel with light-enhanced mobilizable active oxygen species for toluene oxidation *Appl. Surf. Sci.* **484** 479–88

[43] Kubiak A 2023 Comparative study of TiO_2–Fe_3O_4 photocatalysts synthesized by conventional and microwave methods for metronidazole removal *Sci. Rep.* **13** 12075

[44] Li Y, Xu H, Ouyang S and Ye J 2016 Metal–organic frameworks for photocatalysis *Phys. Chem. Chem. Phys.* **18** 7563–72

[45] García-Salcido V, Mercado-Oliva P, Guzmán-Mar J L, Kharisov B I and Hinojosa-Reyes L 2022 MOF-based composites for visible-light-driven heterogeneous photocatalysis: synthesis, characterization and environmental application studies *J. Solid State Chem.* **307** 122801

[46] Navalón S, Dhakshinamoorthy A, Álvaro M, Ferrer B and García H 2023 Metal–organic frameworks as photocatalysts for solar-driven overall water splitting *Chem. Rev.* **123** 445–90

[47] Sun K, Qian Y and Jiang H-L 2023 Metal-organic frameworks for photocatalytic water splitting and CO_2 reduction *Angew. Chem. Int. Ed.* **62** e202217565

[48] DuBose J T and Kamat P V 2022 Efficacy of perovskite photocatalysis: challenges to overcome *ACS Energy Lett.* **7** 1994–2011

[49] Dandia A, Saini P, Sharma R and Parewa V 2020 Visible light driven perovskite-based photocatalysts: a new candidate for green organic synthesis by photochemical protocol *Curr. Res. Green Sustain. Chem.* **3** 100031

[50] Mohd Kaus N H, Ibrahim M L, Imam S S, Mashuri S I S and Kumar Y 2022 Efficient Visible-Light-Driven Perovskites Photocatalysis: Design, Modification and Application *Green Photocatalytic Semiconductors: Recent Advances and Applications* ed S Garg and A Chandra (Cham: Springer International Publishing) 357–98

[51] Sun P, Xing Z, Li Z and Zhou W 2023 Recent advances in quantum dots photocatalysts *Chem. Eng. J.* **458** 141399

[52] Chakraborty I N, Roy P and Pillai P P 2023 Visible light-mediated quantum dot photocatalysis enables olefination reactions at room temperature *ACS Catal.* **13** 7331–8

[53] Li X, Yu J, Wageh S, Al-Ghamdi A A and Xie J 2016 Graphene in photocatalysis: a review *Small* **12** 6640–96

[54] Lu K-Q, Li Y-H, Tang Z-R and Xu Y-J 2021 Roles of graphene oxide in heterogeneous photocatalysis *ACS Materials Au* **1** 37–54

[55] Albero J, Mateo D and García H 2019 Graphene-based materials as efficient photocatalysts for water splitting *Molecules* **24** 906

[56] Suresh R, Mangalaraja R V, Mansilla H D, Santander P and Yáñez J 2020 Reduced Graphene Oxide-Based Photocatalysis *Green Photocatalysts* ed M Naushad *et al* (Cham: Springer International Publishing) 145–66

[57] Singh P P and Srivastava V 2022 Recent advances in visible-light graphitic carbon nitride (g-C_3N_4) photocatalysts for chemical transformations *RSC Adv.* **12** 18245–65

[58] Gao R-H, Ge Q, Jiang N, Cong H, Liu M and Zhang Y-Q 2022 Graphitic carbon nitride (g-C_3N_4)-based photocatalytic materials for hydrogen evolution *Front. Chem.* **10**

[59] Ong W-J, Tan L-L, Ng Y H, Yong S-T and Chai S-P 2016 Graphitic carbon nitride (g-C_3N_4)-based photocatalysts for artificial photosynthesis and environmental remediation: are we a step closer to achieving sustainability? *Chem. Rev.* **116** 7159–329

[60] Gogotsi Y and Anasori B 2019 *The rise of MXenes* (Washington, DC: ACS Publications) 8491–4

[61] Kuang P, Low J, Cheng B, Yu J and Fan J 2020 MXene-based photocatalysts *Journal of Materials Science & Technology* **56** 18–44

[62] Zhong Q, Li Y and Zhang G 2021 Two-dimensional MXene-based and MXene-derived photocatalysts: recent developments and perspectives *Chem. Eng. J.* **409** 128099

[63] Zhang X, Chen Y L, Liu R-S and Tsai D P 2013 Plasmonic photocatalysis *Rep. Prog. Phys.* **76** 046401

[64] Verma R, Belgamwar R and Polshettiwar V 2021 Plasmonic photocatalysis for CO_2 conversion to chemicals and fuels *ACS Mater. Lett.* **3** 574–98

[65] Wang T, Wang H-J, Lin J-S, Yang J-L, Zhang F-L, Lin X-M, Zhang Y-J, Jin S and Li J-F 2023 Plasmonic photocatalysis: mechanism, applications and perspectives *Chinese J. Struct. Chem.* **42** 100066

[66] Díaz F J P, del Río R S and Rodriguez P E D S 2022 Plasmonic Photocatalysts for Water Splitting *Photoelectrochemical Hydrogen Generation: Theory, Materials Advances, and Challenges* ed P Kumar and P Devi (Singapore: Springer Nature Singapore) 117–73

[67] Jiang X, Chen Y-X and Lu C-Z 2020 Bio-inspired materials for photocatalytic hydrogen production *Chinese J. Struct. Chem.* **39** 2123–30

[68] Fu J, Zhu B, You W, Jaroniec M and Yu J 2018 A flexible bio-inspired H_2-production photocatalyst *Appl. Catal.* B **220** 148–60

[69] Masood H, Toe C Y, Teoh W Y, Sethu V and Amal R 2019 Machine learning for accelerated discovery of solar photocatalysts *ACS Catal.* **9** 11774–87

[70] Mai H, Le T C, Chen D, Winkler D A and Caruso R A 2022 Machine learning for electrocatalyst and photocatalyst design and discovery *Chem. Rev.* **122** 13478–515

[71] Keith J A, Vassilev-Galindo V, Cheng B, Chmiela S, Gastegger M, Müller K-R and Tkatchenko A 2021 Combining machine learning and computational chemistry for predictive insights into chemical systems *Chem. Rev.* **121** 9816–72

[72] Li X, Maffettone P M, Che Y, Liu T, Chen L and Cooper A I 2021 Combining machine learning and high-throughput experimentation to discover photocatalytically active organic molecules *Chem. Sci.* **12** 10742–54

[73] Liu C, Huang H, Du X, Zhang T, Tian N, Guo Y and Zhang Y 2015 *In situ* co-crystallization for fabrication of g-C_3N_4/Bi_5O_7I heterojunction for enhanced visible-light photocatalysis *J. Phys. Chem.* C **119** 17156–65

[74] Ágoston Á and Janovák L 2023 Hydrothermal co-crystallization of novel copper tungstate-strontium titanate crystal composite for enhanced photocatalytic activity and increased electron–hole recombination time *Catalysts* **13** 287

[75] Li P, Yan X, He Z, Ji J, Hu J, Li G, Lian K and Zhang W 2016 α-Fe_2O_3 concave and hollow nanocrystals: top-down etching synthesis and their comparative photocatalytic activities *CrystEngComm.* **18** 1752–9

[76] Yusran Y, Li H, Guan X, Li D, Tang L, Xue M, Zhuang Z, Yan Y, Valtchev V and Qiu S 2020 Exfoliated mesoporous 2D covalent organic frameworks for high-rate electrochemical double-layer capacitors *Adv. Mater.* **32** 1907289

[77] Cui L, Liu Y, Fang X, Yin C, Li S, Sun D and Kang S 2018 Scalable and clean exfoliation of graphitic carbon nitride in NaClO solution: enriched surface active sites for enhanced photocatalytic H_2 evolution *Green Chem.* **20** 1354–61

[78] Liu W, Li X, Wang C, Pan H, Liu W, Wang K, Zeng Q, Wang R and Jiang J 2019 A scalable general synthetic approach toward ultrathin imine-linked two-dimensional covalent organic framework nanosheets for photocatalytic CO_2 reduction *J. Am. Chem. Soc.* **141** 17431–40

[79] Tian B, Tian B, Smith B, Scott M, Lei Q, Hua R, Tian Y and Liu Y 2018 Facile bottom-up synthesis of partially oxidized black phosphorus nanosheets as metal-free photocatalyst for hydrogen evolution *Proc. Natl. Acad. Sci.* **115** 4345–50

[80] Liu M, Xing Z, Li Z and Zhou W 2021 Recent advances in core–shell metal organic frame-based photocatalysts for solar energy conversion *Coord. Chem. Rev.* **446** 214123

[81] Chen J, Tang T, Feng W, Liu X, Yin Z, Zhang X, Chen J and Cao S 2021 Large-scale synthesis of p–n heterojunction Bi_2O_3/TiO_2 nanostructures as photocatalysts for removal of antibiotics under visible light *ACS Appl. Nano Mater.* **5** 1296–307

[82] Wang L, Hong Y, Liu E, Wang Z, Chen J, Yang S, Wang J, Lin X and Shi J 2020 Rapid polymerization synthesizing high-crystalline g-C_3N_4 towards boosting solar photocatalytic H_2 generation *Int. J. Hydrogen Energy* **45** 6425–36

[83] Weng S, Chen B, Xie L, Zheng Z and Liu P 2013 Facile *in situ* synthesis of a Bi/BiOCl nanocomposite with high photocatalytic activity *J. Mater. Chem.* A **1** 3068–75

[84] Ge L, Han C and Liu J 2012 *In situ* synthesis and enhanced visible light photocatalytic activities of novel PANI–gC$_3$N$_4$ composite photocatalysts *J. Mater. Chem.* **22** 11843–50

[85] Li G, Zhang K, Li C, Gao R, Cheng Y, Hou L and Wang Y 2019 Solvent-free method to encapsulate polyoxometalate into metal-organic frameworks as efficient and recyclable photocatalyst for harmful sulfamethazine degrading in water *Appl. Catalysis* B **245** 753–9

[86] Xin X, Song Y, Guo S, Zhang Y, Wang B, Yu J and Li X 2020 In-situ growth of high-content 1T phase MoS$_2$ confined in the CuS nanoframe for efficient photocatalytic hydrogen evolution *Appl. Catal.* B **269** 118773

[87] Yi J, Fei T, Li L, Yu Q, Zhang S, Song Y, Lian J, Zhu X, Deng J and Xu H 2021 Large-scale production of ultrathin carbon nitride-based photocatalysts for high-yield hydrogen evolution *Appl. Catalysis* B **281** 119475

[88] Deng S, Liu C, Zhang Y, Ji Y, Mei B, Yao Z and Lin S 2023 Large-scale preparation of ultrathin bimetallic nickel iron sulfides branch nanoflake arrays for enhanced hydrogen evolution reaction *Catalysts* **13** 174

[89] Jahanshahi R, Khazaee A, Sobhani S and Sansano J M 2020 g-C$_3$N$_4$/γ-Fe$_2$O$_3$/TiO$_2$/Pd: a new magnetically separable photocatalyst for visible-light-driven fluoride-free Hiyama and Suzuki–Miyaura cross-coupling reactions at room temperature *New J. Chem.* **44** 11513–26

[90] Izadifard M, Achari G and Langford C H 2013 Application of photocatalysts and LED light sources in drinking water treatment *Catalysts* **3** 726–43

[91] Jo W-K and Tayade R J 2014 New generation energy-efficient light source for photo-catalysis: LEDs for environmental applications *Ind. Eng. Chem. Res.* **53** 2073–84

[92] Sergejevs A, Clarke C, Allsopp D, Marugan J, Jaroenworaluck A, Singhapong W, Manpetch P, Timmers R, Casado C and Bowen C 2017 A calibrated UV-LED based light source for water purification and characterisation of photocatalysis *Photochemical & Photobiological Sciences* **16** 1690–9

[93] Gondal M A, Ali M A, Chang X F, Shen K, Xu Q Y and Yamani Z H 2012 Pulsed laser-induced photocatalytic reduction of greenhouse gas CO$_2$ into methanol: a value-added hydrocarbon product over SiC *J. Environ. Sci. Health, Part* A **47** 1571–6

[94] Bui V K H, Tran V V, Moon J-Y, Park D and Lee Y-C 2020 Titanium dioxide microscale and macroscale structures: a mini-review *Nanomaterials* **10** 1190

[95] Loeb S K, Alvarez P J, Brame J A, Cates E L, Choi W, Crittenden J, Dionysiou D D, Li Q, Li-Puma G and Quan X 2019 The Technology Horizon for Photocatalytic Water Treatment: Sunrise or Sunset? *Environ. Sci. Technol.* **53** 2937–47

[96] Phan D D, Babick F, Trịnh T H T, Nguyen M T, Samhaber W and Stintz M 2018 Investigation of fixed-bed photocatalytic membrane reactors based on submerged ceramic membranes *Chem. Eng. Sci.* **191** 332–42

[97] Karakaş Z K 2022 A comprehensive study on the production and photocatalytic activity of copper ferrite nanoparticles synthesized by microwave-assisted combustion method as an effective photocatalyst *J. Phys. Chem. Solids* **170** 110927

A Classical
Thermodynamics Toolkit

Online at: https://doi.org/10.1088/978-0-7503-6029-6

A Classical Thermodynamics Toolkit

Srinivas Vanapalli
*Faculty of Science and Technology, University of Twente (Netherlands),
Enschede, The Netherlands*

IOP Publishing, Bristol, UK

ISBN   978-0-7503-6029-6 (ebook)
ISBN   978-0-7503-6027-2 (print)
ISBN   978-0-7503-6030-2 (myPrint)
ISBN   978-0-7503-6028-9 (mobi)

DOI   10.1088/978-0-7503-6029-6

Version: 20251201

IOP ebooks

British Library Cataloguing-in-Publication Data: A catalogue record for this book is available from the British Library.

Published by IOP Publishing, wholly owned by The Institute of Physics, London

IOP Publishing, No.2 The Distillery, Glassfields, Avon Street, Bristol, BS2 0GR, UK

US Office: IOP Publishing, Inc., 190 North Independence Mall West, Suite 601, Philadelphia, PA 19106, USA

To the memory of **Professor Miko Elwenspoek**, *my PhD supervisor and lifelong inspiration.*

His unwavering encouragement to stay curious, seek diverse perspectives, and explain complex ideas with clarity has shaped not only this book but also the way I approach science. To me, he was like a Richard Feynman – but one I had the privilege to know, learn from, and walk alongside.

Contents

Preface

The world has changed. We no longer live in the age of steam engines, and even internal combustion engines are fading into the background. Yet, many thermodynamics textbooks continue to reflect the priorities of that era. They are often written for mechanical engineers or physicists, each with their own framing of the subject—and understandably so. But as we move deeper into the 21st century, we must prepare scientists and engineers for a new reality: one that is inherently multidisciplinary.

This book is written with that future in mind.

The idea of making thermodynamics accessible across disciplines was first instilled in me through the Advanced Technology program at the University of Twente—a visionary curriculum launched over two decades ago by my late PhD supervisor, Professor Miko Elwenspoek. He always encouraged me to keep an open mind, to seek different perspectives, and to explain things simply. That spirit runs through every chapter of this book.

Whether you specialize in mechatronics, robotics, semiconductors, chemical processes, aerospace, or applied physics, a foundational understanding of thermodynamics is essential. But thermodynamics is often seen as opaque, filled with abstract definitions and mysterious quantities—none more so than entropy. Many textbooks also fail to clearly distinguish the *system* from its *surroundings*, or they develop ideas in ways that obscure their practical relevance. This book attempts to break through those barriers. It is written with no inherent assumptions, built from the ground up to foster clear, intuitive understanding. My goal is to give every reader—whether a student or a practicing scientist—a platform from which they can take off in any specialized direction.

This book is for **everyone who is curious about Nature**. It requires no advanced mathematics; elementary calculus is helpful but not essential. A first-year bachelor student from any discipline can benefit from it. Practicing engineers and scientists may also find it a useful reference, particularly for revisiting concepts that are too often rushed in traditional education.

The structure of this book has been shaped by more than a decade of teaching and research. When I began teaching thermodynamics at the University of Twente 14 years ago, students would often ask me questions that I didn't always have good answers for. That pushed me to sharpen my own understanding—not just by reading, but by *working through* the ideas until they clicked. I kept notes on the questions that challenged me most, and those notes gradually evolved into this book.

Importantly, many of the insights here also stem from my research work in the lab. It was in the context of real experimental systems that I began to see how the abstract concepts of thermodynamics come to life—how energy flows, how entropy reveals itself, and how to reason about systems rigorously yet simply.

This practical grounding led me to an essential realization: thermodynamics is **universal**. It's not just about gases, engines, or power cycles. It applies just as much to electrical, magnetic, and chemical systems. I remember teaching thermodynamics

to applied physics students who asked, 'Why should we learn about chemical reactions?' My answer was that even boiling water—a phase change—can be interpreted as a kind of chemical transformation. Learning fundamental thermodynamics allows us to translate insights across domains.

And yes, we often start with the classic piston–cylinder system—not because it's outdated, but because it offers an elegant, visual way to grasp foundational ideas like pressure, volume, energy, and entropy. But the tools developed in this book go far beyond that. They provide a language for understanding a wide range of natural and engineered systems, from simple heat exchange to complex technological applications.

Along the way, I've also noticed that many learners confuse thermodynamics with other fields such as heat transfer or kinetics. A helpful analogy is the game of chess: thermodynamics defines the rules—what is allowed, what is forbidden—while kinetics and heat transfer determine how fast the game unfolds. Both are important, but this book is about the rules. Clarifying such distinctions is part of what makes the subject approachable and useful.

This book is the product of many conversations, mistakes, realizations, and refinements—all of which I am grateful for. I hope it serves not only as a learning resource but also as an invitation: to look more closely, to question confidently, and to carry the spirit of curiosity into your own work, wherever it leads.

Srinivas Vanapalli
University of Twente

Acknowledgements

Although my academic background is in electrical engineering, it was Ray Radebaugh (NIST, Boulder, Colorado) who first inspired my deep interest in thermodynamics. His clarity of thought and passion for the subject left a lasting impression on me, and I am deeply grateful for that influence.

At the University of Twente, my teaching experience has been enriched by many people. I want to especially acknowledge Jelle van der Meulen, whose creative style of designing exercises—often like solving puzzles or filling in thought-provoking tables—inspired a more intuitive and engaging approach to teaching. My co-teacher Herman Hemmes has also been a consistent source of support and insight throughout the years.

This book would not have taken its current form without the help of my PhD students, Rick Spijkers and Adam Kovács, who contributed to many of the illustrations and figures. I sincerely thank them for their patience and eye for detail.

I am deeply grateful to the students of Advanced Technology and Applied Physics at the University of Twente. Their curiosity, questions, and willingness to challenge the material pushed me to refine my explanations and clarify my thinking. Much of this book was shaped by those classroom interactions.

I am also thankful to IOP Publishing for their proactive and professional support throughout the publication process. Their clear communication and attention to detail helped turn the manuscript into a well-produced final book.

Lastly, I want to express my heartfelt appreciation to my wife and son for their constant encouragement and patience during this journey. Their support made all the difference.

Author biography

Srinivas Vanapalli

Srinivas Vanapalli is a Professor of Applied Physics at the University of Twente, where he leads the Applied Thermal Sciences (ATS) group. His team is pioneering a revolution in cryogenics—rethinking traditional systems to make them energy-efficient, compact, scalable, and sustainable. They develop next-generation cryocoolers with dramatically reduced power consumption, thermal management solutions like cryogenic circulators and heat switches, and even liquid-hydrogen zero boil-off systems and modular cryogenic platforms.

Originally trained as an electrical engineer, Professor Vanapalli earned his PhD conducting cryocooler research at NIST in Boulder, Colorado. His work has since bridged fundamental science, industrial innovation, and focused entrepreneurship—collaborating with companies and developing novel cryogenic technologies for quantum computing, energy systems, and beyond.

A committed educator, he has taught thermodynamics for over 14 years, known for his clear, intuitive approach shaped by both student interactions and hands-on lab research. He co-designed problem-based exercises that bring concepts to life, and his PhD students have supported the book with technical illustrations.

Professor Vanapalli also serves the global scientific community—he is Chairman of the Cryogenics Society of Europe and Chair of the Cryogenic Engineering Conference (CEC)—working to raise professional standards and foster collaboration across academia and industry.

This book reflects his belief in universal thermodynamics: that the same principles underpin diverse systems—from engines to cryocoolers to hydrogen energy infrastructure—and that clarity and simplicity are the keys to unlocking innovation.

IOP Publishing

A Classical Thermodynamics Toolkit

Srinivas Vanapalli

Chapter 1

Introduction

This chapter introduces the foundational perspectives and principles of classical thermodynamics, focusing on the dual approach of examining systems from both microscopic and macroscopic viewpoints. In thermodynamics, we can describe the behavior of systems at a detailed molecular level or from an averaged perspective. Each approach provides valuable insights into the properties and interactions of matter in various states—solid, liquid, and gas—and how they respond to changes in energy and forces.

We explore key thermodynamic properties such as internal energy, the concept of the system and surroundings, and how systems exchange energy through heat and work. By examining simple models, such as a cube of gas or phase changes of water, we illustrate the essential processes of energy transfer. Additionally, we discuss the ideal gas law and how real gases deviate from ideal behavior under certain conditions, providing a basis for understanding the complexities of more advanced thermodynamic systems.

This chapter also introduces essential tools for studying force balance and property relationships in thermodynamics. Together, these principles set the stage for a deeper examination of thermodynamic processes and the practical applications of energy transformations, as seen in the boundary work of piston–cylinder devices. Through exercises and examples, readers will gain insight into the dynamic exchanges between energy forms and how these interactions define the structure and function of systems in thermodynamic equilibrium and beyond.

1.1 Microscopic and macroscopic viewpoints

The behavior of a system may be investigated from either a microscopic or macroscopic point of view. Let us briefly describe a system from a microscopic point of view. Consider a system consisting of a cube 25 mm on a side and containing a monatomic gas at atmospheric pressure and temperature. This volume contains approximately 10^{20} atoms. We are interested in predicting the properties of

doi:10.1088/978-0-7503-6029-6ch1
1-1

this system. There are two approaches to this problem that reduce the number of equations and variables to a few that can be computed relatively easily. One is the statistical approach, in which, based on statistical considerations and probability theory, we deal with average values for all particles under consideration. This is usually done in connection with a model of the atom under consideration. This is the approach used in the disciplines of kinetic theory and statistical mechanics.

The other approach to reducing the number of variables to a few that can be handled relatively easily involves the macroscopic point of view of classical thermodynamics. As the word macroscopic implies, we are concerned with the gross or average effects of many molecules. These effects can be perceived by our senses and measured by instruments.

However, what we really perceive, and measure, is the time-averaged influence of many molecules. For example, consider the pressure a gas exerts on the walls of its container. This pressure results from the change in momentum of the molecules as they collide with the wall. From a macroscopic point of view, however, we are concerned not with the action of the individual molecules but with the time-averaged force on a given area, which can be measured by a pressure gauge.

1.2 Solids, liquids, and gases

A pure substance is one that has a homogeneous and invariable chemical composition. It may exist in more than one phase, but the chemical composition is the same in all phases. Thus, liquid water, a mixture of liquid water and water vapor (steam), and a mixture of ice and liquid water are all pure substances; every phase has the same chemical composition. In contrast, a mixture of liquid air and gaseous air is not a pure substance.

Consider an amount of ice contained in a weightless piston and cylinder arrangement that maintains a constant atmospheric pressure, as in figure 1.1. Assume that the water starts at −10 °C, in which the state is solid. If ice is slowly heated, the temperature increases and by design in this example the pressure stays constant. When the temperature reaches 0 °C additional heat transfer results in a phase change with the formation of liquid, as indicated in figure 1.1. and the

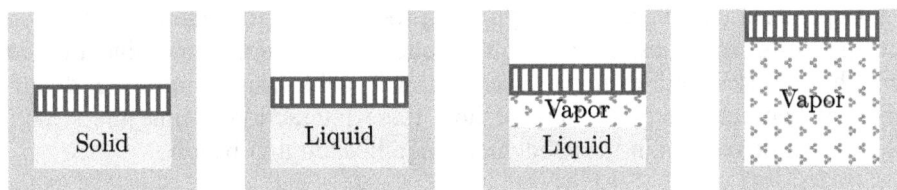

Figure 1.1. Schematic representation of phase changes of a substance. The external pressure exerted on the substance is held constant. When heat is added the solid warms up until its melting point. Further addition of heat melts the solid and the temperature remains constant during melting. After all the solid has melted the addition of heat causes the temperature of the liquid to rise until it reaches the boiling temperature, where the liquid converts to vapor at a constant temperature. After all the liquid is converted to vapor, further addition of heat increases the temperature of the vapor. In the case of water, the density of the solid (ice) is smaller than that of the liquid, therefore the volume of the solid is larger than that of the liquid.

temperature stays constant. Further heating melts all of the solid to a liquid state. If the water is slowly heated, the temperature increases, and the volume increases slightly. When the temperature reaches 99.6 °C additional heat transfer results in a phase change with the formation of some vapor, as indicated in figure 1.1, while the temperature stays constant, and the volume increases considerably. Further heating generates more and more vapor and a large increase in the volume until the last drop of liquid is vaporized. Subsequent heating results in an increase in both the temperature and volume of the vapor, as shown in figure 1.1.

1.3 Important characteristics of substances in various states

1.3.1 The liquid and solid phases

Both liquid and solid states can generally be treated as **incompressible substances** because their specific volume changes very little with temperature. This allows for simpler modeling, as volume is nearly constant across a range of conditions in these phases.

1.3.2 The vapor phase (real gas)

In the vapor phase, molecules experience both attractive and repulsive forces due to neighboring molecules. This molecular interaction makes the behavior of a **real gas** distinct from that of an ideal gas, as intermolecular forces significantly influence its properties.

1.3.3 The ideal gas law

At low densities, gases can often be treated as **ideal gases**, which simplifies the analysis of many thermodynamic processes. The assumptions behind ideal gas behavior are:

1. Molecules do not experience intermolecular forces, meaning they do not attract or repel each other.
2. Molecules are treated as point particles with no volume.

The ideal gas equation, which relates the state variables of an ideal gas, is given by

$$pV = nRT \ \text{ or } \ pV = NkT,$$

where n is the number of moles and N the number of molecules, R is the universal gas constant, and k is Boltzmann's constant.

This equation accurately describes the behavior of gases over a broad range of conditions. However, the ideal gas law significantly differs from equations for liquids and solids (considered incompressible) because an ideal gas's specific volume is highly sensitive to changes in both pressure p and temperature T. Specifically, it varies linearly with temperature and inversely with pressure, reflecting the gas's high compressibility.

The simplicity of the ideal gas law makes it convenient for thermodynamic calculations, although it raises two important questions:

1. What constitutes 'low density' for the ideal gas law to be accurate? In other words, over what density range does this equation of state hold with reasonable precision?
2. How much does a real gas at a specific pressure and temperature deviate from ideal gas behavior?

These questions are essential when assessing whether the ideal gas approximation is valid for a given system.

1.3.4 Internal energy

Internal energy represents the sum of all microscopic forms of energy within a system. It relates to the molecular structure and the level of molecular activity, encompassing the kinetic and potential energies of the molecules.

To better understand internal energy, let us look at a system at the molecular level. In gases, molecules move freely through space, possessing kinetic energy known as **translational energy**. For polyatomic molecules, atoms also rotate around an axis, resulting in **rotational kinetic energy**. Additionally, atoms in polyatomic molecules may vibrate around a common center of mass, which creates **vibrational kinetic energy**. In gases, kinetic energy primarily comes from translational and rotational motion, with vibrational energy becoming significant only at higher temperatures.

On a more detailed level, electrons orbit around an atom's nucleus and possess rotational kinetic energy, with electrons in outer orbits having higher energy. Electrons also spin about their axes, contributing **spin energy**. Other particles, such as neutrons and protons within the nucleus, have similar spin energy. The portion of a system's internal energy associated with the kinetic energies of its molecules is known as **sensible energy**. As temperature increases, molecules move faster and with greater energy, leading to a higher internal energy in the system.

Internal energy also arises from various **binding forces**. These forces exist between molecules within a substance, between atoms within a molecule, and within the particles in an atom's nucleus. Binding forces are strongest in solids and weakest in gases. When a substance absorbs enough energy to overcome these forces, its molecules break free, changing its phase, such as from solid or liquid to gas. This phase-related internal energy is called **latent energy**. Phase changes typically occur without altering the chemical composition of a substance, which is why most practical thermodynamic problems do not require us to consider the forces binding atoms within molecules.

Each molecule contains atoms bound by **chemical energy**, resulting from the bonds between electrons and the nucleus. During chemical reactions, such as combustion, some chemical bonds break while others form, resulting in a change in internal energy. Stronger than these chemical forces are **nuclear forces**, which bind protons and neutrons in the atomic nucleus. The enormous energy associated with nuclear bonds is called **nuclear energy** and only becomes relevant in thermodynamics when considering nuclear reactions, such as fusion or fission. Chemical reactions

involve changes in electron structure, but nuclear reactions change the nucleus itself, resulting in the loss of atomic identity.

Atoms may also possess **electric and magnetic dipole-moment energies** when exposed to external electric or magnetic fields. This is due to the twisting of magnetic dipoles caused by the electric currents of orbiting electrons.

Determining the absolute internal energy of a substance (other than an ideal gas) can be complex and challenging, often requiring advanced calculations. Property databases may show large differences in values for internal energy, enthalpy, and entropy (to be discussed later) because these absolute values are somewhat arbitrary. The specific values of enthalpy, entropy, and internal energy at a single state are generally irrelevant; what matters is the difference in these values between two state points. For convenience, many reference handbooks set an arbitrary state point so that the values of these properties remain positive for most common liquid or gas states.

1.3.5 Internal energy of an ideal gas

We can treat a special case to determine the internal energy of an ideal gas, which relies on the concept of **degrees of freedom**. Each degree of freedom represents an independent way in which molecules can store energy. For example, degrees of freedom include the three possible directions of translational motion (movement along the x-, y-, and z-axes), and in multi-atomic gases, additional rotational or vibrational modes.

For **monatomic gases** (such as helium or neon), each molecule has only three translational degrees of freedom because these atoms are treated as single points with no rotational or vibrational motion. According to the equipartition theorem, each degree of freedom contributes $\frac{1}{2}kT$ (where k is Boltzmann's constant and T is temperature) to the energy per molecule. Thus, the internal energy U of N molecules or n moles of a monatomic ideal gas is

$$U = \frac{3}{2}NkT = \frac{3}{2}nRT.$$

For **diatomic gases** (such as oxygen or nitrogen), there are additional degrees of freedom due to rotational motion around two perpendicular axes. Therefore, diatomic gases have five active degrees of freedom at moderate temperatures (three translational and two rotational). The internal energy of a diatomic ideal gas is then

$$U = \frac{5}{2}NkT = \frac{5}{2}nRT.$$

At higher temperatures, vibrational motion also becomes significant, adding more degrees of freedom and increasing the internal energy further. However, at room temperature, this vibrational energy remains negligible for most diatomic gases.

We can generalize the expression for internal energy for an ideal gas as

$$U = f\frac{1}{2}NkT = f\frac{1}{2}nRT = f\frac{1}{2}pV.$$

1.4 System and surroundings

In thermodynamics, a **system** refers to a specific quantity of matter or a region in space selected for study. The **surroundings** include everything outside the system, and the **boundary** is the real or imaginary surface that separates them (see figure 1.2). This boundary can be fixed or movable and acts as the contact surface between the system and its surroundings. The boundary is typically treated as having zero thickness, meaning it contains no mass or volume.

Systems are categorized as closed or open, depending on whether the mass is fixed or if mass and energy can cross its boundaries.

- **Closed system**: A closed system, also known as a control mass, consists of a fixed amount of mass, meaning no mass can enter or leave the system (figure 1.2). However, energy in the form of heat or work can cross the boundary. The volume of a closed system may change. If no energy transfer occurs across the boundary, the system is called an **isolated system**.

  Example: Consider the gas inside a piston–cylinder device shown in figure 1.2. Here, the gas is the system, and the boundary is formed by the inner surfaces of the piston and cylinder. Since no mass enters or exits, it is a closed system, although energy can cross the boundary, and the piston may move. Everything outside the gas, including the piston and cylinder, is considered the surroundings.

- **Open system**: An open system, also known as a **control volume**, is a selected region in space where both mass and energy can cross the boundary. The boundaries of a control volume are referred to as the **control surface**, which can be real or imaginary (figure 1.3).

  Example: If we analyse the flow of air through a tube connecting two cylinders (the right-hand side of figure 1.3), the tube could be selected as a control volume. The inner surface of the tube forms the real boundary, while the entrance and exit areas represent imaginary boundaries. This control volume allows mass and energy to move in and out and enables specific analysis of energy and work interactions, optimizing system components individually.

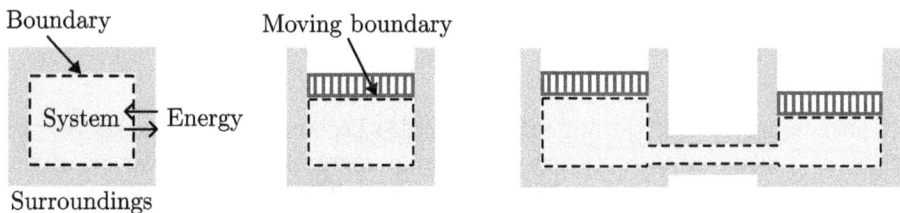

Boundary   Moving boundary

System ⇆ Energy

Surroundings

Figure 1.2. Description of a closed system. A thin line (real or imaginary) boundary separates the system from the surroundings. The boundary may move or is firmly fixed. Energy in the form of work or heat can be transferred between the system and the surroundings.

Control volume

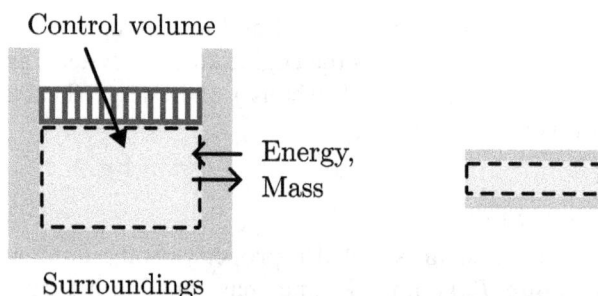

Energy,
Mass

Surroundings

Figure 1.3. Description of a control volume used in the analysis of an open system. Left: Mass can leave the control volume and the control volume may change in shape. In addition to the boundary work due to change in volume of the system, there can be other work, such as shaft or electrical work and heat interaction. Right: A simple tube connecting the two cylinders shown in figure 1.2. By isolating the tube like this, the focus is only on the optimization of this tube. In this case there will be flow in and out and energy interaction with the surroundings. This is the beauty of open system analysis as it allows optimizing individual components of a system.

A control volume may be fixed in size and shape, as with the tube in figure 1.3, or it may involve a moving boundary. Like closed systems, control volumes can experience heat and work interactions, with the added possibility of mass transfer across the boundary.

1.4.1 Examples of closed systems

1. *A sealed water bottle*: When a bottle of water is tightly sealed, it is a closed system. No water or air can enter or leave the bottle, but heat can still transfer across the bottle's surface, affecting the temperature of the water inside.
2. *A pressure cooker*: In operation, a pressure cooker acts as a closed system where no steam or food particles escape (until the pressure relief valve opens). However, heat from the stove transfers through the cooker's walls to raise the temperature inside, cooking the food.
3. *A car battery*: While discharging or charging, a car battery allows electrons to flow through external circuits, transferring energy in the form of electrical work. However, the mass of the battery remains constant, as no material crosses its boundaries, making it a closed system in terms of mass.

1.4.2 Examples of open systems

1. *A boiling pot of water (without a lid)*: When water boils, it allows steam to escape into the air, transferring both mass (water vapor) and energy (heat). The pot is considered an open system because mass and energy cross its boundary.
2. *An air conditioner*: An air conditioner is an open system because it exchanges both mass (air) and energy (cooling effect) with its surroundings. The system intakes warm air, cools it, and releases the cooler air back into the environment.

3. *A car engine*: During operation, fuel and air enter the engine, and exhaust gases are expelled. This makes the engine an open system, as it continuously exchanges mass (fuel, air, and exhaust) and energy (heat and mechanical work) with its surroundings.

1.4.3 Properties of a system

Any characteristic of a system is called a property. Some familiar properties are pressure p, temperature T, volume V, and mass m. The list can be extended to include less familiar ones such as viscosity, thermal conductivity, modulus of elasticity, thermal expansion coefficient, electric resistivity, and even velocity and elevation.

Properties are either intensive or extensive. **Intensive properties** are those that are independent of the mass of a system, such as temperature, pressure, and density. **Extensive properties** are those whose values depend on the size—or extent—of the system. Total mass, total volume, and total momentum are some examples of extensive properties. An easy way to determine whether a property is intensive or extensive is to divide the system into two equal parts with an imaginary partition, as shown in figure 1.4. Each part will have the same value of intensive properties as the original system, but half the value of the extensive properties.

1.4.4 Force balance

Pressure is defined as the normal force exerted by a fluid per unit area. When we discuss pressure, we usually refer to fluids, such as gases or liquids. In solids, the analogous concept is normal stress, although it is important to note that pressure is a scalar quantity, while stress is a tensor. Since pressure measures force per unit area, it is expressed in units of newtons per square meter (N m^{-2}), known as pascals (Pa). This relationship is given by

$$1 \text{ Pa} = 1 \text{ N m}^2.$$

Additionally, 1 bar is equivalent to 10^5 Pa, or 0.1 MPa (100 kPa).

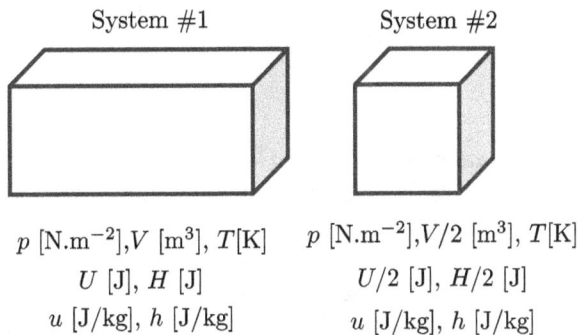

System #1   System #2

p [N.m^{-2}], V [m^3], T[K] p [N.m^{-2}], $V/2$ [m^3], T[K]

U [J], H [J] $U/2$ [J], $H/2$ [J]

u [J/kg], h [J/kg] u [J/kg], h [J/kg]

Figure 1.4. The properties of a system can be distinguished as two types: intensive or extensive. Intensive (or effort variables) do not change with the size of the system whereas extensive variables scale proportionally to the size of the system.

1.4.5 Piston–cylinder set-up

Consider the piston–cylinder system shown in figure 1.5. Here, a piston of mass M encloses a gas within a cylinder. To analyse the pressure inside the cylinder, we can apply a force balance on the piston. The forces acting on the piston include:
1. The downward gravitational force due to the piston's weight, Mg,
2. The atmospheric pressure exerted on the piston's surface, $p_{atm}A_p$, and
3. The internal gas pressure acting upwards, pA, where A is the cross-sectional area of the piston.

Setting up the force balance equation,

$$Mg + p_{atm}A - pA = 0.$$

Solving for p, we find that

$$p = p_{atm} + \frac{Mg}{A}.$$

This result shows that the pressure exerted by the gas inside the cylinder depends on both the atmospheric pressure and the weight of the piston. In figure 1.5, the force exerted on the gas is shown as the sum of the atmospheric pressure and the force due to the piston's weight.

1.4.6 Balloon buoyancy and floating conditions

Archimedes' principle states that any object fully or partially submerged in a fluid experiences an upward buoyant force equal to the weight of the fluid it displaces. For balloons, this principle explains how they achieve lift: the balloon displaces a volume of air as it expands. If the weight of the displaced air (the buoyant force) is greater than the combined weight of the balloon and its contents, the balloon will rise. In hot air balloons, this is achieved by heating the air inside, making it less dense. For gas balloons, using a lighter-than-air gas such as helium or hydrogen allows the balloon to displace a greater weight of air than its own weight, creating lift. Archimedes' principle is foundational for understanding how varying densities and volumes allow balloons to float and reach stable altitudes.

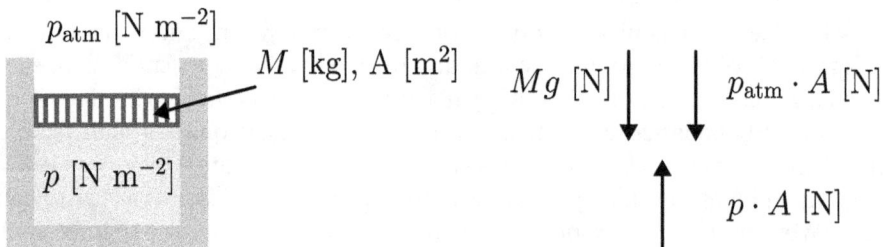

Figure 1.5. Force exerted on the gas is the sum of the force of weight of the piston and the atmosphere.

Let M represent the mass of a balloon at a height z above the ground. This mass includes the gondola, the balloon skin, and any payload, but excludes the gas inside (which could be hot air, hydrogen, or helium). For the balloon to float stably at a height z, the upward buoyant force provided by the displaced air must balance the combined weight of the balloon and the enclosed gas.

This balance condition is given by

$$M + \rho_{in} V = \rho_{air} V,$$

where:
- ρ_{in} is the density of the gas inside the balloon,
- ρ_{air} is the density of the surrounding air, and
- V is the volume of the gas inside the balloon.

If the left side of the equation (representing the total weight) is less than the right side, the balloon will rise; if it is greater, the balloon will descend.

1.4.6.1 Hot air balloons
A hot air balloon operates by heating the air within the balloon, which reduces the density of the air inside compared to the cooler, denser air outside. The balloon is open at the bottom, so the internal pressure equalizes with the external atmospheric pressure. Since warmer air is less dense, we have $T_{in} > T_{air}$, where T_{in} and T_{air} are the internal and external temperatures, respectively, and therefore $\rho_{in} < \rho_{air}$.

1.4.6.2 Gas balloons
Gas balloons, typically filled with hydrogen or helium, operate on a slightly different principle. These gases are naturally less dense than air, allowing the balloon to rise. A gas balloon is initially only partially filled and takes an inverted teardrop shape at ground level. As the balloon ascends, the external air pressure decreases, causing the gas inside to expand. Eventually, the balloon becomes nearly spherical.

If the balloon continues to rise, the internal gas may continue to expand until it reaches the balloon's structural limits, potentially causing it to burst if the balloon material cannot stretch further. For this reason, gas balloons often have mechanisms to release gas or vent it gradually to maintain a stable altitude.

1.4.7 Energy transfer

In classical thermodynamics, energy transfer between a system and its surroundings occurs through two primary modes: heat and work. Consider a system positioned at a certain height above the ground, moving at a constant velocity (figure 1.6). This system, defined by its boundaries, can interact with its surroundings only through these two forms of energy transfer. **Heat transfer** occurs due to a temperature difference between the system and its surroundings, moving energy spontaneously from warmer to cooler regions. **Work transfer**, on the other hand, involves energy transfer due to force applied over a distance, such as when an external force acts on the system or when the system does work on its environment. The energy of the system itself, termed **internal energy**,

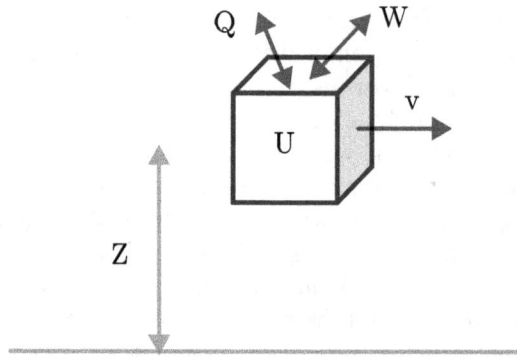

Figure 1.6. Illustration of a system and its interactions with the surroundings. The system, positioned at a height, possesses potential energy due to gravitational forces and kinetic energy due to its velocity. Additionally, it holds internal energy. The system exchanges energy with its surroundings through heat transfer and work transfer.

encapsulates all microscopic kinetic and potential energies within the system and remains distinct from the energy transferred as heat or work. Together, these modes outline the only ways energy can cross the boundary of a system, shaping the fundamental dynamics of thermodynamic processes.

In thermodynamics, work transfer can be broadly categorized into two types: $-pdV$ work and other forms of work transfer. The $-pdV$ work, also known as boundary work, arises specifically from a change in the volume of a system. When a system expands, it performs work on its surroundings, displacing them to create more space. Conversely, when the system compresses or shrinks, the surroundings may do work on the system, exerting force to reduce its volume. Although we often consider the external surroundings as the Earth's atmosphere, they can vary widely; for instance, a system might be in a high-pressure environment or in Martian conditions, where atmospheric pressure is lower than on Earth. On the Moon, with almost no atmosphere, $-pdV$ work would approach zero since there is no significant external pressure. This highlights how $-pdV$ work accounts for work transfer due to volume changes within a system and adapts based on the surrounding pressure, whether atmospheric or otherwise.

In thermodynamics, aside from $-pdV$ work, there are several other types of work transfer modes, each arising from different energy interactions. Here are some of the most common forms:

1. **Shaft work**: Shaft work occurs when a rotating shaft transfers energy into or out of a system, commonly found in engines, turbines, and compressors. For example, in a turbine, the rotating shaft converts fluid energy into mechanical energy, performing work that can drive generators or other machinery. This type of work is calculated based on the torque applied and the angular displacement of the shaft, making it essential in applications where rotational motion is key to energy transfer.

2. **Electrical work**: Electrical work is the transfer of energy due to an electric potential difference. An example would be an electric heater within a system

boundary, where energy is transferred by an electric current to generate heat. The work done is a function of the voltage (potential difference) and charge transfer. This form of work is prevalent in systems where electrical devices operate, such as batteries, electric motors, and circuits.

3. **Magnetic work**: Magnetic work involves the interaction of a system with a magnetic field, where energy is transferred as magnetic field strength aligns or moves magnetic dipoles within a material. An example is a magnetic refrigeration system, where work is done by aligning the magnetic dipoles to cool the system. The work in this case is calculated as the product of magnetic field strength and the magnetic dipole moment.

4. **Dielectric or polarization work**: This type of work occurs when a material's electric dipoles are aligned by an electric field. The interaction between the electric field and the dipole moment can induce work within the system, which is relevant in materials used as capacitors or dielectric insulators. For instance, dielectric heating in microwave ovens uses an alternating electric field to polarize water molecules, resulting in heat production.

5. **Elastic or spring work**: Elastic work, sometimes called spring work, occurs when a force deforms an elastic material, like stretching or compressing a spring. This energy transfer is proportional to the force applied and the displacement, allowing for storage and release of potential energy. It is often seen in applications like suspension systems or mechanical clocks.

Each of these types of work illustrates how diverse energy interactions can facilitate energy transfer across system boundaries.

1.4.8 Moving boundary work

Moving boundary work describes the mechanical work associated with any system that changes its volume, resulting in an energy transfer across its boundaries. This concept applies universally, whether in a piston–cylinder set-up or a more general system where boundaries expand or contract.

Consider a system that, due to internal conditions, begins to expand, increasing its volume. This expansion forces the surroundings to make room for the larger volume, exerting a force on the boundary. As the system volume increases, it performs work on the surroundings by pushing outward against the external pressure. This work transfer, resulting from the volume change, is termed **moving boundary work** and is often expressed as $-pdV$ work, reflecting the product of pressure and the differential change in volume.

Figure 1.7 illustrates two examples of moving boundary work due to volume changes in natural systems:

(a) *A growing tree*: In this scenario, a tree is growing, which means it is gradually increasing in volume. As the tree expands, it displaces the surrounding air, pushing it outward to make space for its growth. This expansion requires the tree (system) to exert a force on the surrounding air (surroundings), effectively performing work on the surroundings. Therefore, in this case, the work transfer occurs from the system (tree) to the surroundings (air).

Figure 1.7. Illustration of boundary work.

(b) *Decaying bananas*: Here, bananas are shown undergoing decay over time, resulting in a gradual reduction in volume. As they decay and shrink, they occupy less space, creating additional volume within the surroundings. Consequently, the surrounding air moves in to fill the space left by the shrinking bananas. In this scenario, the surroundings perform work on the system, as the work transfer is from the surroundings (air) to the system (bananas).

In this book, we adopt a consistent **sign convention for energy transfers**. *Any energy transferred to the system is considered positive*, whether it occurs through heat or work. Conversely, energy transferred from the system to the surroundings is considered negative. This convention simplifies the analysis by clearly indicating whether energy is entering or leaving the system. Since thermodynamics defines only two modes of energy transfer—heat and work—this sign convention applies universally to both. By using this approach, we maintain clarity in interpreting thermodynamic processes and ensure consistency in energy accounting across different scenarios.

In the examples provided, we can use the concept of moving boundary work and our established sign convention to analyse the direction and sign of work transfer.

For the *growing tree* example, our intuition tells us that as the tree expands, it pushes against the surrounding air to make room for its growth. This means that the system (tree) is doing work on the surroundings, transferring energy out of the system. According to our sign convention, this transfer should be negative. The moving boundary work is given by $-pdV$, where p is the atmospheric pressure and dV is the change in volume. In this case, the volume change dV is positive, as the tree's volume increases, while p is constant (atmospheric pressure). Therefore, $-pdV$ yields a negative result, indicating that the work transfer is negative—consistent with our convention for energy leaving the system.

In the *decaying bananas* example, the volume of the system (bananas) decreases over time as they decay, creating additional space that the surrounding air fills. Here, work is done by the surroundings on the system, transferring energy into the system. According to our sign convention, work transferred to the system is positive. In this case, the moving boundary work, $-pdV$, has a negative volume change (dV is

negative) since the system's volume is decreasing. Multiplying a negative dV by the negative sign in $-pdV$ results in a positive work value, showing that work is indeed transferred to the system. This result aligns with our sign convention, as energy is entering the system.

These examples illustrate that the $-pdV$ formulation is consistent with our sign convention, accurately reflecting the direction of energy transfer in each case.

1.4.8.1 Gas enclosed in a cylinder by a piston

A **special case of work transfer** that resembles moving boundary work occurs when a gas expands within a cylinder enclosed by a piston. In this case, the work transfer is between the gas (system) and the surroundings, specifically through the movement of the piston. Although this work transfer has the same mathematical expression, $-pdV$, it is important to note that this case differs conceptually from the traditional moving boundary work discussed earlier.

In this scenario, p represents the **pressure of the gas inside the cylinder**, not the external atmospheric pressure. As the gas expands, it pushes the piston outward, increasing the system's volume (dV is positive). This expansion requires the gas to perform work on the surroundings (the piston), effectively transferring energy out of the system. According to our sign convention, this work is negative, as energy is leaving the gas. Thus, the expression $-pdV$ accurately represents the direction of work transfer.

Conversely, when the piston compresses the gas, the volume change dV is negative. In this case, the surroundings do work on the gas, transferring energy into the system. The expression $-pdV$ then yields a positive result, indicating that work is being added to the gas, in line with our sign convention.

In summary, while the expression for work transfer $-pdV$ applies here as it does for moving boundary work, *the two scenarios differ*. This case specifically describes work transfer between the gas and its surroundings (the piston), and it relies on the *internal gas pressure* rather than the *external atmospheric pressure*, making it a distinct application of the $-pdV$ relationship.

Building on this specific example, we can calculate the differential work transferred to or from the system by considering the force exerted by the gas pressure on the piston. Consider the gas confined within this piston–cylinder device, as illustrated in figure 1.8. The initial pressure of the gas is p, the total volume is V, and the cross-sectional area of the piston is A. When the piston moves a small distance ds, the differential work W (a scalar quantity) transferred to or from the system during this process is given by

$$W = \overrightarrow{F_{\text{ext}}} \cdot d\vec{s} = -\overrightarrow{F_{\text{sys}}} \cdot d\vec{s}.$$

To simplify, we replace the force (a vector quantity) and displacement with system properties. The external force exerted on the piston is the opposite of the force exerted by the system, which is equal to the pressure of the gas multiplied by the piston area:

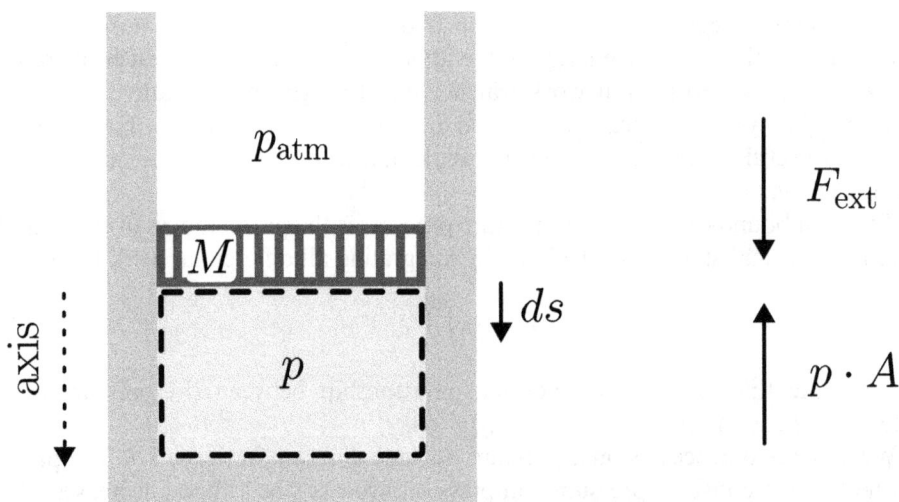

Figure 1.8. Schematic showing the boundary work performed by an external force on the system.

$$\overrightarrow{F_{\text{ext}}} = -\overrightarrow{F_{\text{sys}}} = -\overrightarrow{pA}.$$

Substituting this into the equation, we obtain

$$W = -\overrightarrow{F_{\text{sys}}} \cdot \overrightarrow{ds} = -pA\frac{dV}{A} = -pdV.$$

This transformation elegantly allows us to evaluate the gas work using only system properties (pressure and volume). There is no need to account for the specifics of how the external force is applied. However, it is important to note that the pressure within the system may vary over time. Therefore, to calculate the boundary work over an entire process, we would need to know how the pressure p changes as a function of volume.

It is essential to note that, although the expression $-pdV$ appears the same in both cases, the work calculated for a gas enclosed within a piston–cylinder differs fundamentally from moving boundary work. In traditional moving boundary work, p represents the **external pressure** applied to the system. However, in the specific example of a gas within a piston–cylinder device, p refers to the **internal pressure of the gas** inside the cylinder. This internal gas pressure results from the force exerted by external factors, such as atmospheric pressure and the weight of the piston (as in the above example), but it is an internal property of the system. Thus, while both cases use the $-pdV$ expression, they represent distinct applications, with different interpretations of the pressure term depending on the system's context.

In some books and sources, including online literature, boundary work is defined as pdV without the negative sign. This difference arises from varying sign conventions for energy flows. In this book (and in this course), we adopt the convention that energy flowing *into* the system is positive, while energy flowing *out*

1-15

of the system is negative. This approach is physically intuitive, as positive energy flow increases the system's energy. Some alternative conventions treat heat transfer to the system as positive but work transfer into the system as negative, with work done by the system (energy transferred out) considered positive. This choice is primarily useful when focusing on the work output of a system but can introduce inconsistency.

The total boundary work over an entire process, as the piston moves from the initial state *i* to the final state *f*, is calculated by integrating all differential work terms:

$$W = -\int_i^f pdV.$$

To evaluate this integral, a functional relationship between the pressure *p* and volume *V* throughout the process is required.

We show two processes on a pressure–volume diagram in figure 1.9. The process on the left is a constant pressure compression process (the volume decreases). You might wonder how a gas can be compressed at a constant pressure. This is possible by immersing the cylinder in cold surroundings, leading to transfer of heat from the gas to the cold bath. In this diagram, the differential area dA is equal to pdV. The **differential** work is the *negative* of this area. The total area A under the process curve from an initial state (*i*) to the final state (*f*) is obtained by adding these differential areas.

A gas can follow several different paths as it expands from state *i* to state *f*. In general, each path will have a different area underneath it, and since this area represents the magnitude of the work, the work done will be different for each process. See the process on the right of figure 1.9.

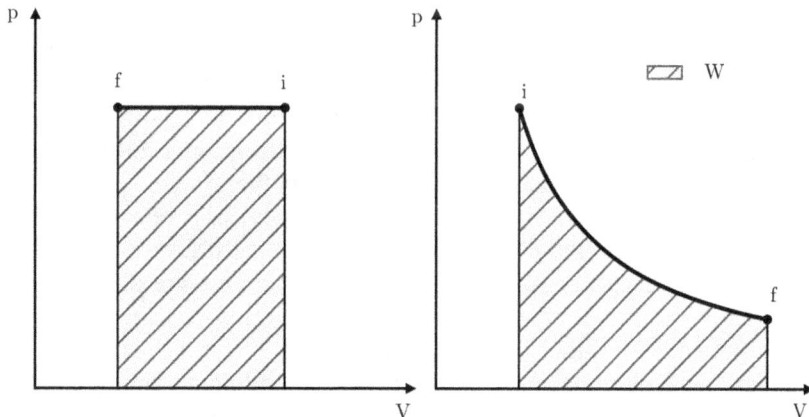

Figure 1.9. A constant pressure (or an isobaric) process (left) and a constant temperature (or an isothermal) process (right) shown on a pressure–volume property diagram. The initial state is indicated by *i* and the final state with *f*. The area under the curve is the negative of the work transferred from the surroundings on the system.

1.5 Summary

In this chapter, we laid the groundwork for classical thermodynamics by exploring fundamental concepts such as energy transfer, system boundaries, and state properties. We examined systems through microscopic and macroscopic lenses, understanding how molecular behaviors aggregate into observable phenomena. Key distinctions between energy in forms such as internal, kinetic, and potential energy were clarified, along with how systems interact with surroundings through heat and work. From the ideal gas law to real gas deviations, we discussed the limitations and applications of these principles in practical scenarios.

Chapter 2

Enthalpy and heat capacity

Chapter 2 introduces the concept of enthalpy as a foundational component in thermodynamic analysis. Starting with enthalpy, defined as the sum of internal energy and the product of pressure and volume, this chapter highlights its utility in evaluating energy transfer, particularly for open systems where mass and energy flow across boundaries. By establishing enthalpy as an energy potential, this chapter sets the stage for deriving the first law of thermodynamics for both open and closed systems, focusing on how energy in the form of heat, work, and mass flow contributes to a system's energy balance.

Following enthalpy, the chapter delves into heat capacity, exploring how different substances require varying amounts of energy to change temperature. By distinguishing between heat capacity at constant pressure and constant volume, readers will gain tools for analysing how energy interacts with substances under different constraints. This framework prepares readers to tackle more complex thermodynamic applications, particularly those involving energy transfer in flowing systems.

2.1 Enthalpy

By definition enthalpy is

$$H = U + pV,$$

where U represents the internal energy of the system, p is the pressure, and V is the volume. Enthalpy serves as a crucial energy potential, especially useful in the analysis of open systems where mass and energy flow in and out of the system.

To illustrate enthalpy's meaning, let us consider a balloon filled with gas. Initially, the balloon's volume is zero. It is gradually filled with a diatomic ideal gas (with degrees of freedom $f = 5$), reaching a final volume of 1 l, as shown in figure 2.1. The gas pressure and the surrounding atmospheric pressure remain constant at 1 bar. The internal energy difference between the final and initial states is 250 J, and the boundary work done by the gas to expand the balloon is 100 J. (*Note*: this work, as

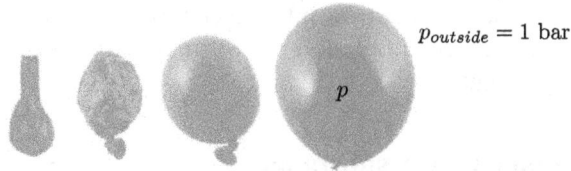

$p_{outside} = 1$ bar

p

Volume, V	0	1 liter
Internal Energy, U	0	$f\dfrac{1}{2}pV = 250$ J
Boundary work performed to create the balloon		$W = \displaystyle\int_0^{1 liter} pdV = 100$ J
Total energy		$U + W = U + pV = H = 350$ J

Figure 2.1. Example illustrating the physical meaning of enthalpy. See the text for details.

discussed earlier, corresponds to $-pdV$ rather than work done *on* the balloon.) Therefore, the total energy required to inflate the balloon is the sum of the internal energy of the gas and the work done to create the space for it, or $U + pV = 350$ J. This value corresponds to the enthalpy of the ideal gas in its final state.

Therefore, in creating objects 'out of thin air', one must account for both the internal energy and the work done to make space for them. This total energy is the enthalpy. On the surface of the Moon, which lacks an atmosphere, enthalpy is effectively equal to internal energy. On Mars, with its atmospheric pressure of about 6.5 mbar, the enthalpy of gases is lower than it would be on Earth.

2.1.1 Special case: enthalpy of an ideal gas

Using expressions of ideal gas, the enthalpy is

$$H = U + pV = \frac{f}{2}nRT + nRT = \left(\frac{f}{2} + 1\right)nRT \ \text{(J)}.$$

$$h_{mol} = \frac{H}{n} = \left(\frac{f}{2} + 1\right)RT \ \text{(J mol}^{-1}\text{)}; \quad h_{mass} = \frac{H}{m} = \left(\frac{f}{2} + 1\right)\frac{RT}{M} \ \text{(J kg}^{-1}\text{)}.$$

2.1.2 Enthalpy in reference databases

Enthalpy data for most substances is available in property databases. It is important to note that absolute values of enthalpy are not necessary for process analysis. In fact, except for ideal gases, determining absolute enthalpy values is cumbersome and generally unnecessary for system analysis. The REFPROP database, for example, uses the melting point of each substance at one atmosphere as its reference point, deriving all other property data from this baseline. When looking up energy

potential values—such as enthalpy, internal energy, Gibbs free energy, or Helmholtz energy—you may notice differences across databases because each may use a distinct reference point.

2.2 Heat capacity of a substance

Experience shows that different substances require different amounts of energy to increase the temperature of identical masses or moles by one degree. This variation leads us to define a property that allows comparison of the energy storage capacities of various substances—**specific heat**. Specific heat is defined as the energy required to raise the temperature of a unit mass or mole of a substance by one degree.

- **Per unit mass**: c_m J (kg K)$^{-1}$.
- **Per unit mole**: c_n J (mol K)$^{-1}$.

The amount of energy needed depends on the process conditions. In thermodynamics, we focus on two main types of specific heats: specific heat at constant volume (c_v) and specific heat at constant pressure (c_p).

- **Constant-volume specific heat** (c_v): This is the energy required to increase the temperature of a unit mass by one degree while keeping volume constant.
- **Constant-pressure specific heat** (c_p): This is the energy needed to raise the temperature of a unit mass by one degree under constant pressure.

Specific heats c_v and c_p can be expressed in terms of energy transferred for any mass or mole of a substance:

energy transferred (J) = mass (kg) $\cdot$ c_m $\cdot$ ΔT (K) = moles (mol) $\cdot$ c_n $\cdot$ ΔT (K).

In specific domains, other definitions for specific heat capacity exist, such as specific heat at a constant magnetic field, however, these are outside our current scope.

Now, let us connect specific heats to other thermodynamic properties. Consider a stationary, closed system with a fixed mass undergoing a constant-volume process (meaning no boundary work, $-pdV = 0$). We express the differential energy change per unit mass or mole of the system as:

energy transferred as work or heat per unit mass or mole, $du = c_v dT$ or, equivalently,

$$c_v = \left(\frac{\partial u}{\partial T} \right)_V.$$

Here, u represents the internal energy per unit mass or mole of the substance, maintaining consistent units throughout. Note that while $-pdV = 0$ in a constant-volume process, in addition to heat transfer, energy can still be transferred by other means, such as shaft work or electric work.

For constant-pressure processes, the expression for c_p can be derived similarly. In this scenario, the energy transferred contributes to both the internal energy and the work required to expand the system boundary. Thus, we have:

energy transferred as work or heat per unit mass or mole, $dh = c_p dT$ or, equivalently,

$$c_p = \left(\frac{\partial h}{\partial T}\right)_p.$$

These expressions illustrate that c_v corresponds to changes in internal energy, while c_p relates to changes in enthalpy. To be precise, c_v can be defined as the change in a substance's internal energy per unit temperature change at constant volume, while c_p measures the change in enthalpy per unit temperature change at constant pressure. Thus, c_v provides insight into how internal energy varies with temperature, and c_p similarly reflects enthalpy's temperature dependence.

For an ideal gas, specific heat at constant volume (c_v) and specific heat at constant pressure (c_p) have a straightforward relationship.

Starting from the basic thermodynamic relationship for enthalpy (H),

$$H = U + pV,$$

and knowing from the ideal gas law that

$$pV = nRT,$$

we can substitute $pV = nRT$ into the enthalpy equation, resulting in

$$H = U + nRT.$$

Differentiating with respect to temperature, we obtain

$$dH = dU + nRdT.$$

From definition,

$$nc_p dT = nc_v dT + nRdT,$$

which simplifies to

$$c_p = c_v + R.$$

For an ideal gas, we can further express c_v in terms of the gas's degrees of freedom f. The internal energy U for an ideal gas is

$$dU = nc_v dT = d\left(\frac{f}{2}nRT\right),$$

which gives

$$\frac{f}{2}nRdT = nc_v dT \Rightarrow c_v = \frac{f}{2}R.$$

For a constant-pressure process, we use the relationship

$$c_p = c_v + R = \frac{f}{2}R + R = \left(\frac{f}{2} + 1\right)R.$$

Thus, c_p for an ideal gas is greater than c_v by R, the gas constant. This difference exists because at constant pressure, energy must also account for the work done in expanding the gas, in addition to increasing its internal energy.

2.2.1 Specific heat of solids and liquids

A substance with a constant specific volume (or density) is considered **incompressible**. For solids and liquids, we typically assume that specific volume or density remains unchanged during a process. This assumption allows us to treat solids and liquids as incompressible substances without significant loss of accuracy. Consequently, for such incompressible substances, the specific heats c_p and c_v can be approximated as equal.

To understand why, let us examine the expression for c_p in terms of enthalpy (h):

$$c_p = \left(\frac{\partial h}{\partial T}\right)_p = \left(\frac{\partial (u + pV)}{\partial T}\right)_p = \left(\frac{\partial u}{\partial T}\right)_p + \left(\frac{\partial pV}{\partial T}\right)_p.$$

Expanding this further, we obtain

$$c_p = \left(\frac{\partial u}{\partial T}\right)_p + p\left(\frac{\partial V}{\partial T}\right)_p + V\left(\frac{\partial p}{\partial T}\right)_p.$$

For incompressible substances, both $\left(\frac{\partial V}{\partial T}\right)_p$ and $\left(\frac{\partial p}{\partial T}\right)_p$ are approximately zero, simplifying the expression to

$$c_p \approx \left(\frac{\partial u}{\partial T}\right)_p \approx c_v.$$

Thus, for solids and liquids, c_p and c_v are effectively the same.

2.2.2 Why is 'heat capacity' not correct?

'Heat capacity' can be confusing and perhaps even misleading, given our modern understanding of thermodynamics. The term originates from a historical context, when 'heat' was not clearly distinguished from other forms of energy transfer. In light of that, 'energy capacity' would arguably be a more accurate term, as it would encompass all forms of energy input, not just heat, that can raise the temperature of a substance.

In fact, reframing 'heat capacity' as 'energy capacity' aligns with the way thermodynamics treats energy transfers in a unified framework. Here's why 'energy capacity' could indeed be a better term:

1. *Inclusive of all energy forms*: Temperature can be increased through various forms of energy transfer, not just heat. For example, both shaft work and microwave radiation can raise a substance's temperature. 'Energy capacity' would recognize that any energy input capable of raising temperature is relevant, whether it is in the form of heat, work, or electromagnetic energy.

2. *Alignment with internal energy*: The concept of internal energy in thermodynamics encompasses both heat and work. Referring to the energy required to change temperature as 'energy capacity' would harmonize this term with the broader understanding of internal energy, rather than implying a special status for heat.

3. *Modern thermodynamics language*: Thermodynamics now distinguishes heat, work, and other forms of energy transfer without giving special emphasis to heat. Using 'energy capacity' could help move away from legacy terms that might reinforce outdated distinctions, making thermodynamics more conceptually accessible.

While tradition maintains 'heat capacity' as a term, a general suggestion to rename it as 'energy capacity' could indeed make thermodynamic concepts clearer and more accurate for learners, aligning with the current understanding that temperature change results from total energy transfer rather than from heat alone.

2.3 Flow work and enthalpy

In open systems, or **control volumes**, mass flows across system boundaries, and work is required to push this mass into or out of the control volume. This work, known as **flow work** (or **flow energy**), is essential to maintain a continuous flow through the control volume.

To derive a relation for flow work, let us consider a small fluid element of volume V (see figure 2.2). The fluid element directly upstream exerts force, pushing this volume into the control volume, much like an imaginary piston would. By assuming the fluid element is sufficiently small, we can treat its properties as uniform.

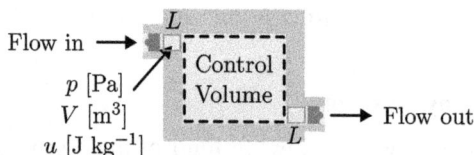

Flow in	Flow out
$W = -\int_{V_{cv}+V}^{V_{cv}} p \, \mathrm{d}V$	$W = -\int_{V_{cv}}^{V_{cv}+V} p \, \mathrm{d}V$
$W = -p\left(V_{cv} - (V_{cv}+V)\right) = pV$ [J]	$W = -p\left((V_{cv}+V) - V_{cv}\right) = pV$ [J]
$W = p\vartheta$ [J kg^{-1}]	$W = -p\vartheta$ [J kg^{-1}]
Energy into the control volume u [J/kg]	Energy into the control volume $-u$ [J/kg]
Total Energy $u + p\vartheta = h$ [J/kg]	Total Energy $-u - p\vartheta = -h$ [J/kg]

Figure 2.2. Flow work during mass transfer across a control volume boundary. The energy associated with the mass entering or leaving the control volume is represented by the enthalpy of that mass.

If the fluid pressure is p and the fluid element has volume V, the flow work can be expressed as the product pV. It is important to note the sign convention here: work is considered positive when pushing fluid into the control volume and negative when the control volume pushes fluid out. Here, ϑ denotes the specific volume ($m^3\ kg^{-1}$).

2.3.1 Total energy of a flowing fluid

The **total energy** of a fluid per unit mass is composed of three primary components,

$$e = u + ke + pe = u + \frac{v^2}{2} + gz,$$

where:

- u is the internal energy,
- $ke = \frac{v^2}{2}$ represents kinetic energy (with v as the velocity), and
- $pe = gz$ is the potential energy (where g is gravitational acceleration and z is the elevation relative to a reference point).

For a flowing fluid, an additional form of energy, **flow energy**, pV, must be included. Thus, the total energy per unit mass for a flowing fluid becomes

$$e_{\text{total}} = u + ke + pe + pv.$$

Since enthalpy h is defined as $h = u + pv$, this relation simplifies to

$$e_{\text{total}} = h + \frac{v^2}{2} + gz.$$

By using enthalpy h instead of internal energy u to represent the energy of a flowing fluid, the flow work pv is automatically included. *This is one of the main reasons enthalpy is defined and commonly used in thermodynamic analysis of open systems.*

2.3.2 Energy transport by mass flow

The total energy associated with a flowing fluid of mass m is given by

$$\text{Total energy transport} = m\left(h + \frac{v^2}{2} + gz\right).$$

For a fluid stream flowing at a mass flow rate $\dot{m}$, the rate of energy transport is

$$\text{Rate of energy transport} = \dot{m}\left(h + \frac{v^2}{2} + gz\right).$$

In many practical situations, the kinetic and potential energy contributions are negligible, allowing these expressions to simplify to mh and $\dot{m}h$ for the total energy transport and rate of energy transport, respectively. This simplification often applies to low-velocity, low-elevation systems where enthalpy alone effectively describes the energy transport.

2.3.3 Examples of flow energy

- *Pumping water into a storage tank*: Flow work is required to pump water up to an elevated storage tank, overcoming both gravitational potential and pressure within the tank.
- *Steam turbines in power plants*: High-pressure steam flows through turbine blades, doing flow work to drive the turbine and generate electricity.
- *Airflow in a building's ventilation system with heat recovery*: Fans push air through ducts, requiring flow work to maintain airflow. Heat recovery units capture thermal energy from outgoing air to warm incoming fresh air, reducing heating needs and increasing efficiency.
- *Compressed air moving through pipes in manufacturing*: Compressed air flows through pipes to power tools and clean surfaces, with flow work enabling continuous supply and pressure maintenance.
- *Fuel flow in a biomass boiler system*: Biomass boilers require fuel to be pushed into the combustion chamber, with flow work ensuring a steady supply of fuel for sustainable heat production.

2.4 Heat transfer definition

To complete our discussion, let us define heat transfer. Heat is the spontaneous flow of energy driven by a temperature difference. There are three main modes of heat transfer—conduction, convection, and radiation—which govern the rate of energy flow:

1. **Conduction**: Transfer of heat through a solid or between solids in contact, where energy moves from molecule to molecule.
2. **Convection**: Transfer of heat by the movement of a fluid, such as air or water, over a surface or within a fluid.
3. **Radiation**: Transfer of heat in the form of electromagnetic waves, which can occur even in a vacuum (for example, the Sun heating the Earth).

Thermodynamics focuses on equilibrium states and establishes the *rules* of energy transfer, while *kinetics*—which we will not cover in this course—determines the *rate* at which energy moves. To illustrate, think of this course as an example: thermodynamics provides the rules for achieving a perfect score, but it is up to you, the student, to study and apply yourself (the kinetics) to achieve that score. How quickly you reach your goal depends on factors such as motivation and focus, just as heat transfer rates depend on the properties of the materials and conditions involved.

2.4.1 First law of thermodynamics

The first law of thermodynamics describes how energy can be transferred to or from a system in three forms: heat, work, and mass flow (figure 2.3). These energy interactions occur at the system boundaries, representing energy gained or lost by the system during a process.

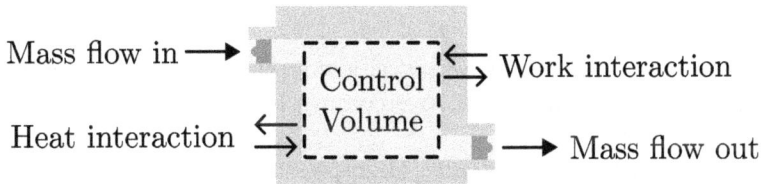

Figure 2.3. The energy of a control volume changes through heat, work, or mass interactions. Multiple inlets or outlets, as well as multiple heat and work interactions, can be present. By convention, energy (heat or work) or mass entering the system is considered positive.

1. **Heat transfer** (Q): Heat transfer to a system (or heat gain) increases the energy of the molecules, raising the internal energy of the system. Conversely, heat transfer from a system (heat loss) decreases the system's energy. Heat transfer occurs when there is a temperature difference, and the energy flows from the higher-temperature area to the lower-temperature area.

2. **Work transfer** (W): Work involves energy interactions that are not caused by a temperature difference between a system and its surroundings. Examples of work include a rising piston, a rotating shaft, or an electric current crossing system boundaries. When work is done on a system (positive work), the system's energy increases. When the system does work on its surroundings (negative work), the system's energy decreases.

3. **Mass flow** (m): Mass flow into and out of a system serves as an additional energy transfer mechanism. When mass enters the system, it brings energy (internal energy) with it, increasing the system's total energy. Conversely, when mass exits the system, it carries energy away, decreasing the system's energy.

2.4.2 Conservation of mass

In control volumes, the **conservation of mass** principle states that the net mass transfer into or out of a control volume over a time interval Δt equals the net change in the total mass within the control volume over Δt:

$$m_{\text{in}} - m_{\text{out}} = \Delta m_{\text{CV}} \ (\text{kg})$$

$$\dot{m}_{\text{in}} - \dot{m}_{\text{out}} = \frac{dm_{\text{CV}}}{dt} \ (\text{kg s}^{-1}).$$

During a **steady-state** process, the total mass in a control volume remains constant ($m_{\text{CV}} = \text{constant}$). In this case, the conservation of mass requires that the total mass entering equals the total mass exiting. For a steady-flow system with multiple inlets and outlets, we write this in rate form as

$$\sum_{\text{in}} \dot{m} = \sum_{\text{out}} \dot{m}.$$

This means the total rate of mass entering the control volume is equal to the total rate of mass leaving it.

2.4.3 First law for open systems

For a control volume (open system), the first law of thermodynamics in rate form is practical when dealing with flows. The rate of change of the total energy within a control volume is equal to the net rate of energy entering the control volume via heat, work, and mass flow:

$$\frac{dE_{CV}}{dt} = \sum \dot{Q} + \sum \dot{W} + \sum_{in} \dot{m}\left(h + \frac{V^2}{2} + gz\right) - \sum_{out} \dot{m}\left(h + \frac{V^2}{2} + gz\right).$$

In a **steady-flow** process, the total energy within the control volume remains constant (E_{CV} = constant), so the change in total energy is zero:

$$\sum \dot{Q} + \sum \dot{W} + \sum_{in} \dot{m}\left(h + \frac{V^2}{2} + gz\right) - \sum_{out} \dot{m}\left(h + \frac{V^2}{2} + gz\right) = 0.$$

If changes in kinetic and potential energy are negligible, this simplifies to

$$\sum \dot{Q} + \sum \dot{W} + \sum_{in} \dot{m}h - \sum_{out} \dot{m}h = 0.$$

2.4.4 First law for a closed system

In a closed system, there is no mass flow across the boundaries, so the energy transfer due to mass flow is zero. The first law for a closed system simplifies to

$$\frac{dE_{CM}}{dt} = \sum \dot{Q} + \sum \dot{W}$$

Neglecting kinetic and potential energy changes (e.g. if the system is at rest), the conservation of energy equation reduces to

$$\frac{dU}{dt} = \sum \dot{Q} + \sum \dot{W}.$$

Over a time interval, this can be expressed as

$$\Delta U = \sum Q + \sum W.$$

2.4.5 Examples

1. **Closed system**: heating water in a rigid tank
   - *System description*: A rigid, sealed tank contains a fixed mass of water. The tank is heated from the outside, increasing the water's temperature.
   - *First law application*: Since the tank is sealed, no mass enters or exits (closed system). The first law simplifies to $\Delta U = Q$ because there is no work done (no boundary movement).

- *Relation to $h = c_p T$*: In this case, we primarily use internal energy U rather than enthalpy h, as the volume is constant and there is no mass flow.
- *Key learning*: This example illustrates that in a closed, constant-volume system, the heat added directly changes the internal energy, with no flow or work done by the system.

2. **Open system:** steam flowing through a turbine
    - *System description*: High-pressure, high-temperature steam flows through a turbine, expanding as it drives the turbine blades and exits at a lower pressure and temperature.
    - *First law application*: For this open system with steady flow, the first law in rate form is $\sum \dot{Q} + \sum \dot{W} + \sum_{\text{in}} \dot{m}h - \sum_{\text{out}} \dot{m}h = 0$. In an adiabatic turbine (no heat transfer, $\dot{Q} = 0$) the energy equation becomes $\dot{W} = \dot{m}(h_{\text{out}} - h_{\text{in}})$.
    - *Relation to $h = c_p T$*: For ideal gases, we can approximate enthalpy with $h = c_p T$, allowing us to use temperature data to determine enthalpy change.
    - *Key learning*: This example demonstrates how the first law applies to an open, steady-flow process with energy extracted as work, and how enthalpy changes directly relate to temperature differences.

3. **Closed system:** compressed air in a piston–cylinder assembly
    - *System description*: A piston–cylinder device contains air, initially at low pressure. As the piston compresses the air, its temperature and pressure increase.
    - *First law application*: Here, no mass enters or exits, so we use the closed system form of the first law: $\Delta U = Q + W$. If the process is adiabatic (no heat transfer, $Q = 0$), then $\Delta U = W$, meaning work done on the system increases internal energy.
    - *Key learning*: This example shows how work done on a closed system (in compression) directly affects internal energy and temperature without involving enthalpy.

4. **Open system:** water flowing through a heat exchanger
    - *System description*: Water flows continuously through a heat exchanger, where it is heated from an initial low temperature T_{in} to a higher temperature T_{out}.
    - *First law application*: For a steady-flow open system, the first law becomes $= \dot{m}(h_{\text{out}} - h_{\text{in}})$, assuming no work is done. This shows that heat added equals the increase in enthalpy of the water.
    - *Relation to $h = c_p T$*: With liquids, we can approximate $h \approx c_p T$ if the specific heat c_p is relatively constant over the temperature range.
    - *Key learning*: This example emphasises how enthalpy change (and thus temperature change) in an open system corresponds to the heat added, ideal for analysing liquid flows in systems such as heating or cooling circuits.

5. **Open system**: compressed air cooling in a nozzle
   - *System description*: Compressed air flows through a nozzle, expanding and cooling as it exits at a higher velocity.
   - *First law application*: In this open system, we apply the first law as $h_{in} + \frac{v_{in}^2}{2} = h_{out} + \frac{v_{out}^2}{2}$, neglecting heat and potential energy changes. This equation shows that any increase in kinetic energy (velocity) results in a drop in enthalpy, or temperature, of the exiting air.
   - *Relation to $h = c_p T$*: For an ideal gas, we can approximate $h = c_p T$, making it easier to relate temperature drop to changes in velocity.
   - *Key learning*: This example highlights the trade-off between kinetic energy and enthalpy in an open, adiabatic system, illustrating how temperature decreases as velocity increases in expansion processes.

2.5 Summary

In this chapter, we explored enthalpy as an energy potential that is particularly advantageous in analysing open systems. By introducing enthalpy as the sum of internal energy and flow work, we gain a powerful tool for understanding how energy moves in systems with mass exchange. This chapter systematically derived the first law of thermodynamics for open and closed systems, emphasizing the conservation of energy across heat, work, and mass flows.

We also examined flow work in the context of open systems, identifying how flow energy combines with internal energy to create a complete picture of energy movement. The application of heat capacity concepts, particularly at constant pressure and constant volume, helped clarify how energy requirements change with system conditions, especially in the context of ideal gases and incompressible substances.

IOP Publishing

A Classical Thermodynamics Toolkit

Srinivas Vanapalli

Chapter 3

Thermodynamic processes in a closed system

In this chapter, we delve into the core principles of thermodynamic processes within a closed system, exploring how systems transition between states and the energy transformations that accompany these changes. At the heart of this study are **state variables**—key properties such as pressure, temperature, and internal energy—that define a system's state independently of the process path taken. We introduce the concept of **thermodynamic processes** as paths connecting initial and final states, with energy exchanged in the form of **heat** and **work**. This chapter covers various types of processes, including **quasistatic** and **non-quasistatic** processes, as well as specialized cases such as **isobaric**, **isothermal**, **isochoric**, **adiabatic**, and **isentropic** processes. By examining these idealized processes, we build a foundation for understanding real-world applications, with particular emphasis on quasistatic processes involving an ideal gas, where we assume the absence of friction and other forms of non-pV work. This chapter provides essential tools and concepts that will enable students to analyse thermodynamic systems under controlled conditions and understand the implications of idealized versus real behaviors.

3.1 State variables

In thermodynamics, **state variables** (or **state properties**) are essential quantities that describe the current condition of a system. These properties are called 'state' variables because they depend solely on the state of the system at a given moment, not on the process or path taken to reach that state. A defining feature of state variables is that once any two independent state variables are known, other properties of the system can often be derived through established thermodynamic relationships.

For example, in the case of an ideal gas, if the temperature and the amount and type of gas (whether monoatomic or diatomic, for instance) are known, the internal energy U can be determined. Furthermore, if the volume V is also known, the

doi:10.1088/978-0-7503-6029-6ch3
3-1

pressure p can be calculated using the ideal gas law, and from these, related properties such as enthalpy H can be deduced.

Common state variables:

- Pressure, p
- Temperature, T
- Volume, V, or density
- Internal energy, U
- Entropy, S
- Enthalpy, H
- Helmholtz free energy, F
- Gibbs free energy, G.

These variables are crucial for analysing the behavior of closed systems under different conditions. Additional state variables may exist, particularly in specialized fields such as electrochemistry or electrical systems, where properties such as electric charge and electric potential are considered.

3.1.1 A note on misconceptions

It is worth mentioning that there is a common misconception where **heat transfer** Q and **work transfer** W are sometimes mistakenly referred to as state properties. This is incorrect, as neither Q nor W are properties of a system; they are forms of energy transfer that depend on the process or path taken during a state change. Unlike state variables, which are path-independent and determined solely by the current state of the system, Q and W cannot be defined by a state alone. Therefore, they are not represented as exact differentials (dQ and dW) but rather as inexact differentials, emphasizing their path-dependent nature. This distinction is crucial for understanding thermodynamic processes and properly analysing energy transfer within a system.

3.2 Thermodynamic processes

A **thermodynamic process** is defined as a path that connects two states of a system, tracing the sequence of changes that the system undergoes as it moves from an initial state to a final state. During this transition, the system may exchange energy with its surroundings in the form of heat (Q) or work (W). The specifics of how the system changes depend on the conditions imposed, such as temperature, pressure, and volume constraints. Each process follows a unique path on a thermodynamic diagram (e.g. p–V or T–S), indicating how state variables evolve over time. The total energy change in a process depends on both the starting and ending states; however, the way in which energy is transferred—whether through heat, work, or a combination of both—depends on the process path.

It is important to note that thermodynamic processes are not limited to hydraulic or gas systems. In other domains, such as electric or magnetic fields, additional variables such as electric potential and magnetic flux can also define the state and

processes of a system, illustrating the broad applicability of thermodynamic principles across various fields of physics and engineering.

3.2.1 Classification of thermodynamic processes

Thermodynamic processes can be classified based on how the system transitions between states, as illustrated in the schematic. Broadly, processes can be categorized as quasistatic or non-quasistatic (figure 3.1). **Quasistatic processes** are those in which the system changes state at a rate much slower than the speed of sound within the system, allowing the system to remain infinitesimally close to equilibrium at each step. This condition ensures that the state variables can be defined consistently throughout the process. These processes can further be divided into reversible and irreversible types. A **reversible process** is an idealized concept where the system and surroundings can be returned to their original states without any net change. **Irreversible processes**, on the other hand, involve inherent dissipative effects such as friction, unrestrained expansion, or heat transfer across a finite temperature difference, making them impossible to reverse without external intervention.

Non-quasistatic processes occur when the system changes at a rate comparable to or faster than the speed of sound, preventing it from maintaining equilibrium during the transition. As a result, significant deviations from a stable state occur throughout the process.

Additionally, while the schematic highlights key classifications, it is important to note that processes can also be categorized as **spontaneous** or **non-spontaneous**. These types are based on the natural tendency of a system to move towards a specific state without external influence. Although spontaneous and non-spontaneous processes are not shown in the current schematic, they are important classifications that will be discussed in greater detail later in the book.

3.2.2 Examples of quasistatic reversible process

1. *Adiabatic expansion of an ideal gas*: A gas confined in a cylinder with a perfectly insulating piston undergoes a slow expansion. The system performs

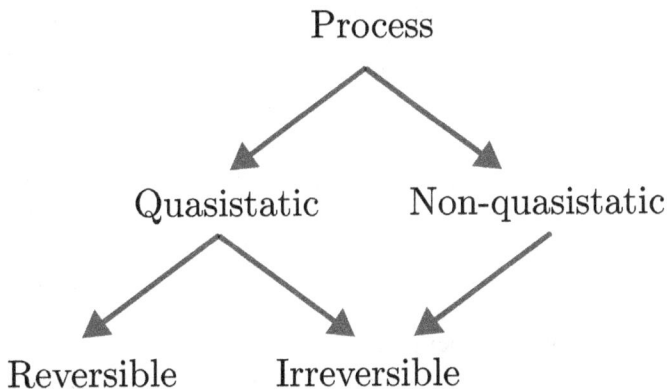

Process

Quasistatic   Non-quasistatic

Reversible   Irreversible

Figure 3.1. Classification of thermodynamic processes.

work on the surroundings without any heat exchange, maintaining a quasistatic and reversible nature due to the controlled, equilibrium-preserving rate of expansion.

2. *Charging of a capacitor*: In an idealized scenario where there is no electrical resistance in the electrical leads, a capacitor connected to a voltage source undergoes a quasistatic, reversible process. The absence of resistance ensures that there is no energy dissipation, allowing the system to stay in a state of electrical equilibrium throughout the charging process.

3. *Compression of a mechanical spring*: An ideal spring being compressed very gradually in a frictionless environment exhibits a quasistatic reversible process. Each incremental compression maintains mechanical equilibrium, and the energy stored in the spring can be fully recovered without loss when the process is reversed.

3.2.3 Examples of quasistatic irreversible processes

1. *Heat transfer across a finite temperature difference*: Consider a system where heat is transferred between a warm body and a cooler body at a controlled, slow rate. Even though the process occurs quasistatically, the finite temperature difference between the system and surroundings means that energy transfer is not perfectly efficient, making the process irreversible.

2. *Isothermal expansion of a gas*: A gas in a cylinder with a piston undergoes an isothermal expansion where it stays at a constant temperature. During this slow, controlled expansion, the system absorbs heat from the surroundings. However, due to the finite temperature difference between the system and the surroundings, the process cannot be perfectly reversed, making it irreversible despite being quasistatic.

3. *Compression of a mechanical lossy spring*: A spring being compressed in a mechanical system with slight internal damping or energy dissipation (such as material hysteresis) represents a quasistatic irreversible process. The slow compression maintains near-equilibrium conditions, but the inherent energy loss within the spring means the process cannot be perfectly reversed.

3.2.4 Examples of non-quasistatic processes

1. *Sudden expansion of gas*: When a gas is released from a high-pressure container into a vacuum or a region of much lower pressure, the process occurs rapidly. This sudden expansion does not allow the system to maintain equilibrium as it changes states, resulting in significant deviations in temperature, pressure, and density throughout the gas. The process is non-quasistatic because the system variables are not uniform during the transition.

2. *Sudden discharge of a capacitor*: When a charged capacitor is suddenly connected across a low-resistance circuit, it discharges almost instantaneously. The rapid flow of current leads to abrupt changes in voltage and current, preventing the system from maintaining equilibrium throughout the

process. The rate of change is so fast that the system's state variables (e.g. electric potential) do not adjust uniformly during the discharge, making it a non-quasistatic process.

3. *Mechanical impact or sudden compression*: When a force is suddenly applied to a mechanical system, such as a hammer hitting a metal surface or a piston being rapidly pushed into a cylinder, the system undergoes abrupt changes in pressure and deformation. The speed of these changes exceeds the rate at which the system can internally equilibrate, resulting in a non-quasistatic process where mechanical and thermodynamic state variables fluctuate unevenly.

3.3 Special cases of processes

In the study of thermodynamics, certain processes are defined by imposing specific constraints on how a system transitions between states. These processes, while idealized, are essential for developing fundamental thermodynamic concepts and analysing the behavior of systems under controlled conditions. Understanding these special cases provides a basis for more complex analyses and helps in recognizing how real systems may approach or deviate from these idealized processes. Below is a summary of common processes and their defining constraints:

- **Isobaric process**: The process occurs at **constant pressure** throughout, often used to study how systems behave when pressure remains unchanged during energy exchanges.
- **Isothermal process**: The process occurs at **constant temperature**, requiring heat transfer to ensure the system temperature does not change despite work being performed.
- **Isochoric process**: The process occurs at **constant volume**, implying that no expansion/compression work is done since the volume of the system does not change.
- **Adiabatic process**: The process involves **no heat transfer** between the system and its surroundings. Energy changes occur solely through work, causing variations in temperature and other state variables.
- **Isentropic process**: The process occurs at **constant entropy**, typically idealized as a reversible adiabatic process where no entropy is generated, allowing for analysis of perfectly efficient systems.

These processes represent idealized models that help establish fundamental thermodynamic principles. However, real systems often exhibit non-ideal behaviors such as friction, heat loss, or imperfect insulation, which can affect the outcomes of these processes. Although we start by studying these idealized processes to build foundational knowledge, later in the book, we will show strategies for analysing real systems that adhere to these constraints but include non-ideal effects. This approach helps bridge the gap between theoretical concepts and practical applications, providing a more complete understanding of how to handle real-world thermodynamic analyses.

3.4 Quasistatic process with a solid

Consider a 100 kg solid with a specific heat capacity $c_s = 0.1$ kJ (kg K)$^{-1}$ initially at a temperature $T_s(t_0) = 500$ K, placed into a liquid reservoir maintained at a constant temperature $T_r = 300$ K. The liquid is assumed to act as an infinite reservoir, implying that its heat capacity is so large that its temperature remains effectively constant throughout the process (figure 3.2).

3.4.1 Example 1: Analysis without considering density change

3.4.1.1 Application of the first law of thermodynamics

To analyse the cooling process of the warm solid when it is immersed in the liquid reservoir, we apply the first law of thermodynamics for a closed system. The first law states that the change in internal energy of a system is equal to the net heat added to the system plus the work done on the system by its surroundings:

$$\Delta U = Q + W.$$

For this initial analysis, we assume there is no significant density change in the solid as it cools. This means any work due to volume change ($W = -p\Delta V$) is negligible, simplifying the equation to

$$\Delta U = Q.$$

3.4.1.2 Analysis of the process
- *Initial condition*:

  ○ The solid has an initial temperature $T_s(t_0) = 500$ K.
  ○ The liquid reservoir, which acts as a constant-temperature heat source, is at $T_r = 300$ K.

- *Assumptions*:

  ○ The process is quasistatic, meaning the heat transfer between the solid and the reservoir is slow enough for the system to remain near thermal equilibrium throughout the process.

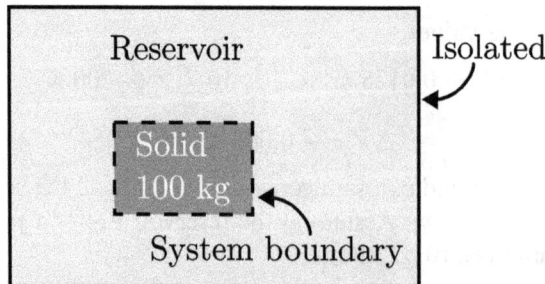

Figure 3.2. Schematic of a hot solid in thermal contact with a reservoir of liquid.

o The reservoir has a very large heat capacity, so its temperature remains constant at 300 K during the process.
o Density changes in the solid with temperature are neglected in this analysis.
o The only form of energy transfer involved is heat exchange (no work interaction).

- *Internal energy change*: The change in internal energy of the solid can be expressed as
  o $\Delta U = mc_v \Delta T$.
  o Substituting the given values
  o $\Delta U = 100 \text{ kg} \times 0.1 \text{ kJ kg K}^{-1} \times (-200 \text{ K}) = -2000 \text{ kJ}$.
- *Heat transfer*: The heat Q transferred out of the solid is equal to the decrease in its internal energy,

$$Q = \Delta U = -2000 \text{ kJ}.$$

The negative value indicates heat flows out of the solid and into the reservoir.

3.4.2 Example 2: Analysis including work due to density change

When the density of the solid changes with temperature, we must consider the work done by the system as it expands or contracts. In this case, the first law of thermodynamics for the solid is

$$\Delta U = Q + W.$$

3.4.2.1 Analysis of the process

- *Initial volume calculation*: To calculate ΔV, we need the initial volume V_0 of the solid. Assuming the density (ρ) of the solid is approximately 8000 kg m^{-3} (a typical value for metals), we can find V_0 as

$$V_0 = \frac{m}{\rho} = \frac{100 \text{ kg}}{8000 \text{ kg m}^{-3}} = 0.0125 \text{ m}^3$$

- *Change in volume*: Using the coefficient of thermal expansion (α),

$$\Delta V = V_0 \alpha \Delta T.$$

Substituting the values,

$$\Delta V = 0.0125 \text{ m}^3 \times 2 \times 10^{-5} \times (-200 \text{ K})$$

$$\Delta V = -0.000\,05 \text{ m}^3.$$

The negative sign indicates a decrease in volume as the solid cools.
- *Work done calculation*: Assuming the reservoir exerts a pressure p of 1 atm (or approximately 101.3 kPa),

$$W = -p\Delta V = -(101.3 \text{ kPa}) \times (-0.000\,05 \text{ m}^3)$$

$$W \approx 0.0051 \text{ kJ}.$$

- *Heat transfer calculation*: Using the first law of thermodynamics,
$$Q = \Delta U - W.$$
Substituting the values,
$$Q = -2000 \text{ kJ} - 0.0051 \text{ kJ}$$

$$Q \approx -2000.0051 \text{ kJ}.$$

The result shows that the heat transferred out of the solid is approximately -2000.0051 kJ. The work term is relatively small compared to the total energy change, but it is included to reflect the slight mechanical energy interaction due to thermal contraction.

Conclusion: In this example, the effect of work due to volume change is minor for the solid, as it undergoes only a small change in volume during cooling. However, including the work term illustrates how the first law of thermodynamics can account for such interactions, providing a more comprehensive analysis. For gases, the work term is typically much larger, as gases expand or contract significantly with temperature changes compared to solids and liquids. This approach becomes especially important for analysing gases or other materials with substantial thermal expansion, as well as for systems under higher pressures where volume changes contribute more prominently to the overall energy exchange.

3.5 Quasistatic process with an ideal gas

In this section, we focus on analysing quasistatic processes involving an ideal gas. These processes are simplified by assuming no additional forms of work, such as electrical or magnetic work, other than pV work. Additionally, we consider idealized conditions where friction is absent, allowing the gas to move smoothly and maintain near-equilibrium at each stage. It is important to note that if heat transfer occurs in any quasistatic process, the process is inherently **irreversible** due to the temperature gradient between the system and its surroundings. The only quasistatic process that remains **reversible** in this context is the **adiabatic process**, where no heat transfer takes place, and the system changes solely through pV work. This is analogous to a **mechanical spring** that compresses and expands without energy loss, provided no other dissipative effects are present. Keeping these assumptions in mind will help clarify the step-by-step analysis presented in this section.

3.6 Isobaric process

In an **isobaric process**, the pressure of an ideal gas remains constant as it undergoes changes in state. Imagine an ideal gas confined in a cylinder with a movable, **frictionless** piston. The pressure on the gas remains constant due to the combined effect of the

atmospheric pressure and the weight of the piston. When heat is transferred to the gas, it expands, pushing the piston upward and increasing its volume.

According to the **first law of thermodynamics**,

$$\Delta U = Q + W.$$

In this case, since the gas expands and does work on the surroundings by moving the piston, the work done by the gas can be expressed as

$$W = -p\Delta V,$$

where p is the constant gas pressure, and ΔV is the change in volume of the gas.

Rearranging the first law, we obtain

$$Q = \Delta U - W.$$

3.6.1 Step-by-step analysis

1. *Work done* (W): Since the pressure is constant, the work done by the gas as it expands from an initial volume V_1 to a final volume V_n is

$$W = -p(V_n - V_1).$$

2. *Internal energy change* (ΔU): For an ideal gas, the change in internal energy depends only on the temperature change and is given by

$$\Delta U = nc_v\Delta T,$$

where:
   - n is the number of moles of gas,
   - c_v is the specific heat at constant volume (J mol^{-1} K^{-1}), and
   - $\Delta T = T_n - T_1$ is the temperature difference between the final and initial states.

3. *Substituting in the first law*: Plugging in the expressions for ΔU and W into the first law:

$$Q = \Delta U - W$$

$$Q = nc_v\Delta T + p\Delta V.$$

4. *Relating $p\Delta V$ to temperature change*: From the ideal gas law, $pV = nRT$, so we can express $p\Delta V$ as

$$p\Delta V = nR\Delta T.$$

5. *Final expression for heat transfer*: Substituting $p\Delta V = nR\Delta T$ back into the equation for Q,

$$Q = nc_v\Delta T + nR\Delta T.$$

Factoring out $n\Delta T$,

$$Q = n(c_v + nR)\Delta T.$$

6. *Defining c_p*: The term $c_p = c_v + R$ is known as the specific heat at constant pressure, so we can write

$$Q = nc_p\Delta T.$$

Thus, the heat transfer in an isobaric process is given by

$$Q = nc_p\Delta T.$$

This formula accounts for both the internal energy change and the work done by the gas as it expands. The specific heat at constant pressure, c_p, includes the extra energy required to perform this work, making it larger than the specific heat at constant volume, c_v.

3.7 Isothermal process

In an **isothermal process**, the temperature of the ideal gas remains constant as the system undergoes changes in state. For an ideal gas, internal energy U depends only on temperature, $U = \frac{f}{2}nRT$. Since the temperature is constant in an isothermal process, there is no change in internal energy, so $\Delta U = 0$.

For an ideal gas at a constant temperature, the first law of thermodynamics is

$$\Delta U = Q + W.$$

Since $\Delta U = 0$ in an isothermal process, this simplifies to

$$Q = -W.$$

This tells us that the heat added to the system is exactly equal (in magnitude) and opposite (in sign) to the work done by the gas.

3.7.1 Step-by-step analysis

1. *Starting with the ideal gas law*: For an ideal gas, the relationship between pressure, volume, and temperature is given by the ideal gas law,

$$pV = nRT.$$

Since the temperature T is constant in an isothermal process, the product pV also remains constant.
2. *Calculating pV work (W)*: As the gas expands or contracts, the pressure and volume change, but the temperature and the product pV remain constant. The work done by the gas when it expands or contracts from an initial volume V_1 to a final volume V_n is given by

$$W = \int_{V_1}^{V_n} -pdV.$$

Using the ideal gas law to express p in terms of V,

$$p = \frac{nRT}{V}.$$

Substituting $p = \frac{nRT}{V}$ into the work integral,

$$W = \int_{V_1}^{V_n} -\frac{nRT}{V} dV.$$

3. *Evaluating the integral*: Since n, R, and T are constants in this process, we can pull them out of the integral

$$W = -nRT \int_{V_1}^{V_n} -\frac{1}{V} dV.$$

The integral of $\frac{1}{V}$ with respect to V is $\ln V$, so

$$W = -nRT(\ln V_n - \ln V_1).$$

Using the logarithmic identity $\ln a - \ln b = \ln \frac{a}{b}$, we obtain

$$W = -nRT \ln \frac{V_n}{V_1}.$$

Therefore, the pV work done by the gas in an isothermal process is

$$W = -nRT \ln \frac{V_n}{V_1}.$$

4. *Calculating heat transfer Q*: From the first law of thermodynamics, we have

$$Q = \Delta U - W.$$

Since $\Delta U = 0$ in an isothermal process, we find

$$Q = -W.$$

Substituting for W

$$Q = nRT \ln \frac{V_n}{V_1}.$$

5. *Final expressions*: For an isothermal process, the heat transferred Q and the work done W are given by

$$W = -nRT \ln \frac{V_n}{V_1}$$

$$Q = nRT \ln \frac{V_n}{V_1}.$$

The positive sign for Q indicates that heat flows into the gas if it is expanding (since $V_n > V_1$), whereas W is negative, meaning the gas does work on the surroundings.

3.8 Isochoric process

In an **isochoric process**, the volume of the ideal gas remains constant as it undergoes changes in state. Since the volume does not change, the gas does no work on the surroundings (and no work is done on the gas) because $W = -p\Delta V$ and $\Delta V = 0$.
According to the **first law of thermodynamics**:

$$\Delta U = Q + W.$$

In this case, since $W = 0$ in an isochoric process, the first law simplifies to

$$\Delta U = Q.$$

This means that any heat transferred to or from the gas goes entirely into changing its internal energy.

3.8.1 Step-by-step analysis

1. *Work done* (W): Since the volume is constant ($V = V_1 = V_n$), the work done by the gas in an isochoric process is zero:

$$W = \int_{V_1}^{V_n} - p\, dV = -p. \ 0 = 0.$$

2. *Internal energy change* (ΔU): For an ideal gas, the internal energy U depends only on temperature. The change in internal energy ΔU as the temperature changes from an initial temperature T_1 to a final temperature T_n is

$$\Delta U = U_n - U_1 = \frac{f}{2}nR(T_n - T_1).$$

This can also be expressed using the specific heat at constant volume $c_v = \frac{f}{2}R$,

$$\Delta U = nc_v\Delta T,$$

where $\Delta T = T_n - T_1$ is the change in temperature.

3. *Calculating heat transfer* (Q): Using the first law of thermodynamics, we have

$$Q = \Delta U - W.$$

Since $W = 0$, we obtain

$$Q = \Delta U.$$

Substituting the expression for ΔU,

$$Q = nc_v\Delta T.$$

4. *Final expressions*: In summary, for an isochoric process, the heat transfer Q and the internal energy change ΔU are given by

$$Q = nc_v \Delta T$$

$$\Delta U = nc_v \Delta T.$$

Since the work done $W = 0$, all the energy transferred as heat is used to change the internal energy of the gas.

3.9 Adiabatic process

In an **adiabatic process**, there is no heat transfer between the system and its surroundings, meaning $Q = 0$. Any energy change in the system results entirely from work done on or by the gas.

According to the first law of thermodynamics,

$$\Delta U = Q + W.$$

Since $Q = 0$ in an adiabatic process, this simplifies to

$$\Delta U = W.$$

This implies that the change in internal energy of the gas is equal to the work done by or on the system.

3.9.1 Step-by-step analysis

1. *Starting with the ideal gas law*: For an ideal gas, the relationship between pressure, volume, and temperature is given by the ideal gas law,

$$pV = nRT.$$

2. *Internal energy change (ΔU)*: For an ideal gas, the internal energy U depends only on temperature. The change in internal energy ΔU as the temperature changes from an initial temperature T_1 to a final temperature T_n is

$$\Delta U = nc_v \Delta T,$$

where $c_v = \frac{f}{2}R$ is the specific heat at constant volume, and $\Delta T = T_n - T_1$ is the change in temperature.

3. *Work done (W)*: Since $Q = 0$, all the change in internal energy comes from the work done. Thus

$$W = \Delta U = nc_v \Delta T.$$

4. *Relationship between p and V*: For an adiabatic process in an ideal gas, there is a specific relationship between pressure and volume that can be derived

using the first law. Using the ideal gas law $pV = nRT$ and the relationship $dU = -pdV$, we find that

$$pV^\gamma = \text{constant},$$

where $\gamma = \frac{c_p}{c_v} = \frac{f+2}{f}$ is the heat capacity ratio, often referred to as the adiabatic index.

5. *Calculating pV work*: The work done in an adiabatic process can also be derived using the pV^γ relationship. Integrating W over the process from an initial volume V_1 to a final volume V_n,

$$W = \int_{V_1}^{V_n} -pdV.$$

Using $pV^\gamma = \text{constant}$, we can substitute for p,

$$W = \int_{V_1}^{V_n} -\frac{C}{V^\gamma}dV = -\frac{C}{1-\gamma}(V_n^{1-\gamma} - V_1^{1-\gamma}).$$

Here, $C = p_1 V_1^\gamma = p_n V_n^\gamma$.

6. *Summary of results*: For an adiabatic process:
   - The change in internal energy $\Delta U = nc_v\Delta T$.
   - The work done $W = \Delta U = nc_v\Delta T$.
   - The relationship between pressure and volume is $pV^\gamma = \text{constant}$.

3.10 Arbitrary quasistatic process

An **arbitrary quasistatic process** is one in which an ideal gas undergoes a transition between two states, following no specific constraint on pressure, volume, or temperature. Unlike isothermal, isobaric, isochoric, or adiabatic processes, an arbitrary process does not adhere to a predefined path. Instead, the gas changes state in a general manner, with its pressure and volume varying according to an unspecified functional relationship. By applying the first law of thermodynamics, we can calculate the changes in internal energy, work, and heat transfer for this process, regardless of the exact path taken. This approach allows for a flexible analysis applicable to any quasistatic transition from an initial to a final state, as long as we have sufficient information about the temperature change and the p–V relationship throughout the process.

3.10.1 Step-by-step analysis

1. *Change in internal energy* (ΔU): For an ideal gas, the internal energy U depends only on the temperature T. The change in internal energy between the initial state (1) and the final state (n) is given by

$$\Delta U = nc_v\Delta T.$$

2. *Work done* (W): In a quasistatic process, the work done by the gas as it expands or contracts from volume V_1 to V_n is given by the area under the curve in a p–V diagram. The pV work is defined as

$$W = \int_{V_1}^{V_n} -p\,dV.$$

To evaluate this integral, we need to know the functional relationship between pressure p and volume V throughout the process. If an explicit expression for p as a function of V is provided (e.g. $p = f(V)$), we can integrate directly. For example:
  - **Linear relationship**: If p varies linearly with V, such as $p = aV + b$, we substitute and integrate.
  - **Other relationships**: For any other relationship, we follow a similar approach, using the given expression for p in terms of V and integrating over the specified limits.

3. *Heat transfer* (Q): Using the first law of thermodynamics, we can rearrange this to solve for heat transfer Q:

$$Q = \Delta U - W.$$

Substituting the expressions we have for ΔU and W:

$$Q = nc_v\Delta T - \int_{V_1}^{V_n} p\,dV$$

This expression gives the amount of heat transferred during the arbitrary process, accounting for both the change in internal energy and the work done by or on the gas.

3.11 Summary

This chapter provides students with a structured understanding of thermodynamic processes in a closed system, focusing on the following key points:
  1. **State variables**: We establish the importance of state variables—properties that depend solely on the system's current state and not on how it reached that state. These variables, such as pressure, temperature, and internal energy, form the basis for analysing changes within a closed system.

  2. **Thermodynamic processes**: We define a thermodynamic process as the path taken between two states of a system, highlighting how energy transfer in the form of heat and work varies depending on the process. We examine different types of processes, from simple transitions at constant pressure or temperature to adiabatic and isentropic processes.

3. **Classification of processes**: Processes are categorized into **quasistatic** and **non-quasistatic** types, with quasistatic processes further divided into **reversible** and **irreversible**. We discuss how reversible processes are idealized, with the ability to return both the system and its surroundings to their original states, while irreversible processes involve real-world dissipative effects.

4. **Quasistatic processes with an ideal gas**: Detailed step-by-step analyses are provided for isobaric, isothermal, isochoric, adiabatic, and arbitrary processes, assuming an ideal gas without friction or additional forms of work. We clarify that only the quasistatic adiabatic process can be fully reversible under ideal conditions.

5. **Special process constraints**: We explore the significance of idealized constraints—constant pressure, temperature, volume, and entropy—in defining specific thermodynamic processes, and we caution readers about the limitations of these idealized cases in practical applications.

6. **Energy transfers and the first law**: Using the first law of thermodynamics, we derive expressions for heat transfer, work done, and changes in internal energy for each type of quasistatic process. The chapter emphasizes that in ideal gases, the internal energy change depends solely on temperature, making the analyses applicable across different processes.

By the end of this chapter, students should have a solid grasp of various thermodynamic processes and the tools needed to analyse energy exchanges within closed systems. This foundation prepares them to extend their understanding to the more complex, real-world thermodynamic applications introduced in later chapters.

IOP Publishing

A Classical Thermodynamics Toolkit

Srinivas Vanapalli

Chapter 4

Entropy

In this chapter, we delve into the concept of entropy, a fundamental quantity in thermodynamics that describes the disorder or randomness within a system. We start by exploring the statistical foundation of entropy, viewing it as a measure of possible configurations (microstates) of a system that contribute to an observable state (macrostate). This statistical perspective not only demystifies entropy but also lays the groundwork for understanding it as a state property that only depends on the current condition of a system, independent of its history.

From this foundation, we proceed to define entropy changes under various thermodynamic processes—such as isobaric, isochoric, isothermal, and adiabatic—and examine how each type of process uniquely impacts entropy. By calculating entropy changes in specific, controlled set-ups, we aim to clarify the relationships among heat, work, and entropy in idealized systems. The chapter also discusses arbitrary processes and the conditions required for reversibility and irreversibility, highlighting how entropy dictates the direction and feasibility of natural processes.

Finally, we address the often-misunderstood notion of irreversibility, emphasizing that for a process to be irreversible, the *total* entropy change—accounting for both the system and surroundings—must be positive. This comprehensive view brings us closer to the second law of thermodynamics, which governs the behavior of entropy in all processes.

4.1 Statistical definition of entropy

To understand entropy in statistical terms, let us begin with a simple analogy that introduces the concepts of the microstates and macrostates of a system. Imagine a railway car with three wagons ($N = 3$) transporting a total of three sugar beets ($q = 3$).

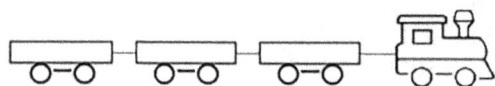

In this scenario, the beets can be arranged in the three wagons in multiple ways. Each unique arrangement of beets in the wagons is a **microstate** (Ω), and each configuration has an equal probability of occurring. The **macrostate** of the system, on the other hand, is defined by the overall distribution of beets across the wagons—for example, three beets in one wagon, two beets in one wagon and one in another, or one beet in each wagon.

Table 4.1 shows all possible arrangements of the beets among the wagons:
- Three beets in a single wagon—three possible microstates.
- Two beets in one wagon, one in another—six possible microstates.
- One beet in each wagon—one possible microstate.

Thus, we observe three macrostates with different probabilities, as the number of microstates varies for each one. Although each microstate is equally likely, the likelihood of observing a specific macrostate depends on the number of microstates that correspond to it.

For a system with N wagons and q beets, the total number of possible microstates is given by

$$\Omega = \frac{(q + N - 1)!}{q!(N - 1)!}.$$

Table 4.1. All possible arrangements of the beets among the wagons.

	Wagon #1	Wagon #2	Wagon #3
	3	0	0
3 beets in one wagon	0	3	0
	0	0	3
	2	1	0
	2	0	1
2 beets in one wagon	1	2	0
	1	0	2
	0	2	1
	0	1	2
1 beet in one wagon	1	1	1

N_{osc}	q	multiplicity Ω
3	6	28
10	50	6.3×10^{10}
100	1000	10^{144}
10^6	10^7	$10^{1.5 \cdot 10^6}$
$6 \cdot 10^{23}$	10^{26}	$10^{1.6 \cdot 10^{24}}$

Figure 4.1. An example of the astronomical number of microstates in real materials.

Substituting $q = 3$, we calculate $\Omega = 10$, meaning there are ten distinct microstates for this arrangement. Extending this analogy, we can imagine each beet as a unit of energy (or energy quantum) and each wagon as an oscillator within a solid. In real materials, the numbers of oscillators and energy quanta are typically vast, leading to astronomical numbers of microstates, as illustrated in figure 4.1.

4.1.1 Key points from this example

This analogy highlights two crucial aspects of statistical entropy:
1. A system can be described by a large number of equally probable microstates.
2. Observable properties (e.g. counting beets in each wagon) represent macrostates, which differ in likelihood due to varying numbers of corresponding microstates.

4.2 Defining entropy

Entropy (S) is defined as

$$S = k \ln \Omega,$$

where k is the Boltzmann constant, and Ω is the number of microstates. The logarithmic function is used due to the typically immense size of Ω, and the Boltzmann constant serves to scale the units of entropy to joules per kelvin (J K^{-1}).

This statistical view of entropy reveals it as a measure of a system's multiplicity or the number of ways energy can be distributed among particles, reflecting the underlying probability of various macrostates.

4.3 Temperature as a measure of entropy change

Temperature in thermodynamics can be understood through the relationship between **internal energy** and **entropy** when two systems (or bodies) at different initial temperatures come into thermal contact.

Consider two bodies, A and B, each with their own internal energies, U_A and U_B, and entropies, S_A and S_B. When these bodies are brought into thermal contact, energy flows from the body with higher temperature to the one with lower

temperature until thermal equilibrium is reached. At equilibrium, no further net energy transfer occurs, and both bodies are at the same temperature.

During this process, the **total entropy** of the combined system, $S_{\text{total}} = S_A + S_B$, changes. Since entropy tends to increase until equilibrium, the process of energy exchange between the two bodies is governed by the maximization of total entropy. The amount of energy exchanged depends on how much each body's entropy changes with a small change in its internal energy.

The change in entropy with respect to internal energy for a body i is given by the derivative $\partial S_i / \partial U_i$. This quantity is fundamental to the definition of temperature. For each body, we define **temperature** as

$$\frac{1}{T_i} = \frac{\partial S_i}{\partial U_i},$$

where:
- T_i is the absolute temperature of body i,
- $\partial S_i / \partial U_i$ is the rate of change of entropy with respect to internal energy.

This relationship means that temperature is a measure of how a body's entropy responds to changes in its internal energy. A system with a high temperature has a relatively small increase in entropy for a given increase in internal energy, while a low-temperature system has a larger entropy change for the same energy addition.

4.4 Interaction and equilibrium

When two bodies A and B interact, energy flows between them in such a way as to increase the total entropy S_{total} until equilibrium is reached. At equilibrium,

$$\frac{\partial S_A}{\partial U_A} = \frac{\partial S_B}{\partial U_B}.$$

By substituting $\frac{1}{T} = \frac{\partial S}{\partial U}$, we find that

$$\frac{1}{T_A} = \frac{1}{T_B}$$

or equivalently, $T_A = T_B$. This condition defines thermal equilibrium: two bodies have the same temperature when they no longer exchange energy and their entropies change in sync with each other's internal energy.

Thus, **temperature** can be defined as the parameter that equalizes the entropy changes per unit of energy between interacting bodies at equilibrium, reflecting a balance in the tendency to disperse energy between them. This statistical view of temperature highlights its role as the factor that regulates entropy exchange in response to energy changes.

4.5 Mechanical equilibrium and the definition of pressure

In thermodynamics, equilibrium is central to defining the measurable properties of a system. In the previous section, we introduced thermal equilibrium as the basis for defining temperature. Now, we extend this concept to mechanical equilibrium, which provides the foundation for defining pressure in a similar way.

4.5.1 Understanding mechanical equilibrium

A system is in **mechanical equilibrium** when no net force acts to change its volume. This means that, under mechanical equilibrium, any infinitesimal change in volume does not lead to a redistribution of energy within the system or between the system and its surroundings. When two systems are in mechanical contact, equilibrium is achieved when they exert equal pressures on each other, preventing spontaneous volume changes.

4.5.2 The definition of pressure

Analogous to the way temperature arises from thermal equilibrium, **pressure** is defined in terms of mechanical equilibrium. Pressure can be understood as a measure of how entropy S changes with volume V at a fixed internal energy U. We define pressure as

$$p = T\left(\frac{\partial S}{\partial V}\right)_U.$$

In this expression:
- p is the pressure of the system,
- T is the temperature, and
- $\left(\frac{\partial S}{\partial V}\right)_U$ represents the rate of entropy change with respect to volume at constant energy.

This relationship shows that pressure is a force per unit area exerted by the system due to its internal molecular motion and interactions. It implies that pressure is a measure of how entropy would increase if the volume were allowed to expand at constant energy.

4.5.3 Interdependence of temperature and pressure

Together, temperature and pressure are both fundamentally related to the system's entropy. Temperature reflects the entropy's sensitivity to changes in energy, while pressure reflects its sensitivity to changes in volume. This unified view of temperature and pressure through entropy provides a powerful framework for understanding equilibrium and thermodynamic processes.

4.6 Entropy is a state property

Entropy, like pressure p, temperature T, volume V, and internal energy U, is a **state property**, meaning it depends only on the current state of the system and not on the specific path taken to reach that state. For an idealized system, entropy S can be expressed as a function of internal energy U, volume V, and particle number N:

$$S = S(U, V, N).$$

This relationship highlights that entropy reflects the possible ways (microstates) in which energy can be distributed within the system at given values of U, V, and N. When any of these variables changes, the number of accessible microstates—and thus the entropy—increases.

4.6.1 Path-independence of entropy changes

Since entropy is a state property, the entropy S of a system depends solely on its current state and not on the process used to reach it. Consequently, the change in entropy ΔS between any two states depends only on the initial and final states, regardless of whether the process was reversible, irreversible, fast, or slow. This **path-independence** is a defining characteristic of state properties in thermodynamics.

Path-independence means that, for any transition between states, the entropy change ΔS remains the same regardless of the process path. Unlike process-dependent quantities such as work and heat, entropy change reflects only the start and end states, which makes it a reliable measure of a system's internal distribution of energy.

This independence from process details is a powerful concept in thermodynamics, making entropy a fundamental property that consistently describes energy dispersal within the system.

4.6.2 Entropy as a function of energy, volume, and particle number

Entropy's dependency on the state variables U, V, and N allows us to understand how changes in these quantities affect entropy.

1. **Energy** (U): Adding energy to a system generally increases entropy, as it allows the particles more possible configurations, increasing the system's **multiplicity**—the number of ways energy can be distributed.
2. **Volume** (V): Increasing the system's volume also raises entropy by allowing more spatial configurations. For example, in a gas, increasing volume gives particles more room to spread out, which expands the number of accessible microstates.

4.6.3 Example: Phases of water—ice, liquid, and vapor

Consider the entropy of 1 kg of water in different phases: ice, liquid, and vapor. For each phase, both the energy U and volume V differ, which directly affects entropy.

- **Ice** has the lowest entropy, as water molecules are tightly organized in a structured lattice, with few accessible microstates.

- **Liquid water** has a higher entropy than ice, as the molecular structure breaks down, allowing for more movement and more possible configurations.
- **Water vapor** has the highest entropy, with molecules moving freely in a large volume, maximizing the multiplicity.

4.6.4 Anomaly of water near freezing point

Water exhibits a unique behavior near 0 °C, where it expands slightly upon freezing. This expansion, caused by hydrogen bonding, increases the volume and slightly increases entropy, even though temperature is decreasing. This anomaly, due to the density difference between ice and liquid water, is crucial for environmental systems.

4.7 Energy and entropy

A fundamental distinction in thermodynamics is that **energy is conserved**, while **entropy is not**. This means that in any process, the total energy change of the Universe—the sum of energy changes in both the system and its surroundings—is always zero, adhering to the **first law of thermodynamics**. Energy can be transferred or transformed, but it cannot be created or destroyed. In contrast, entropy does not follow a conservation law. The **second law of thermodynamics** tells us that the total entropy change of the universe in any process is either zero or positive. This is because, for natural processes, there is an inherent tendency toward increased disorder or greater energy dispersal.

When a process is **reversible**, the entropy change of the Universe is zero; the system and surroundings can return to their initial states without any net change in entropy. However, in **irreversible** processes—which are more common in real-world conditions—entropy increases as energy spontaneously disperses and becomes less available for work. This distinction is crucial: energy is a conserved quantity, strictly balanced in all interactions, while entropy reflects the direction and feasibility of processes, increasing whenever the energy distribution becomes more randomized or less organized. Understanding this difference is essential for grasping why some processes are irreversible and for appreciating the role of entropy in defining the arrow of time.

4.8 Second law of thermodynamics (entropy change in a process)

The second law of thermodynamics describes the change in entropy of a substance during a process. It states that the change in entropy is equal to the sum of the ratio of heat transfer to the system's temperature and the irreversible entropy generated within the system. Mathematically, this is expressed as

$$\Delta S = \int \frac{Q}{T} + S_{\text{irr}},$$

where Q is the heat transfer, T is the temperature of the system, and S_{irr} represents the entropy generated due to irreversibilities in the process.

In differential form, this becomes

$$dS = \frac{Q}{T} + S_{irr}.$$

In this differential notation, Q represents the infinitesimal heat transfer at each stage of the process. The second law of thermodynamics is a powerful tool for analysing processes, setting thermodynamic limits on various energy conversion processes that we will encounter later in this book.

When the **irreversible entropy** S_{irr} is zero, meaning the process is reversible, we can calculate the entropy change of a substance by using the heat capacity, $c(T)$. For a substance with mass m, the entropy change ΔS is given by

$$\Delta S = \int \frac{Q}{T} = \int \frac{mc(T)dT}{T} = m \int \frac{c(T)dT}{T}.$$

Here, Q represents the heat transferred to the substance, and T is the temperature at each point in the process.

The **heat capacity** $c(T)$ varies depending on the conditions under which heat is added. At **constant volume**, the heat capacity c_v describes how much heat is required to raise the temperature of the substance without changing its volume. Conversely, at **constant pressure**, the heat capacity c_p defines the heat needed to raise the temperature without altering pressure. For ideal gases, c_p is typically greater than c_v because some energy goes into doing work against the surroundings as the gas expands.

This relationship allows us to compute the entropy change by integrating the heat capacity over the temperature range of interest, provided the process is reversible. It is a useful approach for processes where the temperature-dependent behavior of heat capacity, $c(T)$, is known.

4.9 Cooling of a warm solid in a liquid reservoir

Consider a solid with a mass $m_s = 100$ kg and a specific heat capacity $c = 0.1$ kJ kg^{-1} K^{-1} initially at a temperature $T_s(t_0) = 500$ K. This solid is immersed in a large liquid reservoir at a constant temperature $T_r = 300$ K. We assume that the liquid reservoir has an extremely large heat capacity, allowing us to treat its temperature as constant throughout the process.

In this scenario, the solid will transfer heat to the reservoir as it cools down to the reservoir's temperature T_r. This process, where energy flows from a higher temperature to a lower temperature, is inherently irreversible.

Step 1. Heat transfer from the solid to the reservoir

The total heat Q_s transferred from the solid to the reservoir as it cools from $T_s(t_0)$ to T_r can be calculated by

$$Q_s = m_s c(T_r - T_s(t_0)).$$

Substituting the values,

$$Q_s = 100 \text{ kg} \times 0.1 \text{ kJ kg}^{-1} \text{K}^{-1} \times (300 \text{ K} - 500 \text{ K})$$

$$Q_s = -2000 \text{ kJ}.$$

The negative sign indicates that heat is leaving the solid.

Step 2. Entropy change of the solid (ΔS_s)

The entropy change of the solid as it cools from $T_s(t_0)$ to T_r can be calculated by integrating the infinitesimal entropy changes $dS = \frac{Q}{T}$:

$$\Delta S_s = \int_{T_s(t_0)}^{T_r} \frac{m_s c_s \, dT}{T} = m_s c_s \ln \frac{T_r}{T_s(t_0)}.$$

Substituting in the values,

$$\Delta S_s = 100 \times 0.1 \times \ln \frac{300}{500}$$

$$\Delta S_s = -5.108 \text{ kJ K}^{-1}.$$

The negative entropy change reflects the decrease in entropy of the solid as it loses heat.

Step 3. Entropy change of the reservoir (ΔS_r)

Since the liquid reservoir's temperature remains constant at $T_r = 300$ K, its entropy change can be calculated using

$$\Delta S_r = \frac{Q_r}{T_r} = \frac{-Q_s}{T_r}.$$

Substituting in $Q_s = -2000$ kJ and $T_r = 300$ K,

$$\Delta S_r = \frac{2000}{300} = 6.667 \text{ kJ K}^{-1}.$$

The positive sign indicates an increase in entropy of the reservoir as it absorbs heat from the solid.

Step 4. Total entropy change (ΔS_{total})

The total entropy change of the universe for this process is the sum of the entropy changes of the solid and the reservoir:

$$\Delta S_{\text{total}} = \Delta S_s + \Delta S_r.$$

Substituting the values,

$$\Delta S_{\text{total}} = -5.108 + 6.667 = 1.559 \text{ kJ K}^{-1}.$$

Since ΔS_{total} is positive, the process is irreversible. This increase in total entropy confirms that the spontaneous flow of heat from the warmer solid to the cooler reservoir cannot naturally reverse.

Key insights:

1. **Conservation of energy**: While energy is conserved (the heat lost by the solid equals the heat gained by the reservoir), entropy behaves differently. The total entropy of the universe increases in this process, indicating irreversibility.
2. **Irreversibility**: The positive entropy change for the universe signifies that this is an irreversible process, as entropy has been generated in the transfer of heat from a higher to a lower temperature.

3. **Temperature gradient and irreversibility**: The irreversibility arises due to the finite temperature difference between the solid and the reservoir. If the solid and the reservoir had the same temperature initially, there would be no heat transfer, and the entropy change would be zero.

This example illustrates a crucial concept in thermodynamics: although energy is conserved in every process, entropy is not. The second law of thermodynamics states that in an irreversible process, the total entropy of the universe always increases, pointing to the fundamental direction of spontaneous processes.

4.10 Entropy production due to heat diffusion in a bar

In this example (figure 4.2), we have a solid bar connecting two thermal reservoirs at different temperatures, with one end at a high temperature $T_H = 400$ K and the other end at a low temperature $T_c = 300$ K. Heat flows through the bar from the hot reservoir to the cold reservoir, maintaining a constant temperature difference. We will calculate the total entropy generation during this process, including the entropy changes of the system (the bar) and the surroundings (the reservoirs).
Assumptions:
1. *Steady-state heat transfer*: The bar operates at steady state, meaning the heat transfer rate $\dot{Q}$ through the bar is constant.
2. *No change in system entropy*: The entropy of the bar itself does not change over time because it is in a steady state, with energy entering and leaving at a constant rate and no accumulation within the bar.

4.10.1 Step-by-step analysis

Step 1. Calculate the rate of heat transfer $\dot{Q}$ using Fourier's law
From Fourier's law of heat conduction, the rate of heat transfer $\dot{Q}$ through the bar is given by

$$\dot{Q} = -\frac{kA}{L}\Delta T,$$

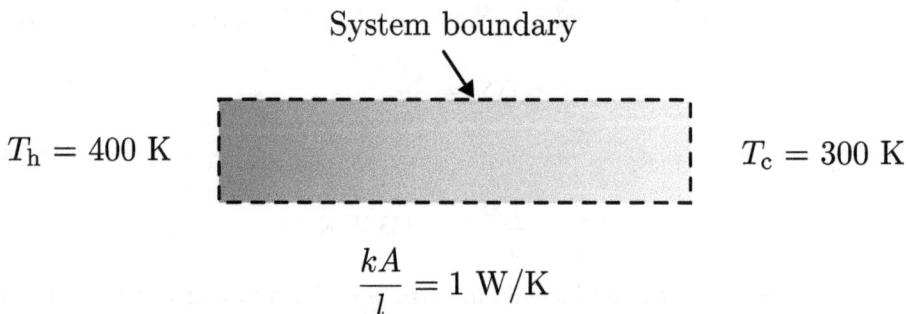

System boundary

$T_h = 400$ K $\qquad\qquad$ $T_c = 300$ K

$$\frac{kA}{l} = 1 \text{ W/K}$$

Figure 4.2. A solid bar connecting two thermal reservoirs at different temperatures.

where:
- k is the thermal conductivity of the bar,
- A is the cross-sectional area,
- L is the length of the bar, and
- $\Delta T = T_C - T_H$ is the temperature difference between the ends of the bar.

Given that $\frac{kA}{L} = 1$ (a simplified unit for this problem) and $\Delta T = -100$ K (since heat flows from hot to cold), we find

$$\dot{Q} = -1 \times (-100) = 100 \text{ W}.$$

This means that heat is transferred at a constant rate of 100 W from the hot reservoir to the cold reservoir.

Step 2. Entropy change of the system (bar)

Since the bar is in steady state, there is no net change in its entropy. Heat enters at one end and leaves at the other end, resulting in no entropy accumulation within the bar. Therefore, the entropy change of the system (the bar) is

$$\Delta S_{\text{bar}} = 0.$$

Step 3. Entropy change of the hot reservoir (ΔS_H)

The hot reservoir loses heat at a rate of $\dot{Q} = 100$ W. Since the temperature of the hot reservoir remains constant at $T_H = 400$ K, the entropy change of the hot reservoir per second is

$$\Delta S_H = \frac{\dot{Q}}{T_H}.$$

Substituting the values

$$\Delta S_H = -\frac{100}{400} = -0.25 \text{ W K}^{-1}.$$

The negative sign indicates a decrease in entropy in the hot reservoir as it loses energy.

Step 4. Entropy change of the cold reservoir (ΔS_c)

The cold reservoir gains heat at the same rate, $\dot{Q} = 100$ W, and its temperature is constant at $T_C = 300$ K. The entropy change of the cold reservoir per second is

$$\Delta S_C = \frac{\dot{Q}}{T_C}.$$

Substituting the values,

$$\Delta S_C = \frac{100}{300} = 0.333 \text{ W K}^{-1}.$$

The positive sign indicates an increase in entropy in the cold reservoir as it absorbs energy.

Step 5. Total entropy generation (ΔS_{total})

The total entropy generation in the process is the sum of the entropy changes of the hot and cold reservoirs:

$$\Delta S_{\text{total}} = \Delta S_H + \Delta S_C.$$

Substituting the values from steps 3 and 4:

$$\Delta S_{\text{total}} = -0.25 + 0.333 = 0.083 \text{ W K}^{-1}.$$

Since ΔS_{total} is positive, this confirms that the process is irreversible due to the finite temperature difference between the two reservoirs.

4.10.2 Summary

- **System entropy change (bar)**: The entropy change of the bar is zero because it is in steady state with energy entering and leaving at the same rate.
- **Surroundings entropy change**: The hot reservoir loses entropy, while the cold reservoir gains entropy. The net entropy change of the surroundings (reservoirs) is positive.
- **Irreversibility and entropy generation**: The total entropy generation rate of 0.083 W K^{-1} reflects the irreversibility of heat transfer across a finite temperature difference, consistent with the second law of thermodynamics.

This example illustrates that while energy transfer occurs at a constant rate, entropy is not conserved; it increases due to the irreversible nature of heat conduction through a temperature gradient.

While the example above focuses on Fourier heat conduction, the principle is universally applicable to all modes of heat transfer, including convection and radiation. Heat transfer inherently involves moving energy from a region of higher temperature to a region of lower temperature, which always leads to an increase in the total entropy of the Universe—an indication of irreversibility. This irreversibility holds true regardless of the specific mechanism of heat transfer, as it fundamentally relies on a temperature gradient.

It is worth noting that some authors in the literature describe heat transfer across a very small temperature difference as 'reversible'. While a small temperature difference may indeed result in a very small (or negligible) total entropy change, the process is still fundamentally irreversible. Conceptually, calling it reversible is misleading, as it implies that no entropy is generated. This can confuse readers by suggesting that near-equilibrium heat transfer is exempt from the principles of irreversibility. In reality, even the slightest finite temperature gradient causes entropy production, meaning that all heat transfer processes—no matter how close they are to equilibrium—are irreversible.

4.11 Entropy change in a process: ideal gas in a quasistatic process

In this section, we examine entropy changes during thermodynamic processes, specifically focusing on an ideal gas undergoing a quasistatic transformation. This set-up offers a simplified, clear view of how entropy behaves when no additional work

or losses are involved. In a quasistatic process—where each step occurs in perfect control—we gain insight into how entropy shifts under these ideal conditions.

By focusing on the example of an ideal gas, we will see how the system's internal state dictates changes in entropy within this specific set-up. Working in this idealized setting allows us to keep calculations straightforward while connecting physical intuition with mathematical expression. We will break down each part of the process, showing how entropy changes in this carefully controlled transformation.

4.11.1 Isobaric process (constant pressure)

In an isobaric process, where pressure remains constant, we can derive a relationship between temperature and entropy. To do this, we start with the second law of thermodynamics in its differential form, expressing the heat transfer in terms of temperature:

$$dS = \frac{Q}{T} = \frac{nc_p dT}{T}.$$

Integrating this expression yields

$$S = \int \frac{nc_p dT}{T} = nc_p \ln T + C,$$

where C is the integration constant. We can now express temperature as a function of entropy,

$$T = Ae^{S/(nc_p)},$$

where A is a constant that incorporates initial conditions.

If we are interested in the entropy difference between any two states, provided the temperatures of these states are known, we use

$$\Delta S = \int_1^n \frac{Q}{T} = \int_1^n \frac{nc_p dT}{T}.$$

Assuming that the heat capacity, c_p, is independent of temperature, this simplifies to

$$\Delta S = nc_p \ln \frac{T_n}{T_1}.$$

This expression gives us the entropy change for a quasistatic isobaric process between states 1 and n.

In the property diagrams in figure 4.3 illustrate the isobaric quasistatic process between states 1 and n: a pressure–volume (p–V) diagram on the left and a temperature–entropy (T–S) diagram on the right.

4.11.2 Isochoric process (constant volume)

For an isochoric process, where the volume remains constant, we can calculate the entropy change in a similar manner to the isobaric process. Starting with the second

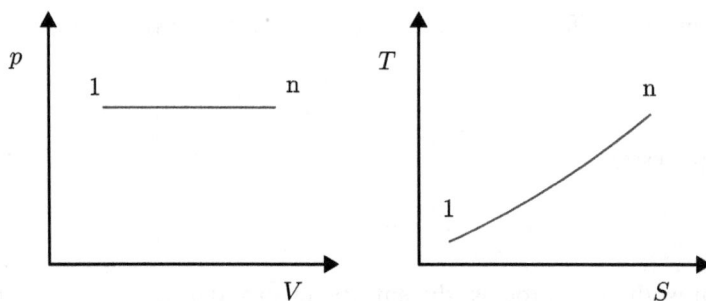

Figure 4.3. Property diagrams for a quasistatic isobaric process between states 1 and n. Left: Pressure–volume (p–V) diagram showing constant pressure. Right: temperature–entropy (T–S) diagram illustrating the entropy increase as temperature rises.

law of thermodynamics in differential form, we express the heat transfer in terms of temperature:

$$dS = \frac{Q}{T} = \frac{nc_v dT}{T}.$$

Integrating this, we find

$$S = nc_v \ln T + C,$$

which allows us to express temperature as a function of entropy,

$$T = Ae^{S/(nc_v)},$$

where A is a constant. For the entropy difference between two states, if temperatures are known and c_V is constant, we have

$$\Delta S = nc_v \ln \frac{T_n}{T_1}.$$

This gives us the entropy change for a quasistatic isochoric process between two states.

4.11.3 Isothermal process (constant temperature)

In an isothermal process, where the temperature remains constant, the change in internal energy is zero. This means that any heat transfer to or from the system is equal to the work done by or on the system, assuming no other types of work are present.

Starting with the second law of thermodynamics in differential form, we have

$$dS = \frac{Q}{T} = \frac{-W}{T} = \frac{-(-pdV)}{T} = \frac{pdV}{T} = \frac{nRdV}{V}.$$

To find the entropy difference between two states with known volumes, we integrate

$$\Delta S = \int_1^n \frac{nRdV}{V}.$$

This simplifies to

$$\Delta S = nR \ln \frac{V_n}{V_1}.$$

Thus, for an isothermal process, the entropy change depends solely on the volume ratio between the two states.

4.11.4 Adiabatic process

In an adiabatic process, there is no heat transfer into or out of the system, and if no other work is done, the entropy change is zero. Since the system is perfectly insulated from its surroundings, no entropy is added or removed, resulting in

$$\Delta S = 0.$$

This means that an adiabatic process is also an isentropic process, where the entropy remains constant throughout the transformation.

4.11.5 Arbitrary process

Entropy, like internal energy, is a state property. This means that the entropy change between two states depends only on the initial and final states, not on the specific path taken. To calculate the entropy difference between any two states, we can follow a similar approach as we did for internal energy by considering a combination of known paths.

In this case, we denote the initial state as 1, the final state as n, and introduce an intermediate state i, as illustrated in figure 4.4.

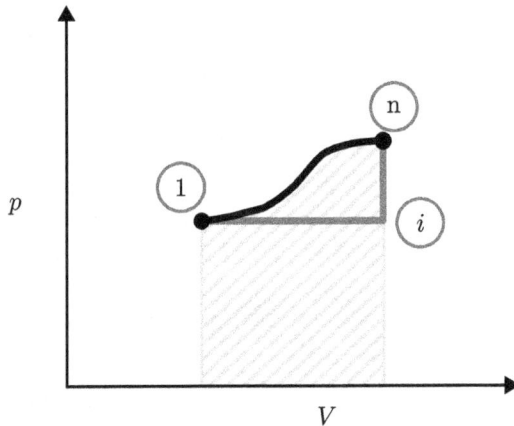

Figure 4.4. Pressure–volume (P–V) diagram of a quasistatic arbitrary process between states 1 and n.

The total entropy change between states 1 and n is the sum of the entropy changes along each segment:

$$\Delta S_{1n} = \Delta S_{1i} + \Delta S_{in}.$$

Using the expressions for entropy change along each segment, we have

$$\Delta S_{1n} = nc_p \ln \frac{T_i}{T_1} + nc_v \ln \frac{T_n}{T_i}.$$

The temperature T_i of the intermediate state, given the pressure p_1 and volume V_n, can be determined using the ideal gas law.

4.12 Condition for an irreversible process

An irreversible process is characterized by a positive *total* entropy change, which includes both the entropy change of the system, ΔS_s, and the entropy change of the surroundings, ΔS_r. This distinction is crucial, as it is sometimes oversimplified in the literature by stating that a process is irreversible when the system's entropy change is positive. However, this is not enough to determine irreversibility. To conclude that a process is irreversible, the *total* entropy change of the combined system and surroundings must be greater than zero:

$$\Delta S_{\text{total}} = \Delta S_s + \Delta S_r > 0.$$

Entropy in a system alone can increase, stay the same, or even decrease during a process, but this does not provide a full picture. We need to consider the surroundings as well, which may also experience an increase, no change, or a decrease in entropy. Only by evaluating the total entropy change of the system and surroundings together can we accurately assess whether a process is truly irreversible. Importantly, the total entropy change can be zero (for a reversible process) or positive, but it can never be negative. This fundamental principle underscores the direction of natural processes and is essential to the correct understanding of entropy and irreversibility.

4.13 Summary

In this chapter, we established entropy as a statistical measure of disorder, introduced its role as a state property, and discussed its calculation across different thermodynamic processes. Key takeaways include:

1. **Statistical definition of entropy**: By connecting microstates and macrostates, we defined entropy as a measure of a system's multiplicity and the probability of various states.
2. **Entropy as a state property**: Like temperature and internal energy, entropy depends solely on a system's state, making its change path-independent.
3. **Entropy change in idealized processes**: We calculated entropy changes in isobaric, isochoric, isothermal, and adiabatic processes, showing how different constraints affect entropy.

4. **Irreversibility and the second law**: We clarified that irreversibility requires a positive total entropy change for the combined system and surroundings, cementing the concept that entropy increases in natural processes.

5. **Scope of idealized systems**: In this chapter, we focused on idealized processes and did not consider non-ideal situations, such as entropy generation due to work dissipation and other irreversible effects.

IOP Publishing

A Classical Thermodynamics Toolkit

Srinivas Vanapalli

Chapter 5

Quasistatic processes with irreversible work transfer and non-quasistatic processes

In this chapter, we delve into the complexities of quasistatic processes with irreversible work transfer and introduce the concept of non-quasistatic processes. Traditional thermodynamic analysis often assumes idealized, reversible conditions, where changes occur slowly enough to maintain equilibrium at each infinitesimal step. However, real-world processes frequently involve irreversible effects—even within quasistatic conditions—that lead to energy dissipation and entropy generation.

We begin by exploring how irreversible work transfer occurs within quasistatic processes, such as frictional and dissipative forces in mechanical systems, and analyse their effects on energy distribution. By separating the reversible and irreversible components of work, we reveal the impact of dissipative forces on internal energy and entropy. To illustrate these effects, we examine a range of ideal gas processes—isochoric, isobaric, isothermal, and adiabatic—that involve irreversibilities, providing insight into entropy generation and the limits of energy recoverability.

The chapter then shifts to non-quasistatic processes, where equilibrium is not maintained, complicating the analysis. Here, classical thermodynamic relationships must be re-examined to account for rapid changes and non-equilibrium conditions. We provide examples, including gas expansion in a vacuum and expansion against constant external pressure, to demonstrate the unique characteristics of non-quasistatic processes and their implications for energy and entropy changes.

Finally, we derive the thermodynamic identity, a fundamental relation that connects internal energy, entropy, temperature, pressure, and volume. This derivation emphasizes a comprehensive approach, respecting the contributions of both heat transfer and irreversibilities, distinguishing it from common simplified treatments. By establishing the thermodynamic identity in this broader context, we reinforce its universal applicability to both reversible and irreversible processes.

doi:10.1088/978-0-7503-6029-6ch5

5.1 Introduction to quasistatic processes with irreversible work transfer

In thermodynamics, a **quasistatic process** is one that occurs slowly enough for the system to remain in a state of equilibrium at each infinitesimal step. For a process to be **reversible**, however, the total entropy change of the combined system and surroundings must be zero, meaning there is no net entropy generation in the Universe. In a reversible process, both the system and surroundings can, in principle, be returned to their initial states without any net impact on the Universe.

This reversibility is an idealization that assumes the absence of any dissipative effects, such as friction or unrestrained expansion, which would otherwise cause irreversibilities and increase the total entropy of the system and surroundings. In a purely reversible quasistatic process, the entropy of the system may increase, decrease, or remain constant, as long as any changes are exactly balanced by corresponding changes in the surroundings.

Quasistatic, reversible processes are foundational in thermodynamics, serving as a benchmark for ideal energy transformations and allowing us to explore fundamental relationships and limits within thermodynamic systems.

However, in real-world scenarios, processes often contain irreversibilities even when they are slow enough to be considered quasistatic. This introduces the concept of irreversible work transfer within a quasistatic process. Here, the process may still be slow enough to maintain equilibrium at each step, but certain irreversible effects —such as friction, inelastic deformations, or viscous dissipation—prevent the process from being fully reversible. These irreversibilities introduce additional energy losses, which cannot be recovered by simply reversing the process.

Irreversible work transfer has a significant impact on energy distribution within these processes, primarily by causing temperature increases due to dissipative effects such as friction. For example, friction within a piston–cylinder set-up causes mechanical work to raise the temperature of the gas or material, which may lead to heat transfer if the system is not adiabatic. However, in an adiabatic system, this temperature increase would remain contained within the substance without necessarily converting mechanical work directly into thermal energy transferred to the surroundings.

An illustrative example of irreversible work transfer can be seen in a mechanical shaft rotating within a liquid. In this scenario, the shaft's mechanical energy is dissipated as it encounters resistance from the fluid. This interaction increases the temperature of the liquid due to frictional forces, making the work done by the shaft irreversible; it cannot be fully recovered to drive the shaft in reverse. This irreversibility leads to an increase in the total entropy of the system and surroundings, reflecting the energy loss associated with irreversible work transfer.

These examples highlight that even in quasistatic processes, irreversibilities are present, leading to entropy generation and irreversible energy loss within the system and surroundings.

5.2 Energy transfer in quasistatic processes with irreversible work

In a quasistatic process with irreversible work transfer, energy distribution within the system consists of both recoverable and irrecoverable components. While the process is slow enough to assume that equilibrium is maintained at each step, certain irreversible effects—such as friction, resistance, and other non-conservative forces—dissipate a portion of the work. This dissipated work cannot be fully recovered and contributes to entropy generation, setting it apart from purely reversible work in an ideal quasistatic process.

To clarify this, we separate the **reversible** and **irreversible components** of work. The reversible component represents the idealized, recoverable work done by the system that could, in principle, be fully recovered. In contrast, the irreversible component stems from dissipative forces like friction, viscosity, or electrical resistance, resulting in internal energy dissipation and entropy generation within the system.

5.2.1 Examples of reversible and irreversible work

Consider a **battery** as an example of reversible work transfer. In an ideal scenario, the battery converts chemical energy into electrical work, which could be recovered by recharging, making it a reversible process. However, in real batteries, **Joule heating** due to internal resistance and **electrode losses** are irreversible components, as they result in entropy generation and energy loss. These effects cannot be fully reversed, even if the battery is recharged.

Another example is a **fuel cell** that combines hydrogen and oxygen to produce electricity and water. Ideally, this process is reversible, as electricity could be used to split water back into hydrogen and oxygen, allowing complete recovery of the input energy. However, in practice, **Joule heating**, **electrode degradation**, and **formation of gas bubbles** within the electrolyte lead to irreversible energy losses that cannot be fully recovered, making part of the work irreversible.

A third example is the **expansion of a gas within a piston–cylinder set-up**. In an ideal scenario, the work done by the gas expanding against the piston could be recovered completely if there were no frictional losses. However, with a **lossy piston**, frictional effects dissipate some of the work, which increases the internal energy of the gas or piston material. This irreversibility results in an increase in the total entropy of the system and surroundings, and the dissipated work cannot be recovered.

5.2.2 Mathematical treatment of reversible and irreversible work

Using the first law of thermodynamics as defined in this book, we can write

$$\Delta U = Q + W,$$

where ΔU represents the change in internal energy of the system, Q is the heat added, and W is the total work done on the system.

In a quasistatic process with irreversibilities, we can separate W into the reversible work, W_{rev}, and irreversible work, W_{irr}, components:

$$W = W_{rev} + W_{irr}.$$

The **reversible work**, W_{rev}, represents the ideal component of work that could, in principle, be fully recovered. The **irreversible work**, W_{irr}, on the other hand, arises due to dissipative forces such as friction or resistance, leading to energy dissipation within the system.

In the modified energy equation, the presence of W_{irr} reflects the energy lost due to irreversibilities:

$$\Delta U = Q + W_{rev} + W_{irr}.$$

In a purely reversible process, W_{irr} would be zero, meaning there would be no net entropy generation and all the work could theoretically be recovered. However, when W_{irr} is non-zero due to dissipative forces, there is an irreversible loss of energy, contributing to an increase in total entropy.

By separating W_{rev} and W_{irr} in the energy equation, we gain insight into how energy transfers are affected by irreversibilities in quasistatic processes. These examples and equations demonstrate the distinct roles of reversible and irreversible work, and underscore the impact of irreversibilities on entropy and energy transfer within thermodynamic systems.

5.3 Entropy generation in quasistatic processes with irreversible work

In thermodynamic processes, **entropy generation** is a direct indicator of irreversibility. Whenever a process involves dissipative effects such as friction, electrical resistance, or inelastic deformation, it becomes irreversible, leading to a net increase in the total entropy of the system and surroundings. This increase in entropy, or entropy generation, is the hallmark of irreversible work within a thermodynamic process, including those that are quasistatic.

5.3.1 Entropy generation due to irreversibilities

In an **idealized reversible process**, the total entropy change of the system and surroundings is zero, meaning no net entropy is generated, and the process could theoretically be reversed without any loss of energy or increase in entropy. However, in a **quasistatic process with irreversible work**, dissipative forces introduce irreversibilities that generate entropy. This entropy generation reflects the portion of the energy that cannot be recovered, as it has been dissipated in ways that increase the randomness or disorder of the system.

5.3.2 Relationship between irreversible work and entropy production

The presence of irreversible work in a quasistatic process directly contributes to entropy generation. For instance, consider a gas expanding in a piston–cylinder

arrangement with friction. The work done by the gas against the piston includes an irreversible component due to the frictional forces between the piston and cylinder wall. This irreversible work, W_{irr}, results in increased temperature and disorder within the system, which cannot be reversed without adding energy from an external source. Consequently, this irreversible work leads to a **positive entropy generation** term, representing the loss of available energy in the system.

The relationship between entropy generation S_{gen} and irreversible work W_{irr} can be expressed as

$$S_{gen} \propto W_{irr}/T,$$

where T is the absolute temperature at which the irreversibility occurs. This relation implies that the greater the irreversible work, the more entropy is generated. The entropy generation term, S_{gen}, measures the degree of irreversibility in the process and signifies the portion of the energy transformation that is irrecoverable.

5.3.3 Entropy in reversible versus irreversible quasistatic processes

In a **purely reversible quasistatic process**, there are no dissipative forces, and as a result, there is no entropy generation. Any entropy change within the system is exactly balanced by an equal and opposite change in the surroundings, leading to zero net entropy change in the Universe. This reversibility implies that all work performed can be fully recovered, with no energy lost to irreversible effects.

In contrast, in a **quasistatic process with irreversible work**, entropy is generated within the system or surroundings (or both), leading to a net positive entropy change in the Universe. This irreversibility reduces the potential for energy recovery, as part of the work has been dissipated through mechanisms that increase entropy, such as frictional heating or resistive heating. The resulting entropy generation thus distinguishes quasistatic processes with irreversible work from ideal reversible processes, underscoring the fundamental difference in energy efficiency and recoverability.

In summary, the presence of entropy generation in a quasistatic process with irreversible work signals the loss of usable energy and highlights the impact of irreversibilities. This entropy generation reflects the inherent inefficiencies in real-world processes, contrasting sharply with the idealized behavior of purely reversible systems.

5.4 Quasistatic irreversible ideal gas processes

This section explores various types of ideal gas processes that are quasistatic but involve irreversibilities. In each case, the process is slow enough to maintain equilibrium at every step, yet friction, heat transfer resistance, or other dissipative effects introduce irreversibilities, resulting in entropy generation.

5.4.1 Isochoric process with irreversibility (constant volume)

In an isochoric process, the volume of the gas remains constant, so no boundary work is done ($W = 0$). We will analyse two cases where the system undergoes a

change in internal energy due to heat transfer, and in one case, additional work through a rotating shaft. In both cases, we assume the gas is ideal, and the system goes from the same initial state to the same final state.

Let T_{sur} be the temperature of the surroundings, which remains constant during the process.

Case 1: Isochoric process with only heat transfer
In this first case, the gas undergoes an isochoric process where heat Q_1 is transferred from the surroundings to the system. According to the first law of thermodynamics as defined in this book,

$$\Delta U = Q_1,$$

where ΔU is the change in internal energy of the gas. Given that volume is constant, all the heat transfer goes into changing the internal energy.

For an ideal gas undergoing an isochoric process, the entropy change of the system is

$$\Delta S_{\text{system}} = nc_v \ln\left(\frac{T_f}{T_i}\right),$$

where T_i and T_f are the initial and final temperatures of the gas, n is the number of moles, and c_v is the molar heat capacity at constant volume.

Since this process is irreversible (likely due to a temperature gradient between the surroundings and system), there is entropy generation in the surroundings, denoted $S_{\text{gen},1}$. For the surroundings, the entropy change due to heat transfer Q_1 at temperature T_{sur} is

$$\Delta S_{\text{surroundings},1} = -\frac{Q_1}{T_{\text{sur}}}.$$

The total entropy generation for case 1, combining the system and surroundings, is

$$S_{\text{gen},1} = \Delta S_{\text{system}} + \Delta S_{\text{surroundings},1} = nc_v \ln\left(\frac{T_f}{T_i}\right) - \frac{Q_1}{T_{\text{sur}}}.$$

Case 2: Isochoric process with heat transfer and shaft work
In the second scenario, in addition to heat transfer, a shaft rotates within the gas, performing mechanical work W_{shaft} on the gas. This work is dissipated entirely as heat within the system, contributing to the internal energy change and reducing the amount of heat Q_2 needed from the surroundings.

The total change in internal energy is the same as in case 1:

$$\Delta U = Q_2 + W_{\text{shaft}}.$$

Since both cases start and end at the same states, we have

$$Q_1 = Q_2 + W_{\text{shaft}}.$$

Therefore, $Q_2 < Q_1$ due to the additional energy input from the shaft work.

For the system, the entropy change remains the same as in case 1, as it only depends on the initial and final states:

$$\Delta S_{\text{system}} = nc_v \ln\left(\frac{T_f}{T_i}\right).$$

For the surroundings, the entropy change due to the heat transfer Q_2 is

$$\Delta S_{\text{surroundings},2} = -\frac{Q_2}{T_{\text{sur}}}.$$

Comparison of entropy generation
The total entropy generation for case 2, combining the system and surroundings, is

$$S_{\text{gen},2} = \Delta S_{\text{system}} + \Delta S_{\text{surroundings},2} = nc_v \ln\left(\frac{T_f}{T_i}\right) - \frac{Q_2}{T_{\text{sur}}}.$$

Since $Q_2 < Q_1$, we have

$$-\frac{Q_2}{T_{\text{sur}}} > -\frac{Q_1}{T_{\text{sur}}}.$$

Therefore,

$$S_{\text{gen},2} > S_{\text{gen},1}.$$

This shows that the additional shaft work in case 2 introduces more irreversibility, leading to a greater entropy generation than in case 1. Although less heat is needed from the surroundings in case 2, the total irreversibility of the process is higher due to the dissipative work done by the shaft. This example highlights how work dissipation can increase entropy generation, even when reducing external heat transfer.

5.4.2 Isobaric process with irreversibility (constant pressure)

In an isobaric process, the pressure of the gas remains constant, which allows for boundary work as the gas expands or contracts. We will consider two cases where the system undergoes an internal energy change due to heat transfer, and in one case, additional work is provided through a rotating shaft. The gas is assumed to be ideal, and both cases go from the same initial to the same final state.

Let T_{sur} represent the temperature of the surroundings, which remains constant throughout the process.

Case 1: Isobaric process with only heat transfer
In the first scenario, heat Q_1 is transferred to the system from the surroundings at constant pressure, causing the gas to expand. According to the first law of thermodynamics as defined in this book,

$$\Delta U = Q_1 + W,$$

where ΔU is the change in internal energy, Q_1 is the heat transfer, and W is the work done by the system on the surroundings. For an isobaric process, the work done by the gas on the surroundings is given by

$$W = -p\Delta V.$$

Since ΔU depends only on the initial and final states, the heat transfer Q_1 is required to account for both the internal energy change and the work done,

$$Q_1 = \Delta U + pdV.$$

The entropy change of the system, given it is an ideal gas, can be expressed as

$$\Delta S_{\text{system}} = nc_p \ln\left(\frac{T_f}{T_i}\right),$$

where T_i and T_f are the initial and final temperatures, n is the number of moles, and c_p is the molar heat capacity at constant pressure.

Because this process is irreversible (possibly due to a temperature gradient between the surroundings and the system), there is entropy generation in the surroundings, which we denote as $S_{\text{gen},1}$. For the surroundings, the entropy change due to heat transfer Q_1 at temperature T_{sur} is

$$\Delta S_{\text{surroundings},1} = -\frac{Q_1}{T_{\text{sur}}}.$$

The total entropy generation for case 1, combining the system and surroundings, is

$$S_{\text{gen},1} = \Delta S_{\text{system}} + \Delta S_{\text{surroundings},1} = nc_p \ln\left(\frac{T_f}{T_i}\right) - \frac{Q_1}{T_{\text{sur}}}.$$

Case 2: Isobaric process with heat transfer and shaft work
In the second scenario, in addition to heat transfer, a shaft is rotating within the gas, performing mechanical work W_{shaft} on the gas. This work is fully dissipated as frictional heat within the system, supplementing the heat input required from the surroundings.

The total change in internal energy remains the same:

$$\Delta U = Q_2 + W_{\text{shaft}} + W.$$

Since both cases start and end at the same states, we have

$$Q_1 = Q_2 + W_{\text{shaft}}.$$

Thus, in this case, the heat transfer Q_2 from the surroundings is lower than Q_1 as part of the energy input is provided by the shaft work W_{shaft}.

The entropy change of the system, given it is an ideal gas and depends only on the initial and final states, remains the same as in case 1:

$$\Delta S_{\text{system}} = nc_p \ln\left(\frac{T_f}{T_i}\right).$$

The entropy change of the surroundings, due to the heat transfer Q_2, is

$$\Delta S_{\text{surroundings,2}} = -\frac{Q_2}{T_{\text{sur}}}.$$

Comparison of entropy generation
The total entropy generation for case 2, combining the system and surroundings, is

$$S_{\text{gen,2}} = \Delta S_{\text{system}} + \Delta S_{\text{surroundings,2}} = nc_p \ln\left(\frac{T_f}{T_i}\right) - \frac{Q_2}{T_{\text{sur}}}.$$

Since $Q_2 < Q_1$, we have

$$-\frac{Q_2}{T_{\text{sur}}} > -\frac{Q_1}{T_{\text{sur}}}.$$

Therefore,

$$S_{\text{gen,2}} > S_{\text{gen,1}}.$$

This analysis shows that the additional shaft work in case 2 increases the total entropy generation. While less heat is required from the surroundings, the irreversibility associated with the dissipative shaft work results in greater entropy generation, highlighting the impact of energy dissipation on entropy production in an isobaric process.

5.4.3 Isothermal process with irreversibility (constant temperature)

In an isothermal process, the temperature of the gas remains constant, so the internal energy change is zero for an ideal gas ($\Delta U = 0$). For this analysis, we assume that the surroundings are at a slightly higher temperature $T_{\text{sur}} = T + \Delta T$ than the system's temperature T. This slight temperature difference drives heat transfer from the surroundings to the system, allowing the gas to perform work while maintaining an isothermal state.

Case 1: Isothermal process with only heat transfer
In the first case, the gas undergoes an isothermal expansion where the work done on the surroundings is entirely supplied by heat transfer Q_1 from the surroundings. According to the first law of thermodynamics,

$$\Delta U = Q_1 + W.$$

Since $\Delta U = 0$ for an isothermal process, we have

$$Q_1 = -W.$$

The work done by the system during an isothermal expansion from an initial volume V_i to a final volume V_f at temperature T is given by

$$W = -nRT \ln(V_f/V_i).$$

Therefore, the heat transfer required is

$$Q_1 = nRT \ln(V_f/V_i).$$

The entropy change for the system, due to the isothermal heat transfer, is

$$\Delta S_{\text{system}} = \frac{Q_1}{T} = nR \ln(V_f/V_i).$$

Since this process is irreversible due to the temperature difference ΔT between the system and surroundings, entropy is generated. For the surroundings, the entropy change due to the heat transfer Q_1 at temperature T_{sur} is

$$\Delta S_{\text{surroundings},1} = -\frac{Q_1}{T_{\text{sur}}} = -\frac{nRT \ln(V_f/V_i)}{T + \Delta T}.$$

The total entropy generation for case 1, combining the system and surroundings, is

$$S_{\text{gen},1} = \Delta S_{\text{system}} + \Delta S_{\text{surroundings},1} = nR \ln(V_f/V_i) - \frac{nRT \ln(V_f/V_i)}{T + \Delta T}.$$

Since $T < T + \Delta T$, the entropy generation $S_{\text{gen},1}$ is positive, reflecting the irreversibility due to the temperature difference.

Case 2: Isothermal process with heat transfer and shaft work
In the second scenario, in addition to heat transfer, a shaft rotates within the gas, performing mechanical work W_{shaft} on the gas, which is fully dissipated as frictional heat within the system. This dissipative work supplements the heat required from the surroundings.

According to the first law of thermodynamics, we have

$$\Delta U = Q_2 + W + W_{\text{shaft}} = 0.$$

Thus

$$Q_2 = -(W + W_{\text{shaft}}) = nRT \ln\left(\frac{V_f}{V_i}\right) - W_{\text{shaft}}.$$

Since W_{shaft} is positive, $Q_2 < Q_1$, meaning less heat is required from the surroundings due to the additional work from the shaft.

The entropy change for the system remains the same:

$$\Delta S_{\text{system}} = nR \ln\left(V_f/V_i\right).$$

For the surroundings, the entropy change due to the reduced heat transfer Q_2 is

$$\Delta S_{\text{surroundings},2} = -\frac{Q_2}{T_{\text{sur}}} = -\frac{nRT \ln\left(\frac{V_f}{V_i}\right) - W_{\text{shaft}}}{T + \Delta T}.$$

Comparison of entropy generation
The total entropy generation for case 2 is

$$S_{\text{gen},2} = \Delta S_{\text{system}} + \Delta S_{\text{surroundings},2} = nR \ln(V_f/V_i) - \frac{nRT \ln\left(\frac{V_f}{V_i}\right) - W_{\text{shaft}}}{T + \Delta T}.$$

Since $Q_2 < Q_1$, we find

$$-\frac{Q_2}{T_{\text{sur}}} > -\frac{Q_1}{T_{\text{sur}}}.$$

Thus

$$S_{\text{gen},2} > S_{\text{gen},1}.$$

This analysis demonstrates that the inclusion of shaft work in case 2 results in greater total entropy generation compared to the heat-only scenario in case 1, due to additional irreversibilities introduced by the dissipative work. By assuming the surroundings are at a slightly higher temperature than the system, the entropy generation is more accurately represented.

5.4.4 Adiabatic process with irreversibility (no heat transfer)

In an adiabatic process, there is no heat exchange between the system and its surroundings ($Q = 0$). Any change in the system's internal energy must come solely from the work done on or by the system. We will examine two cases involving an ideal gas: (i) an ideal, reversible adiabatic process with no entropy generation, and (ii) an adiabatic process with irreversibilities introduced by a dissipative piston.

Case 1: Ideal (reversible) adiabatic process with only expansion/compression work
In the first case, the gas undergoes an ideal, reversible adiabatic process where the work W_1 done on or by the system is the sole contributor to the change in internal energy. According to the first law of thermodynamics:

$$\Delta U = Q + W_1.$$

Since the process is adiabatic ($Q = 0$), this simplifies to

$$\Delta U = W_1.$$

For an ideal gas undergoing an adiabatic process, the relationship between pressure and volume can be described by

$$pV^\gamma = \text{constant},$$

where $\gamma = c_p/c_v$ is the heat capacity ratio. For an adiabatic expansion, the work W_1 done by the gas is negative, leading to a decrease in internal energy and a drop in temperature. For an adiabatic compression, the work W_1 done on the gas is positive, increasing the internal energy and raising the gas temperature.

Since this process is reversible, there are no irreversibilities, and the total entropy change for the system and surroundings is zero:

$$\Delta S_{\text{system}} = 0 \text{ and } \Delta S_{\text{total}} = 0.$$

With no heat exchange, the surroundings experience no entropy change, and the total entropy generation for this process is zero.

Case 2: Adiabatic process with a dissipative piston (irreversible work)
In the second scenario, the gas undergoes an adiabatic expansion or compression, but with a dissipative piston that introduces friction as it moves. This frictional force dissipates part of the work $W_{\text{dissipative}}$ done by or on the gas, causing an *increase in the internal energy* of the gas, which leads to entropy generation within the system and represents an irreversible process.

According to the first law of thermodynamics,

$$\Delta U = Q + W_2 + W_{\text{dissipative}}.$$

Since $Q = 0$ for an adiabatic process, this simplifies to

$$\Delta U = W_2 + W_{\text{dissipative}}.$$

The dissipative work $W_{\text{dissipative}}$ effectively adds to the internal energy change of the gas. For an adiabatic expansion with a dissipative piston, the temperature of the gas does not drop as much as it would in the reversible case because the dissipative work partially offsets the cooling effect of the expansion. Conversely, for an adiabatic compression with a dissipative piston, the gas temperature increases more than in the reversible case due to the additional frictional contribution to internal energy.

The key point here is that, because of the irreversibility introduced by the dissipative piston, the *end state in case 2 cannot match the end state in case 1* for the same amount of work done by or on the gas. The presence of friction changes the internal energy distribution, resulting in a different temperature and pressure compared to the reversible adiabatic process.

The entropy change of the system in case 2 is now positive, as the dissipative piston introduces irreversibilities,

$$\Delta S_{\text{system}} = S_{\text{gen},2} > 0.$$

Since there is no heat exchange, the surroundings experience no entropy change, and the total entropy generation is entirely within the system:

$$\Delta S_{\text{total}} = S_{\text{gen},2} > 0.$$

Comparison of entropy generation

In case 1, the total entropy change ΔS_{total} is zero, as it represents an ideal, reversible adiabatic process with no irreversibilities. In case 2, the additional dissipative work due to friction in the piston results in greater entropy generation within the system, represented by $S_{\text{gen},2}$. Thus

$$S_{\text{gen},2} > 0.$$

Furthermore, because of this dissipative work, *the final state in case 2 differs from that in case 1*. The irreversibilities prevent reaching the same pressure, volume, and temperature as in the ideal, reversible case, even if the initial states were identical.

This example illustrates how introducing dissipative work in an adiabatic process increases total entropy generation and results in a different final state than in the ideal, reversible case. Readers are encouraged to plot the adiabatic expansion and compression processes on p-V and T-S diagrams to visualize the impact of irreversibilities on the process path and final state.

Additionally, in the isobaric and isothermal examples discussed earlier, a similar analysis can be conducted by considering a lossy piston instead of shaft work, as the friction in the piston would also introduce irreversibilities. While we have primarily analysed expansion processes, readers are encouraged to apply a similar approach to compression processes to explore how irreversibilities affect entropy generation and final states in those scenarios as well.

5.5 Transition to non-quasistatic processes

In the study of thermodynamics, we have focused primarily on **quasistatic processes** —those that proceed slowly enough for the system to remain in equilibrium at every step. In a quasistatic process, the state variables (pressure, volume, temperature) are well-defined at each instant, allowing the system to follow a continuous path of equilibrium states. This assumption of equilibrium simplifies the analysis, as we can use precise thermodynamic relationships to describe the process.

However, in real-world applications, many processes are inherently **non-quasistatic**, or **non-equilibrium**, as they occur too rapidly or involve too many internal interactions for the system to maintain equilibrium throughout. In a non-quasistatic process, variables such as pressure and temperature can vary significantly within different parts of the system, making it impossible to describe the system with a single, uniform state at any given moment. These processes introduce new complexities: they require us to account for internal gradients, rapid changes, and potentially chaotic interactions, and they often result in irreversible energy dissipation and entropy generation.

In non-quasistatic processes, the departure from equilibrium means that classical thermodynamic relationships—such as those for work, heat, and entropy—cannot always be applied in their standard forms. Instead, analysing non-quasistatic processes often requires more advanced approaches, such as non-equilibrium thermodynamics.

As we transition to examining non-quasistatic processes, we will explore the unique challenges they pose and investigate the principles that govern thermodynamic behavior in far-from-equilibrium conditions. This shift offers a broader, more realistic view of how thermodynamic systems operate in nature and technology, expanding our understanding beyond the idealized, reversible framework of quasistatic processes.

5.6 Introduction to non-quasistatic processes and moving boundary work transfer for an ideal gas

In previous chapters, we discussed **quasistatic processes**—those in which the system remains in mechanical equilibrium with its surroundings at each step. This equilibrium allows us to define the system's pressure throughout the process. However, in a **non-quasistatic process**, the system is not in mechanical equilibrium with its surroundings and, as a result, the pressure within the system varies in ways that are not well-defined at each instant. Consequently, non-quasistatic processes cannot be represented on standard thermodynamic property diagrams, such as the pressure–volume (p–V) or temperature–entropy (T–S) diagrams. However, we can still determine changes in state variables, such as internal energy and entropy, by comparing the initial and final states, as these are state functions and do not depend on the process path.

5.6.1 Moving boundary work transfer in a non-quasistatic process

In a non-quasistatic process, **moving boundary work** occurs when an external force acts to change the system's volume (i.e. it moves the system boundary). The work done on the system due to this external force is

$$W = \vec{F}_{\text{ext}} \cdot d\vec{s} = (p_{\text{ext}} \cdot A) d\left(\frac{V}{A}\right) = -p_{\text{ext}} dV,$$

where:
- p_{ext} is the **external pressure**, defined as F_{ext}/A,
- dV is the **volume change** equal to $\vec{s} \cdot A$, and
- the minus sign accounts for the convention that work done on the system is positive when the volume decreases.

In a quasistatic process, the system is always in equilibrium with its surroundings, meaning that the system pressure p equals the external pressure p_{ext} at all points during the process. Thus, the work expression simplifies to

Quasistatic process: $W = -p_{\text{ext}} dV = -p dV$.

However, in a non-quasistatic process, the system is not in mechanical equilibrium with its surroundings, so the external pressure p_{ext} differs from the system pressure p. Therefore, for non-quasistatic processes, the work transferred to the system is

Non-quasistatic process: $W = -p_{\text{ext}} dV$.

In this case, we must use the external pressure p_{ext} rather than the system pressure p to calculate the boundary work, as the system's pressure is not well-defined during the non-equilibrium process.

5.6.1.1 Examples of moving boundary work in non-quasistatic processes

1. **Gas expansion in a vacuum (external pressure zero)**

   Consider an ideal gas enclosed in a cylinder with a weightless piston, initially held in place by stops (figure 5.1). The initial pressure and volume of the gas are p_1 and V_1, respectively. When the stops are removed, the gas expands until it reaches a second set of stops, increasing to volume V_2 with a final pressure p_2. Since the external pressure $p_{ext} = 0$ (a vacuum),

   $$W = -\int p_{ext}\, dV = 0.$$

   Although the gas volume increases, no work is done by the system to expand, as there is no opposing external force.

2. **Gas expansion against a constant external pressure**

   In this scenario, a mass m is placed on a weightless piston, creating a constant external pressure $p_{ext} = mg/A$, where A is the piston area (figure 5.2). When the gas expands, the moving boundary work is

   $$W = -\int_{V_1}^{V_2} p_{ext}\, dV = -\int_{V_1}^{V_2} \frac{mg}{A}\, dV = -\frac{mg}{A}(V_2 - V_1).$$

   If the mass m is chosen such that $mg = p_2 A$, the piston reaches the second set of stops, with the work done by the gas given by

   $$W = -p_2(V_2 - V_1).$$

   This area on a p-V diagram represents the work performed by the gas (figure 5.2 (bottom)).

3. **Multi-step expansion with incremental mass changes**

   Now, suppose we add another mass m on the piston (figure 5.3), creating a higher external pressure p_3. After removing the stops, the gas expands to volume V_3. Then, we remove one mass, allowing the gas to further expand to

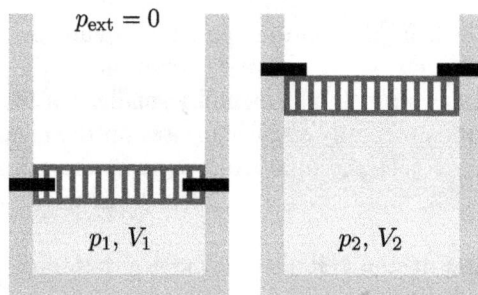

Figure 5.1. An ideal gas enclosed in a cylinder with a weightless piston. Left: The initial state with the piston held by stops. Right: The final state with the stops removed and the gas expanded.

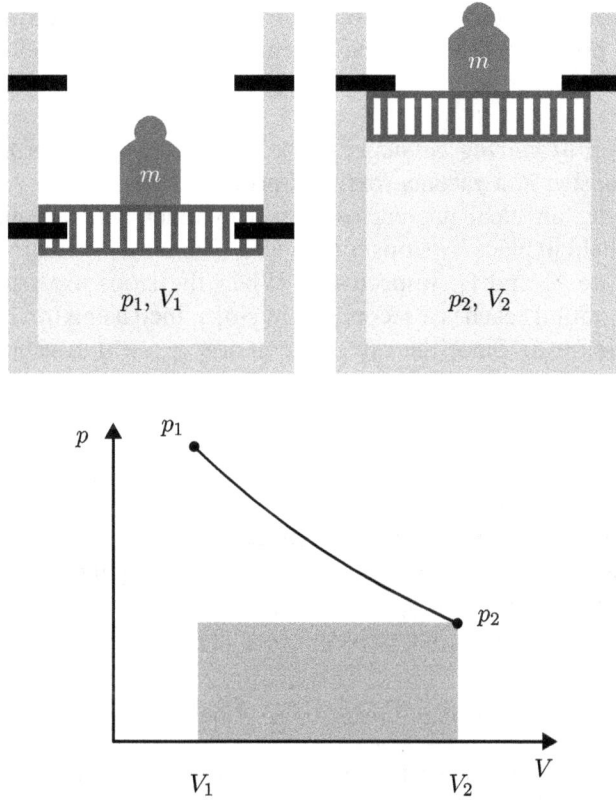

Figure 5.2. Top left: The initial state, a mass m placed on a weightless piston held by stops. Top right: The final state after the bottom stops are removed and the gas has expanded. Bottom: The pV diagram.

volume V_2. The work done during this process can be broken down as follows:

$$W = -\int_{V_1}^{V_3} p_{\text{ext}}\, dV - \int_{V_3}^{V_2} p_{\text{ext}}\, dV = -\int_{V_1}^{V_3} p_3\, dV - \int_{V_3}^{V_2} p_2\, dV$$
$$= -p_3(V_3 - V_1) - p_2(V_2 - V_3).$$

The shaded areas in the bottom panel of figure 5.3 represent the work performed by the gas, which is greater than that with a single mass.

4. **Continuous expansion with incrementally smaller masses**

    If we place incrementally smaller masses on the piston and remove them one by one (figure 5.4), the total work done by the gas approaches that of a quasistatic process, represented by the bold line in the p-V diagram.

To conclude: In the limit of small changes to the external force, the work performed by the gas converges to that of a quasistatic process.

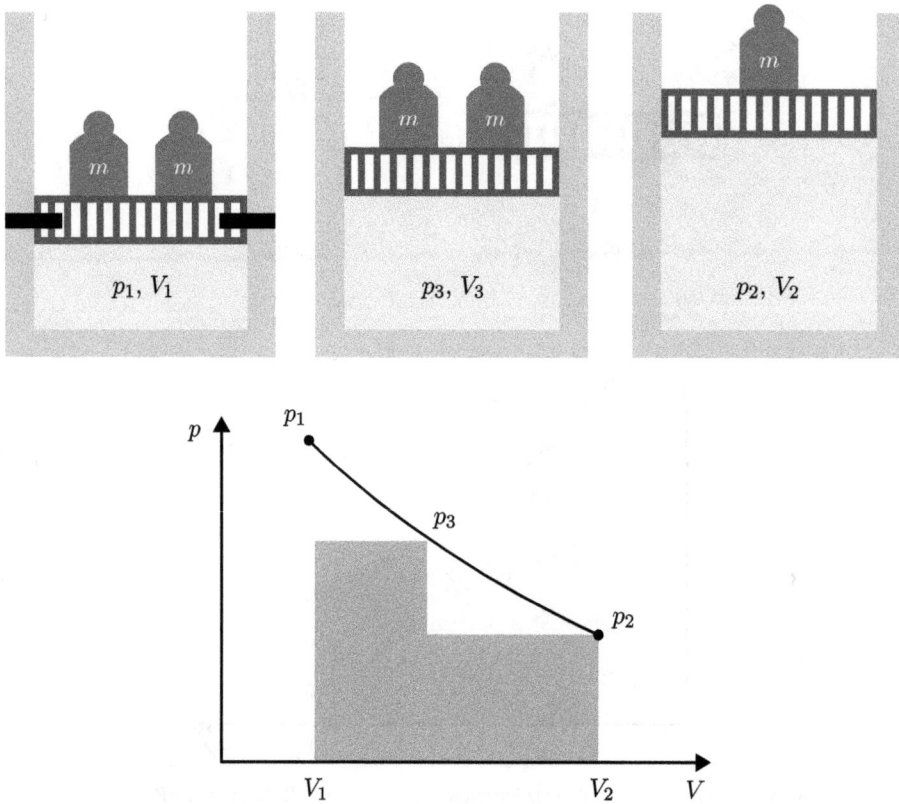

Figure 5.3. Top left: The inital state with two masses *m* on the piston which is held in place by stops. Top middle: The second state after the stops are removed. Top right: The final state with one mass removed. Bottom: The *pV* diagram.

5.6.1.2 Suggested analysis for compression processes

The above examples illustrate moving boundary work during gas expansion under various external conditions. Readers are encouraged to conduct a similar analysis for compression processes, starting from volume V_2 to V_1. This analysis could involve a single mass, two masses, or a series of smaller masses to observe the effects of different external pressures during compression. By applying the same systematic approach, readers can explore how external forces influence work in non-quasistatic compression processes.

5.6.2 Non-quasistatic processes

This section introduces a methodology for calculating changes in internal energy and entropy in non-quasistatic processes, focusing on ideal gas behavior without additional forms of work transfer. The first and second laws of thermodynamics still apply to non-quasistatic processes. For clarity, we restate them here:

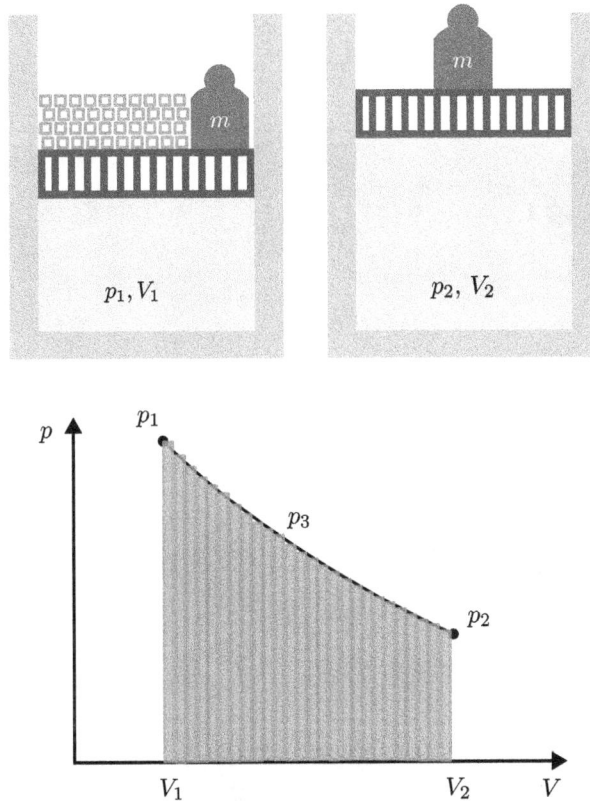

Figure 5.4. Top: Incrementally smaller masses on the piston. Bottom: The pV diagram.

- **First law** (with only pV work):

$$\Delta U = Q + W.$$

- **Second law**:

$$\Delta S = \int \frac{Q}{T} + S_{\text{irr}},$$

where S_{irr} represents the irreversible entropy generated within the system due to irreversibilities.

We will analyse two types of non-quasistatic processes for an ideal gas: an adiabatic expansion and an isothermal compression, and we will compare each with its quasistatic counterpart.

5.6.2.1 Adiabatic expansion—quasistatic versus non-quasistatic
Consider an ideal gas initially at pressure p_1 and volume V_1, enclosed in a cylinder with a weightless piston (figure 5.5).

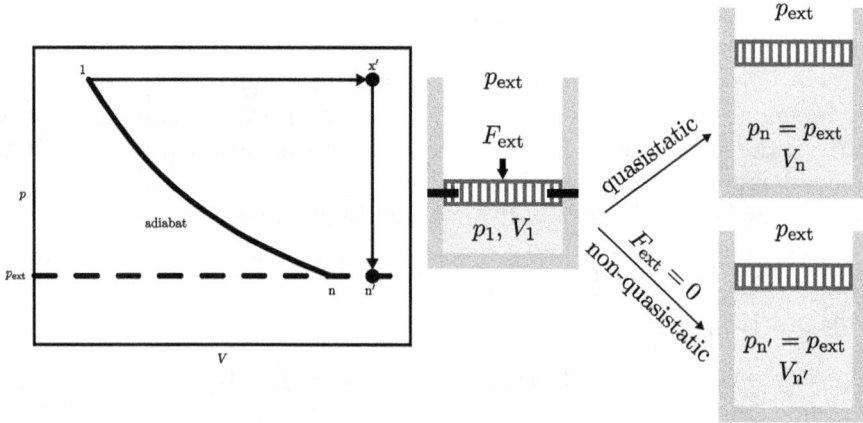

Figure 5.5. Schematic representation of a quasistatic versus non-quasistatic adiabatic expansion of an ideal gas in a piston–cylinder arrangement. In the quasistatic process, the external force is reduced gradually, allowing the system to follow an equilibrium adiabat. In contrast, the non-quasistatic expansion occurs rapidly against a constant external pressure, leading to irreversibilities and greater final volume due to entropy generation.

Quasistatic adiabatic expansion

In a quasistatic adiabatic expansion, an external force F_{ext} on the piston is gradually reduced, allowing the gas to expand along an adiabat. At the final state n, the gas pressure equals the external pressure. For a quasistatic adiabatic process:

$$pV^\gamma = p_1 V_1^\gamma = p_n V_n^\gamma = \text{constant}$$

Where γ is the heat capacity ratio $\frac{c_p}{c_v}$. The ideal gas law applies throughout:

$$\frac{pV}{T} = \frac{p_1 V_1}{T_1} = \frac{p_n V_n}{T_n} = \text{constant}$$

Since this is an adiabatic process, $Q = 0$. Applying the first law:

$$\Delta U = W$$

With no irreversible entropy generation ($S_{irr} = 0$), the entropy change is:

$$\Delta S = \int \frac{Q}{T} + S_{irr} = 0$$

Consider an ideal diatomic gas initially at pressure $p_1 = 10$ bar, volume $V_1 = 1$ L, and temperature $T_1 = 300$ K. The gas is expanded to a pressure of 1 bar. For this quasistatic expansion from state 1 to state n, we calculate:

- Change in internal energy: $\Delta U = -1205$ J.
- Work done: $W = -1205$ J.
- Entropy change: $\Delta S = 0$.

Non-quasistatic adiabatic expansion

Now consider a non-quasistatic expansion starting from the same initial state. In this process, the gas expands rapidly against the external pressure p_{ext}, so the system is not in equilibrium with its surroundings, and intermediate states are not well-defined. At the final state n', the gas pressure equals the external pressure, but the volume differs from the quasistatic case.

The moving boundary work in the non-quasistatic process is:

$$W = -p_{ext}(V_{n'} - V_1)$$

With $Q = 0$, applying the first law yields:

$$\Delta U = W = U_{n'} - U_1 = -p_{ext}(V_{n'} - V_1)$$

To evaluate the final volume $V_{n'}$, we use the thermodynamic expression for internal energy of an ideal gas:

$$U = \frac{f}{2}nRT = \frac{f}{2}pV$$

This relation allows us to express the change in internal energy as:

$$\Delta U = \frac{f}{2}\left(p_{n'}V_{n'} - p_1 V_1\right)$$

Equating this with the expression from the first law:

$$\frac{f}{2}\left(p_{n'}V_{n'} - p_1 V_1\right) = -p_{ext}(V_{n'} - V_1)$$

Rearranging terms gives:

$$\left(\frac{f}{2}p_{n'} + p_{ext}\right)V_{n'} = \frac{f}{2}p_1 V_1 + p_{ext}V_1$$

Substituting the known values:
- $p_1 = 10\text{ bar} = 10^6\text{Pa}$
- $V_1 = 1L = 10^{-3}\text{m}^3$
- $p'_n = p_{ext} = 1\text{ bar} = 10^5$
- $f = 5$

$$\left(\frac{5}{2} \cdot 10^5 + 10^5\right)V_{n'} = \frac{5}{2} \cdot 10^6 \cdot 10^{-3} + 10^5 \cdot 10^{-3}$$

$$\left(2.5 \cdot 10^5 + 10^5\right)V_{n'} = (2.5 + 0.1) \cdot 10^3$$

$$\left(3.5 \cdot 10^5\right)V_{n'} = (2.6) \cdot 10^3$$

$$V_{n'} = \frac{2.6 \cdot 10^3}{3.5 \cdot 10^5} = 7.43 \cdot 10^{-3} \text{m}^3 = 7.43\text{L}$$

Now that we have the final volume, we can compute the internal energy:

$$U_{n'} = \frac{f}{2} p_{n'} V_{n'} = \frac{5}{2} \cdot 10^5 \cdot 7.43 \cdot 10^{-3} = 1857\text{J}$$

Since the initial internal energy was $U_1 = 2500$ J, we find:

$$\Delta U = U_{n'} - U_1 = -643\text{J}$$

and hence:

$$W = -643\text{J}$$

To compute the entropy change, we consider a hypothetical reversible path from state 1 to n' consisting of:

1. An **isobaric** process from 1 to x': $p = p_1$ $V_1 \rightarrow V_{n'}$
2. An **isochoric** process from x' to n': $V = V_{n'}$, $p_1 \rightarrow p_{n'}$

Using:

$$\Delta S = nc_p \ln\left(\frac{T_{x'}}{T_1}\right) + nc_v \ln\left(\frac{T_{n'}}{T_{x'}}\right)$$

From the ideal gas law:

$$T_{x'} = \frac{p_1 V_{n'}}{nR}, \quad T_{n'} = \frac{p_{n'} V_{n'}}{nR}$$

Substituting these gives the total entropy change, which evaluates numerically to:

$$\Delta S = 4.2 \frac{\text{J}}{\text{K}}$$

This value represents the entropy generated due to irreversibility, since the actual adiabatic path is not reversible. The increase in entropy and volume $V_{n'} > V_n$ highlights the thermodynamic consequences of performing the expansion non-quasistatically.

Table 5.1. Thermodynamic state variables for an ideal diatomic gas undergoing adiabatic expansion—comparison between quasistatic and non-quasistatic processes. The table lists pressure, volume, internal energy, and temperature for initial state, final states (quasistatic and non-quasistatic), and intermediate reference states used for entropy calculations.

State	State 1	State n	State n'	State x'
p (bar)	10	1	1	10
V (L)	1	5.18	7.43	7.43
U (J)	2500	1295	1857	18571
T (K)	300	155	223	2229

Main findings for adiabatic expansion comparison
- **Work comparison**: The magnitude of W in the non-quasistatic expansion process is lower than in the quasistatic process.
- **Volume difference**: In the non-quasistatic process, the final volume $V_{n'}$ is greater than the quasistatic final volume V_n, as irreversible entropy generation requires a larger volume change.
- **Entropy generation**: The entropy change in the non-quasistatic process is positive due to irreversible entropy generation. In contrast, the entropy change for the quasistatic process is zero, making it reversible.

This example clarifies how entropy generation and work differ in quasistatic versus non-quasistatic adiabatic expansions. By following alternative paths (involving states x and x') for entropy calculations, we can compare these processes with a more systematic approach. To deepen your understanding, you are encouraged to conduct a similar analysis for an **adiabatic compression process**, considering both quasistatic and non-quasistatic scenarios. Start with the same initial conditions and examine how changes in internal energy, work done, and entropy differ between the two cases. By following a similar methodology—perhaps using an alternative path with intermediate states as we did with states x and x'—you will gain insights into the effects of irreversibility on entropy generation and final state differences during compression. This exercise will further illustrate the unique characteristics of non-quasistatic processes.

5.6.2.2 Isothermal compression—quasistatic versus non-quasistatic
Now let us examine an ideal gas compression process under isothermal conditions, contrasting a quasistatic process with a non-quasistatic one.

5.6.2.2.1 Quasistatic isothermal compression
The gas is initially at a pressure of 1 bar, a volume of 1.0 l, and a temperature of 300 K. It is compressed isothermally to a final volume of 0.1 l, with the surroundings as a constant-temperature reservoir at 290 K.

The work done in the quasistatic isothermal process is

$$W = -\int_1^n p\,dV = -nRT_0 \ln \frac{V_f}{V_i} = -p_i V_i \ln \frac{V_f}{V_i} = 100 \ln 10 \text{ J}.$$

With no change in internal energy ($\Delta U = 0$) for an isothermal process, the heat transfer is

$$Q = -W = -100 \ln 10 \text{ J}.$$

The entropy change of the system is

$$\Delta S_{\text{system}} = \int \frac{Q}{T} = \frac{-100 \ln 10}{300} = -\frac{\ln 10}{3} \text{ J K}^{-1}.$$

Since the surroundings act as a heat reservoir, the entropy change is

$$\Delta S_{\text{surroundings}} = \frac{-Q}{T_{\text{reservoir}}} = \frac{100 \ln 10}{290} = \frac{300 \ln 10}{3 \times 290} \text{ J K}^{-1}.$$

Thus, the total entropy change is

$$\Delta S_{\text{total}} = \Delta S_{\text{system}} + \Delta S_{\text{surroundings}} = \frac{\ln 10}{3} \left(\frac{300}{290} - 1 \right) > 0.$$

5.6.2.2.2 *Non-quasistatic isothermal compression*

In this case, an external force exerts a constant pressure of 10 bar on the gas, causing an instantaneous compression to 0.1 l. The final temperature is maintained at 300 K.

The work done in the non-quasistatic process is

$$W = -p_{\text{ext}} (V_f - V_i) = -10 \text{ bar} \times (0.9 \text{ l}) = 900 \text{ J}.$$

With no change in internal energy ($\Delta U = 0$), the heat transfer is

$$Q = -W = -900 \text{ J}.$$

The entropy change of the system is the same as in the quasistatic process:

$$\Delta S_{\text{system}} = -\frac{\ln 10}{3} \text{ J K}^{-1}.$$

However, the entropy change of the surroundings differs due to the larger heat transfer:

$$\Delta S_{\text{surroundings}} = \frac{-Q}{T_{\text{reservoir}}} = \frac{900}{290} \text{ J K}^{-1}.$$

The total entropy change for the non-quasistatic process is

$$\Delta S_{\text{total}} = \Delta S_{\text{system}} + \Delta S_{\text{surroundings}} = -\frac{\ln 10}{3} + \frac{900}{290} > 0.$$

This larger total entropy change in the non-quasistatic process indicates higher irreversibility than in the quasistatic case.

5.6.2.2.3 *Main findings*

- **Work comparison**: The magnitude of W in non-quasistatic compression processes is generally higher than in quasistatic processes.
- **Entropy generation**: Entropy change in the non-quasistatic process is positive due to irreversible entropy generation, whereas the total entropy change for the quasistatic process is zero, making it a reversible process.

In summary, both examples demonstrate that non-quasistatic processes lead to higher entropy generation, indicating greater irreversibility compared to their quasistatic counterparts. Readers are encouraged to apply this approach to other

scenarios, such as non-quasistatic adiabatic compression, as well as to explore arbitrary variations in external conditions.

It is important to note that there are countless ways in which a non-quasistatic process can unfold, as the system is not constrained to maintain equilibrium with its surroundings. Because of this, analysing a non-quasistatic process requires specific information about how the process occurred or detailed data about the states involved. Only with this information—whether through measurement, experimental data, or clear process definitions—can an accurate analysis of the system's behavior and energy transformations be performed.

5.7 Thermodynamic identity

In this final section, we introduce the **thermodynamic identity**, a fundamental relation that links internal energy, entropy, temperature, pressure, and volume in a concise mathematical form. This identity encapsulates the changes in internal energy in terms of entropy and volume, providing a powerful tool for analysing thermodynamic processes.

5.7.1 Derivation of the thermodynamic identity

We begin by considering entropy S as a function of internal energy U, volume V, and particle number N. For simplicity, we will focus on a fixed number of particles, so entropy becomes a function of U and V alone:

$$S = S(U, V).$$

Taking the differential of entropy S, we have

$$dS = \left(\frac{\partial S}{\partial U}\right)_V dU + \left(\frac{\partial S}{\partial V}\right)_U dV.$$

From our previous discussions on thermal and mechanical equilibrium, we defined **temperature** T and **pressure** p in terms of these partial derivatives:

$$\frac{1}{T} = \left(\frac{\partial S}{\partial U}\right)_V$$

$$p = T\left(\frac{\partial S}{\partial V}\right)_U.$$

Substituting these definitions into the differential of entropy, we obtain

$$dS = \frac{1}{T}dU + \frac{p}{T}dV.$$

Rearranging, we can multiply through by T to isolate dU:

$$TdS = dU + pdV.$$

Rearranging further, we arrive at the **thermodynamic identity**:

$$dU = TdS - pdV.$$

This identity provides an essential link between state variables, relating changes in **internal energy** U to changes in **entropy** S and **volume** V. It reveals that, for a small change in a system's internal energy, the amount of energy is partitioned between the change in entropy (multiplied by temperature) and the change in volume (multiplied by pressure).

5.7.2 Significance of the thermodynamic identity

The thermodynamic identity captures the fundamental interplay of energy, entropy, and volume in thermodynamic systems. Each term in the identity—TdS and $-pdV$—represents a component of the system's internal energy change:

- **Entropy-related energy change**: The term TdS represents the portion of internal energy change associated with entropy change. This includes contributions from both heat transfer and irreversible processes, such as frictional effects or dissipative work, which increase entropy within the system.
- **Work**: The term $-pdV$ corresponds to the work associated with changes in the system's volume. This is often called boundary work, as it reflects the energy required for the system to expand or compress against an external pressure.

Together, these terms describe how internal energy changes in response to both entropy and volume variations, making the thermodynamic identity a versatile tool for analysing any thermodynamic process, whether reversible or irreversible.

This identity is foundational because it holds for any infinitesimal process, whether it is reversible or irreversible, as long as it involves small changes. It allows us to analyse complex processes by breaking them down into contributions from heat and work, thereby deepening our understanding of how energy is conserved and transformed in thermodynamic systems.

5.7.3 Thermodynamic identity clarification

In deriving the thermodynamic identity, many authors take a simplified approach, starting directly from the first law of thermodynamics $dU = Q + W$ and substituting $Q = TdS$ and $W = -pdV$. While this approach is often convenient, it overlooks an important aspect of real thermodynamic systems—namely, that other forms of work transfer, such as friction and shaft work, can occur, especially in irreversible processes.

Our derivation, however, takes a more comprehensive route by incorporating the idea that *entropy change results from both heat transfer and irreversibilities*. Specifically, we start with the expression $dS = \frac{Q}{T} + S_{irr}$, where S_{irr} represents the entropy generated due to irreversible effects within the system. Thus, the internal energy change can be more completely expressed as

$$dU = Q + W_{pV} + W_{other},$$

where:
- Q represents heat transfer, leading to a change in system entropy,
- $W_{pV} = -pdV$ is the work associated with volume change (boundary work), and
- W_{other} represents additional forms of work, such as frictional or shaft work, which can contribute to entropy generation.

This perspective respects the contributions of all energy transfers and entropy changes in real-world systems, offering a more complete picture. The entropy change in this approach accounts for both the reversible entropy change from heat transfer and the irreversible contribution from other forms of work.

5.7.4 The thermodynamic identity in any process

Through this derivation, we obtain the thermodynamic identity

$$dU = TdS - pdV.$$

This identity holds universally for any infinitesimal process, regardless of whether it is reversible or irreversible, because it shows the interplay of **state variables**—internal energy U, entropy S, temperature T, pressure p, and volume V—that define the system. By deriving it through the fundamental principles of entropy and energy transfer, we ensure that it remains applicable even when additional irreversibilities and forms of work transfer are present.

This approach not only honors the first law but also provides a fuller understanding of energy and entropy changes, making it a powerful tool for analysing any thermodynamic process, no matter the complexity.

It is often a limiting perspective that, when first introduced to thermodynamics, readers may come away with the impression that it applies only to mechanical or pneumatic systems—those involving heat transfer, expansion work, and similar phenomena. But this is incorrect. The principles of thermodynamics are universal and extend well beyond mechanical systems. In many physical domains—such as electrical, magnetic, and chemical systems—additional forms of work must be considered. For instance, electrical systems may involve terms like ϕdQ (electric potential times charge), magnetic systems include HdM (magnetic field times change in magnetization), and chemical systems require terms like μdN (chemical potential times particle number). These additional contributions modify the thermodynamic identity and show that entropy can change through more than just heat transfer. Nevertheless, even in such cases, the mechanical work term pdV does not vanish. As long as the system undergoes volume changes under ambient conditions, work is done by or against the surrounding atmosphere, and this must be accounted for in the total energy balance.

5.8 Summary

This chapter provides a detailed examination of thermodynamic processes that extend beyond idealized, reversible assumptions. Through an analysis of quasistatic

processes with irreversible work transfer, we highlight the essential role of dissipative forces—such as friction and inelastic deformation—in generating entropy and causing energy losses that cannot be fully recovered. By distinguishing between reversible and irreversible work components, we gain a clearer understanding of how energy distribution is affected by irreversibilities in real-world systems.

In discussing non-quasistatic processes, the chapter underscores the limitations of equilibrium-based thermodynamic analysis and introduces methodologies for evaluating systems far from equilibrium. Importantly, we have shown a clear strategy for determining changes in internal energy and entropy in non-quasistatic processes. Since these quantities are state functions, their values depend only on the initial and final states—not the path taken. This allows one to calculate entropy or internal energy changes by conceptually constructing alternate quasistatic paths, even if the actual process is highly irreversible or lacks intermediate equilibrium states.

The chapter concludes with the derivation of the thermodynamic identity, incorporating both heat transfer and irreversibilities. This identity provides a unified framework for analysing energy changes in thermodynamic systems, applicable to any infinitesimal process, whether reversible or irreversible. Through this comprehensive approach, we establish the thermodynamic identity as a powerful tool for understanding the interplay of state variables, reinforcing its foundational role in thermodynamics.

Chapter 6

Thermodynamic cycles

6.1 Introduction

Thermodynamic cycles are at the core of many energy transformation systems, from engines that produce mechanical work to coolers that maintain low temperatures. These cycles represent closed-loop sequences of thermodynamic processes through which a working substance interacts with its surroundings. A key concept in understanding thermodynamic cycles is the distinction between a process, where a system transitions between states, and a cycle, where the system returns to its initial state.

We will begin this chapter by distinguishing between a thermodynamic process and a thermodynamic cycle. This distinction will be illustrated using simple real-world examples. We will then explore why thermodynamic cycles are so important in practice, especially in energy conversion systems. From there, we will introduce the classification of thermodynamic cycles into power cycles, refrigeration cycles, and heat pump cycles. We will focus primarily on closed systems and pure substances, although the same principles apply more broadly. Finally, we will examine a few representative idealized cycles to illustrate the underlying principles of thermodynamic analysis.

This chapter bridges the theoretical framework of thermodynamic processes discussed earlier with real-world systems.

6.2 The difference between a process and a cycle

In thermodynamics, it is essential to understand the difference between a *process* and a *cycle* to grasp how energy transformation and efficiency work within systems. This distinction is fundamental because processes and cycles serve unique roles in managing energy flow, achieving specific goals such as cooling, heating, or work generation.

doi:10.1088/978-0-7503-6029-6ch6

6.2.1 The definition of a thermodynamic process

A **thermodynamic process** refers to a change in the state of a system from an initial state to a final state. This can involve changes in variables such as pressure, volume, and temperature, which occur as the system undergoes energy interactions (heat transfer or work). Importantly, a process is not necessarily closed or repeatable—it can happen only once, and the system might not return to its original state.

6.2.2 Definition of a thermodynamic cycle

A **thermodynamic cycle**, on the other hand, is a sequence of processes that ultimately returns a system to its original state. In other words, in a cycle, the initial and final states are the same. This closed-loop sequence allows for continuous operation, making cycles essential in applications such as engines and refrigeration systems where consistent performance over time is required. Cycles enable a system to repeatedly transform energy, often converting heat into work or vice versa.

6.2.3 Examples to differentiate a process and a cycle

- **Air conditioning in an airplane (process)**: Air conditioning in an airplane is an example of an **open system** that undergoes a thermodynamic process but does not complete a closed cycle. The air conditioning system relies on compressed air to cool the cabin, especially when the aircraft is on the ground. Here is how the process typically works:
    1. *Compressed air supply*: When the aircraft is on the ground, an auxiliary power unit (APU) provides compressed air, which is directed into an air conditioning unit to initiate the cooling process.
    2. *Primary heat exchange*: The high-pressure, high-temperature air flows through a primary heat exchanger, where it loses heat to ambient air, cooling down but remaining under relatively high pressure.
    3. *Expansion for cooling*: This cooled, compressed air then passes through a turbine, where it expands, causing a sharp drop in temperature. This expansion produces the cold air needed for cabin cooling.
    4. *Continuous one-way flow*: This cold air is then distributed throughout the cabin and eventually vented out. Because this air is not recycled back to its original state, the air conditioning system is considered an open system performing a thermodynamic *process* rather than a cycle.
- **Brayton cycle (closed cycle)**: In contrast, the **Brayton cycle** is a thermodynamic cycle commonly used in closed-loop applications, such as gas turbines and some advanced refrigeration systems. The closed Brayton cycle consists of a series of processes—compression, heat exchange, and expansion—that allow the system to return to its initial state by the end of each cycle. Here is how it operates:
    1. *Compression*: The working fluid (often air or a gas) is compressed, which raises its pressure and temperature.

2. *Heat exchange*: The compressed air then flows through a heat exchanger, where it releases some of its heat to the surroundings. This stage reduces the temperature of the air while it remains under high pressure.
3. *Expansion*: The cooled, compressed air is then expanded through a turbine, where it drops in both temperature and pressure, generating the desired cooling effect.
4. *Closed-loop recirculation*: The low-pressure, cold air is recirculated back to the compressor, where it repeats the cycle. In this closed-loop configuration, the Brayton cycle can continuously perform cooling or power generation, depending on the set-up.

This closed Brayton cycle is distinct from the one-way flow in an airplane air conditioning system because the working fluid undergoes a continuous loop, returning to its original state after each cycle. This closed-loop operation allows for sustained energy transformation, highlighting the difference between an ongoing **process** and a repeatable thermodynamic **cycle**.

6.3 Why do we need cycles?

In thermodynamics, cycles serve as a foundational concept for understanding how energy is transformed and utilized efficiently. The primary purpose of cycles is to facilitate the conversion of energy from one form to another. This is particularly important in applications where heat, a disorganized form of energy, needs to be transformed into useful work or transferred efficiently for heating or cooling purposes.

A thermodynamic cycle consists of a sequence of processes that return a system to its original state. By completing this cycle, energy can be extracted or supplied repeatedly without permanently altering the working substance. This is critical in engineering systems that need continuous operation, such as engines, refrigerators, and power plants.

The essence of a cycle lies in its ability to exploit the inherent properties of materials, such as gases, liquids, or mixtures, to achieve this transformation. For instance:

- In **power generation**, cycles such as the Brayton or Rankine cycles convert thermal energy from fuel combustion or solar heating into mechanical work, which is then turned into electricity.
- In **refrigeration and heat pumps**, cycles such as the vapor-compression cycle utilize work input to transfer heat from a cooler region to a hotter one, enabling cooling or heating, respectively.
- Advanced technologies, such as **magnetic refrigerators** or **dilution fridges**, employ specialized cycles to achieve extremely low temperatures for scientific or industrial purposes.

Cycles are not only essential for energy conversion but also for optimizing efficiency. They allow engineers to design systems that maximize the useful output from a given energy input while minimizing losses. The Carnot cycle, for example, provides an idealized benchmark for the maximum efficiency a heat engine can achieve between two temperatures.

6.4 Examples of thermodynamic cycles across applications

Thermodynamic cycles are the backbone of numerous technologies, enabling energy conversion and heat transfer for a wide range of applications. Below are examples that illustrate how different cycles are utilized across various domains.

6.4.1 Solar power plant: using cycles for renewable energy generation

In solar power plants, thermodynamic cycles such as the **Rankine cycle** or **Brayton cycle** play a crucial role in converting solar energy into electricity. Concentrated solar power (CSP) systems, for example, use mirrors or lenses to focus sunlight onto a heat transfer fluid. The thermal energy is then used to produce steam, driving a turbine connected to an electric generator. These cycles ensure efficient energy conversion while promoting sustainability by utilizing an abundant and clean energy source.

6.4.2 Home refrigerator: an example of a refrigeration cycle for cooling

The **vapor-compression refrigeration cycle** is fundamental to modern refrigeration systems found in homes. This cycle transfers heat from the interior of the refrigerator to the surrounding environment, keeping food and beverages cool. Key processes involve compression, condensation, expansion, and evaporation, where a refrigerant circulates through the system, absorbing heat inside the refrigerator and releasing it outside. The cycle operates efficiently to maintain the desired temperature with minimal energy consumption.

6.4.3 Heat pump: cycle application for heating purposes

A **heat pump** is essentially a refrigeration cycle in reverse, designed to move heat from a cooler region to a warmer one. In cold climates, heat pumps efficiently extract heat from outdoor air (or other sources such as the ground) and deliver it indoors for heating purposes. By using a thermodynamic cycle similar to the vapor-compression cycle, heat pumps provide a sustainable and energy-efficient alternative to traditional heating methods, reducing greenhouse gas emissions.

6.4.4 Magnetic refrigerator: non-gas-based cooling technology

Magnetic refrigeration relies on the **magnetocaloric effect** instead of conventional gases. In this cycle, a magnetic material is cyclically magnetized and demagnetized, causing changes in its temperature. The material absorbs heat from the space to be cooled during demagnetization and releases heat during magnetization. Magnetic

refrigerators offer a promising alternative to traditional refrigeration systems, with the potential for higher efficiency and the elimination of harmful refrigerants.

6.4.5 Dilution refrigerator: used in ultra-low-temperature physics

The **dilution refrigerator** is a specialized device used in scientific research to achieve temperatures close to absolute zero. This technology operates on the principle of phase separation in a mixture of helium-3 and helium-4 isotopes. The cycle involves continuous dilution of helium-3 into helium-4 at ultra-low temperatures, absorbing heat and maintaining the desired low temperature. Dilution refrigerators are indispensable for experiments in quantum computing, superconductivity, and other fields requiring precise control over extreme cryogenic conditions.

6.5 Classification of thermodynamic cycles

Thermodynamic cycles can be broadly categorized based on their primary function. While the underlying principles remain similar, the objectives of these cycles determine their design and operational parameters. Two major categories are power cycles and refrigeration cycles.

6.5.1 Power cycles: designed to produce work

Power cycles are engineered to convert heat into mechanical work, which is often further transformed into electrical energy. These cycles are integral to the functioning of power plants and other energy generation systems.

Key examples include:

- **Rankine cycle**: This is the backbone of most thermal power plants, where steam is generated by heating water, used to drive a turbine, and then condensed back into water to complete the cycle. It is commonly employed in coal, nuclear, and solar thermal power plants.
- **Brayton cycle**: Found in gas turbines and jet engines, this cycle involves compressing air, heating it by fuel combustion, and then expanding the hot gases through a turbine to produce work.
- **Otto and Diesel cycles**: These are internal combustion engine cycles used in automobiles. The Otto cycle is characteristic of gasoline engines, while the Diesel cycle is used in diesel engines.

Power cycles are designed to maximize efficiency, often operating between high- and low-temperature reservoirs. The efficiency is fundamentally limited by the laws of thermodynamics, with the Carnot cycle serving as an idealized benchmark.

6.5.2 Refrigeration cycles: designed to transfer heat

Refrigeration cycles, in contrast, focus on transferring heat from a lower-temperature region to a higher-temperature one, often against the natural direction of heat flow. These cycles rely on work input to achieve their objectives, making them essential for cooling and heating applications.

Key examples include:

- **Vapor-compression refrigeration cycle**: This is the most widely used refrigeration cycle, found in home refrigerators, air conditioners, and heat pumps. It employs a refrigerant that circulates through compression, condensation, expansion, and evaporation processes to move heat.
- **Absorption refrigeration cycle**: This cycle uses a heat source instead of mechanical work for driving the refrigeration process. It is often employed in industrial and off-grid applications.
- **Reverse Brayton cycle**: A variation of the Brayton cycle used for cooling, especially in cryogenic systems and aerospace applications.

While power cycles are aimed at extracting useful work from heat, refrigeration cycles are designed to control thermal conditions by absorbing or rejecting heat. Together, these two classifications encompass the majority of practical thermodynamic cycles, enabling diverse energy and thermal management applications.

6.6 Cycle performance

In a thermodynamic cycle, the working substance—the system—undergoes a series of processes that eventually return it to its initial state. Previously, we examined processes and energy interactions between the system and its surroundings. In cycles, multiple processes can occur simultaneously across different parts of the working substance. For instance, while one part of the substance absorbs heat, another part may be rejecting heat or performing work. This simultaneous occurrence is characteristic of most thermodynamic cycles, where the system as a whole continuously evolves through the cycle. However, in time-separated cycles, such as magnetocaloric cycles, the entire system undergoes one process at a time in sequence. This distinction helps us understand how the working substance interacts with surroundings throughout a cycle, which we will explore in detail in this section.

We will use this simple four process in a cycle shown on a property diagram (figure 6.1) to discuss the thermodynamics of a cycle.

In a thermodynamic cycle, the working fluid undergoes a series of processes and ultimately returns to its initial state. This cyclic behavior ensures that the net change in state properties, such as pressure, volume, internal energy, and entropy, over a complete cycle is zero:

$$\Delta h_{12} + \Delta h_{23} + \Delta h_{34} + \Delta h_{41} = 0, \quad \Delta s_{12} + \Delta s_{23} + \Delta s_{34} + \Delta s_{41} = 0.$$

From the first law of thermodynamic for a closed system, the total energy change over the cycle is

$$\Delta h_{12} + \Delta h_{23} + \Delta h_{34} + \Delta h_{41} = \{Q_{12} + W_{12}\} + \{Q_{23} + W_{23}\}$$
$$+ \{Q_{34} + W_{34}\} + \{Q_{41} + W_{41}\} = 0$$

$$\sum \Delta h = \sum Q + \sum W = 0.$$

Therefore, the sum of heat and work transfer in a cycle is zero.

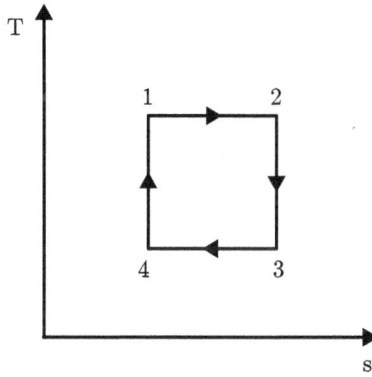

Figure 6.1. A property diagram of a simple four process forming a thermodynamic cycle.

The second law of thermodynamics introduces the concept of entropy, stating that for a reversible cycle,

$$0 = \sum \frac{\dot{Q}}{T} + s_{\text{irr}}.$$

These principles are used to analyse engines, coolers, and heat pumps, each of which involves specific processes of heat absorption, heat rejection, and work input/output.

The performance of a thermodynamic cycle is expressed as the ratio of the desired output to the input effort:

$$\text{performance merit} = \frac{\text{desired}}{\text{effort}}.$$

6.6.1 Engine performance

An engine converts heat absorbed from a high-temperature reservoir (Q_h) into work (W) while rejecting a portion of the heat to a low-temperature reservoir (Q_c).

Figure 6.2 presents a thermodynamic diagram illustrating the global energy flows within a system undergoing a thermodynamic cycle. Unlike the idealized piston–gas systems studied earlier, where the working substance is assumed to undergo the same process at all points, real systems used for energy conversion involve more complex behaviors. In practical systems, heat input (Q_h) and heat output (Q_c) occur at specific locations within the system, corresponding to the hot region and cold region of the working substance. These regions are distinct and physically localized, and they drive the thermodynamic cycle by maintaining a temperature difference within the system.

It is crucial to note that the heat absorption and rejection do not take place uniformly across the system, but at these specific regions. The 'Hot' and 'Cold' labels in the figure represent the temperatures of the working substance in the hot and cold regions of the system, not abstract reservoirs. In the literature, many authors derive performance expressions assuming interactions between the engine and external hot

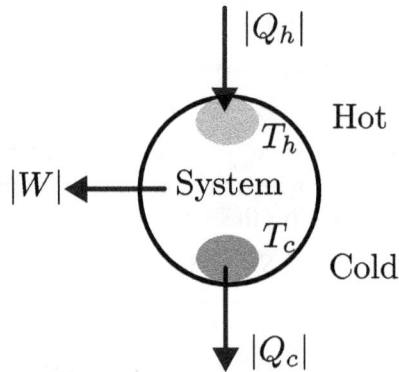

Figure 6.2. Schematic representation of energy flows in a heat engine cycle. The system absorbs heat $|Q_h|$ at a high internal temperature $|T_h|$, rejects heat $|Q_c|$ at a lower internal temperature $|T_c|$, and produces net work output $|W|$. The labels 'Hot' and 'Cold' refer to the relative temperatures within the system during different parts of the cycle.

and cold reservoirs. However, this approach is fundamentally flawed because reservoir temperatures are not system properties. Instead, the performance of the cycle depends on the temperatures of the working substance at these specific locations within the system.

The efficiency (η) of the engine is

$$\text{Efficiency, } \eta = \frac{\text{work output (J or W)}}{\text{heat input (J or W)}} = \frac{|W|}{|Q_h|}.$$

Entropy changes during the heat transfer processes balance out over a cycle. For an engine,

$$\frac{|Q_h|}{T_h} - \frac{|Q_c|}{T_c} + s_{\text{irr}} = 0.$$

We will derive the performance expression for an engine using the T–S diagram provided. This derivation assumes that the *irreversible entropy generation within the system is zero*, meaning that all entropy changes in the system arise solely due to *heat transfer* between the working substance and its surroundings. This does not imply that the cycle is reversible, as heat transfer processes are inherently irreversible. However, we focus on systems where the internal processes (such as friction, turbulence, or mixing) do not contribute additional entropy generation. The entropy changes associated with heat transfer are explicitly accounted for, and these form the basis of the derivation.

6.6.1.1 Heat transfers in the cycle
1. **Heat transfer at the hot side** (Q_h): Heat is absorbed by the working substance during the process $1 \rightarrow 2$, where the working substance is at a temperature T_h. The entropy increase during this process is

$$\Delta S_{12} = S_2 - S_1.$$

The heat absorbed is given by

$$Q_H = T_H \, \Delta S_{12} = T_H \, (S_2 - S_1).$$

2. **Heat transfer at the cold side** (Q_c): Heat is rejected by the working substance during the process $3 \rightarrow 4$, where the working substance is at a temperature T_c. The entropy decrease during this process is

$$\Delta S_{34} = S_4 - S_3.$$

The heat rejected is

$$Q_c = T_c \, \Delta S_{34} = T_c \, (S_4 - S_3).$$

6.6.1.2 Entropy balance in the cycle
While heat transfer processes are irreversible, the assumption here is that no additional entropy is generated within the system beyond what is transferred during heat exchange. Thus, the total entropy change in the cycle is zero:

$$\sum \Delta S = \Delta S_{12} + \Delta S_{23} + \Delta S_{34} + \Delta S_{41} = 0.$$

In the cycle:
- $\Delta S_{23} = 0$ (isentropic process),
- $\Delta S_{41} = 0$ (isentropic process).

This simplifies the entropy balance to

$$\Delta S_{12} + \Delta S_{34} = 0.$$

Substituting the expressions for entropy changes during heat transfer:

$$\frac{Q_h}{T_h} + \frac{Q_c}{T_c} = 0 \ \text{ or } \ \frac{|Q_h|}{T_h} = \frac{|Q_c|}{T_c}.$$

6.6.1.3 Efficiency of the engine
From the first law of thermodynamics, the net work produced is

$$W = Q_h - Q_c.$$

Substituting this into the efficiency expression,

$$\eta = \frac{W}{Q_h} = \frac{Q_h - Q_c}{Q_h} = 1 - \frac{Q_c}{Q_h}.$$

Using the entropy relation $\left(\frac{|Q_h|}{T_h} = \frac{|Q_c|}{T_c} \right)$ the maximum efficiency of an engine is

$$\eta_{\max} = 1 - \frac{T_c}{T_h}.$$

6.6.1.4 Key insights and clarifications

1. **Zero internal irreversibility**: The assumption is that internal sources of entropy generation, such as friction or turbulence, are negligible. Hence, the only entropy changes are due to heat transfer. Therefore, the derived expression gives the maximum efficiency of an engine operating between two temperatures.
2. **Hot and cold regions**: The heat input (Q_h) occurs at a specific hot region with temperature T_h, and heat rejection (Q_c) occurs at a specific cold region with temperature T_c. These are *system properties* and not external reservoirs.
3. **Practical engine representation**: The cycle represents a realistic engine system with physically meaningful locations for heat input and output, ensuring that the derived efficiency is grounded in the system's actual behavior.

6.6.2 Cooler and heat pump performance

The analysis of a cooler or a heat pump follows a similar thermodynamic framework as the engine. However, the key difference lies in the direction of the cycle and the purpose of energy conversion. In a cooler or a heat pump, the cycle traverses in the *opposite direction* compared to an engine. Instead of converting heat into work, as in an engine, these devices *convert work* (or *electricity*) into thermal energy—either to extract heat from a cold region or to supply heat to a warm region (figure 6.3).

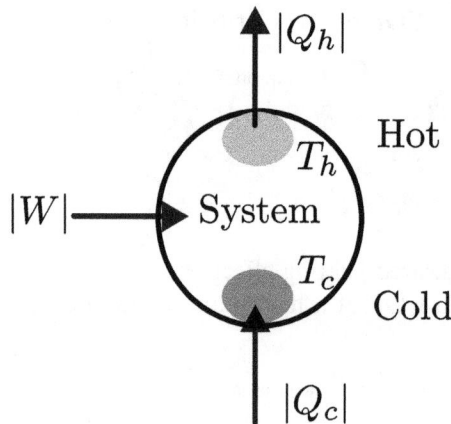

Figure 6.3. Schematic representation of a refrigeration or heat pump cycle. The system extracts heat Q_c at a lower internal temperature T_c, receives input work W, and rejects heat Q_h at a higher internal temperature T_h. The labels 'Cold' and 'Hot' refer to the system's internal temperatures during these interactions.

6.6.3 Key similarities and differences between a cooler and a heat pump

From a thermodynamic perspective, coolers and heat pumps operate on the *same principles* and share identical cycles. The distinction lies in the *desired energy output*:

1. **Cooler**:
   - The primary goal is *cooling*.
   - Heat (Q_c) is extracted from the cold region at temperature T_c. The rejected heat (Q_h) at temperature T_h is considered a by-product.
   - Performance is evaluated using the **coefficient of performance (COP)**, which measures the cooling effect relative to the work input.

2. **Heat pump**:
   - The primary goal is *heating*.
   - Heat (Q_h) supplied to the warm region at temperature T_h is the energy of interest. The extracted heat (Q_c) from the cold region is a means to achieve this.
   - Performance is also evaluated using the COP, but in this case, it measures the heating effect relative to the work input.

The cycle on a T–S diagram is reversed compared to an engine, reflecting the reverse flow of energy.

Coefficient of performance (COP)

The COP is a measure of the cycle's performance, expressed as the ratio of the desired energy output to the work input.

- **COP for a cooler**:

$$\mathrm{COP}_{\text{cooler}} = \frac{|Q_c|}{|W|}.$$

Here, Q_c is the desired cooling effect.
The maximum COP of a cooler (without internal irreversibilities) is

$$\mathrm{COP}_{\text{cooler,max}} = \frac{T_c}{T_h - T_c}.$$

- **COP for a heat pump**:

$$\mathrm{COP}_{\text{heat pump}} = \frac{|Q_h|}{|W|}.$$

Here, Q_h is the desired heating effect.
The maximum COP of a heat pump (without internal irreversibilities) is

$$\mathrm{COP}_{\text{heat pump,max}} = \frac{T_h}{T_h - T_c}.$$

6.7 Ideal versus real cycles

Thermodynamic cycles can be categorized as **ideal cycles** and **real cycles**, with ideal cycles serving as theoretical benchmarks for performance. While real cycles represent practical systems, ideal cycles provide insights into the maximum efficiency a system can achieve under perfect conditions.

The cycle discussed earlier in the T–S diagram is a **Carnot cycle**, characterized by $s_{irr} = 0$. This means that the cycle is free of internal irreversibilities, and all entropy changes are solely due to heat transfer to or from the working substance. The Carnot cycle is the epitome of an ideal cycle, serving as the benchmark for all practical thermodynamic systems. However, it is important to note that the Carnot cycle is purely a thought experiment and cannot be realized in practice. Its assumptions, such as perfectly reversible processes and no heat transfer losses, are idealized and unattainable in real systems. Nonetheless, the Carnot cycle provides a theoretical upper limit for the performance of any thermodynamic system operating between two temperature limits.

6.7.1 Entropy generation and cycle irreversibility

The distinction between ideal and real cycles rests primarily on whether or not irreversible entropy generation is present. In an ideal cycle, all processes are quasistatic and reversible. Entropy is transferred through the system by heat exchange alone, and the net entropy generation within the system is zero. As a result, such cycles achieve maximum theoretical performance. However, these are idealizations.

In practical, real-world systems, some steps in the cycle inevitably deviate from reversibility. This deviation may be due to friction, rapid (non-quasistatic) compression or expansion, turbulent mixing, viscous flow, or uncontrolled heat transfer across finite temperature differences. All these effects lead to irreversible entropy generation within the system.

To visualize the consequences, consider the schematic in figure 6.4. Both panels represent cooling cycles, where the objective is to extract heat $\left|Q_c\right|$ at a low temperature T_c and reject it at a higher temperature T_h, supported by mechanical work $\left|W\right|$. In the left panel, the processes are quasistatic: there is no internal entropy generation, and all entropy delivered to the hot side originates from the cold-side heat absorption. Since entropy is a state function, the system's entropy returns to its initial value after a complete cycle. Thus, in this ideal case, the entropy flow out of the system (via heat at T_h) equals the entropy flow into it (via heat at T_c).

In contrast, the right panel shows a more realistic situation where some steps are non-quasistatic. While the system still performs the same cooling function, the presence of irreversible entropy generation leads to an excess of entropy that must be discharged to the hot side. This results in two important consequences. First, for the same heat absorption $\left|Q_c\right|$ at the cold end, the heat rejected $\left|Q_h\right|$ to the hot side is now larger. Second, more work input $\left|W\right|$ is required to drive the cycle. The internal

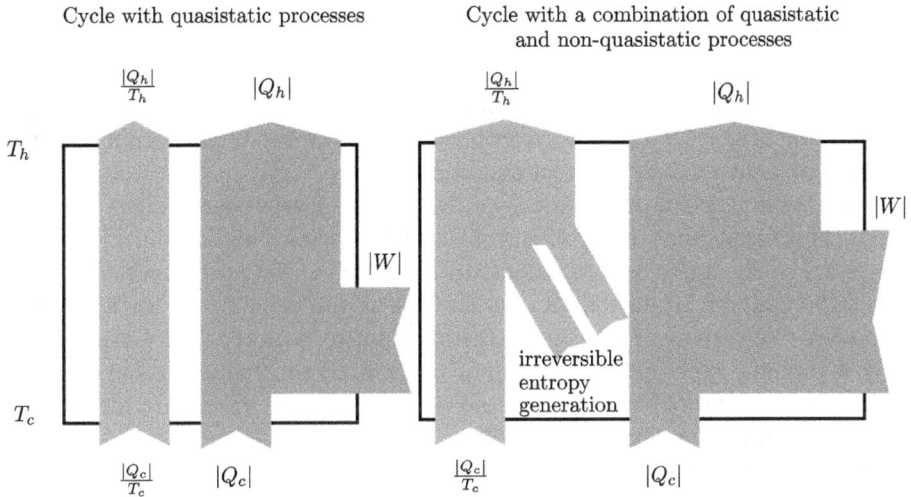

Figure 6.4. Comparison between a quasistatic and a non-quasistatic cooling cycle. In both cases, the system absorbs heat $|Q_c|$ at a lower temperature T_c, receives work input $|W|$, and rejects heat $|Q_h|$ at a higher temperature T_h. In the quasistatic case (left), all entropy transfer occurs via heat exchange, with no internal generation. In the non-quasistatic case (right), irreversible processes within the system generate additional entropy, increasing both the rejected heat and the required work input.

generation of entropy effectively degrades the performance, either by increasing energy input requirements or reducing the net cooling achieved.

The key message here is that the total entropy change of the system over one complete cycle is always zero, regardless of whether the cycle is reversible or not. But the entropy exported to the environment differs significantly between reversible and irreversible cycles, and this directly impacts the efficiency of real devices. Readers are encouraged to sketch a similar energy–entropy flow diagram for a heat engine cycle, identifying the entropy exchange and possible irreversible contributions in that case.

Key characteristics of ideal cycles

1. **No irreversible entropy generation**: Internal irreversibilities, such as friction or turbulence, are absent. The entropy changes observed are due only to heat transfer.
2. **Quasistatic processes**: All processes occur infinitely slowly to maintain equilibrium at every stage.
3. **Perfect heat transfer**: Heat transfer occurs only at the specific hot and cold regions in the system.
4. **Idealized working substance**: The working substance behaves ideally, following simple equations of state without deviations at extreme conditions.

6.8 Classification of cooling cycles: space-separated versus time-separated

The concept of refrigeration is not limited to a specific working fluid or mechanism. At its core, any system that can cyclically absorb heat at a low temperature and

reject it at a higher temperature—while undergoing changes in entropy—qualifies as a refrigeration cycle. This universality allows the design of cooling systems based on gases, liquids, solids, or even fields (magnetic, electric). The essential requirement is that the working substance must undergo a thermodynamic cycle in which its entropy changes during interactions with heat reservoirs at different temperatures.

Cooling systems, particularly those operating on thermodynamic cycles, can be broadly classified into two categories based on how and where heat exchange occurs relative to the working material or system. This classification—**time-separated** versus **space-separated** cooling systems—is helpful in understanding the operational principles of different cooling technologies.

6.8.1 Time-separated cooling systems

In time-separated cooling systems, the processes of heat absorption (from the cold reservoir) and heat rejection (to the hot reservoir) occur at different times, often involving the same physical location or material. That is, the working material undergoes thermodynamic transformations in place, and at different stages in time, it is exposed alternately to cold and hot reservoirs to absorb or reject heat. These systems generally require intermittent thermal contact with the reservoirs, often realized through components like heat switches.

A good example of this class includes magnetocaloric, electrocaloric, and elastocaloric systems. In such systems, the working material exhibits a change in entropy and temperature under an applied field—magnetic, electric, or mechanical stress, respectively. For instance, in a magnetocaloric refrigerator:

- The material is first magnetized adiabatically, increasing its temperature.
- It is then thermally connected to a hot reservoir to reject heat.
- Afterward, the material is demagnetized, causing it to cool.
- It is then brought into contact with the cold reservoir to absorb heat.

This sequence repeats cyclically, but all heat exchange occurs at the same spatial location, while the timing and control of thermal contact (via switches or motion) define the cycle. The physical system itself does not move, but its thermodynamic state evolves over time in synchronization with the reservoir connections.

6.8.2 Space-separated cooling systems

In contrast, space-separated cooling systems involve movement of the working fluid or working material through physically distinct regions for heat absorption and heat rejection. The heat exchange occurs simultaneously but in different spatial locations —as the working substance flows through or oscillates between designated hot and cold zones.

A classical example is the vapor-compression refrigeration cycle (discussed in the next chapter), used in household refrigerators:

- The working fluid evaporates in the evaporator, located at the cold side, absorbing heat from the refrigerated space.

- It is then compressed and directed to the condenser, located in a different part of the system, where it rejects heat to the surroundings.
- The fluid cycles back after passing through an expansion valve, and the process repeats.

Another important example is the Stirling cryocooler, where the working gas oscillates between hot and cold ends of a cylinder. Each location is thermally anchored to a different reservoir, and the cycle is driven by pistons or displacers.

Thermoelectric devices, although solid-state, fall under the space-separated category. Here, electrical current flows through different regions of a thermoelectric element, causing heat to be absorbed at one junction and rejected at another. The two sides are spatially distinct—meaning no heat switch is required, and heat flows continuously while the device is powered.

6.9 Thermoelectric coolers: a non-gas-based cycle

The Peltier effect occurs when an electric current flows through a circuit made of two different materials (usually semiconductors). The current causes one junction to absorb heat (cooling) and the other junction to reject heat (heating). This heat transfer is a direct result of the charge carriers (electrons or holes) moving through the materials and carrying energy from one side of the system to the other.

Interestingly, the electrons in a thermoelectric circuit behave somewhat like a working fluid in traditional thermodynamic cycles. They are 'pumped' around the circuit, undergoing processes that involve energy absorption and rejection, akin to the heating and cooling processes of gases or liquids. However, instead of changes in pressure or volume, the electrons experience variations in energy states as they move through the different materials and across temperature gradients.

The basic process in a thermoelectric cooler involves the following steps:

1. **Heat absorption** (cooling side): Electric current passes through the thermoelectric module, causing heat to be absorbed from the cooler side (low temperature). This process creates a cooling effect at the interface, drawing heat away from the object being cooled.
2. **Heat rejection** (hot side): The heat absorbed from the cold side, along with the energy input from the electrical work, is rejected to the surroundings at the hot side (high temperature).
3. **Work input** (electric current): The energy required to move heat from the cold side to the hot side is provided by the electric current passing through the system.

This analogy of electrons acting as the 'working fluid' provides a bridge between the behavior of thermoelectric coolers and traditional thermodynamic systems. It emphasizes how energy transfer and work interaction are universal principles, even when no physical fluid is involved.

6.10 External temperatures and maximum performance

Thermodynamic systems, such as engines and coolers, interact with their surroundings to exchange heat and perform work. While we derived performance metrics in earlier sections based on the **internal system properties**—specifically the temperatures of the working substance at hot and cold regions—the reality is that these systems operate in conjunction with external surroundings. The temperatures of these surroundings, referred to as **external temperatures** ($T_{h,\text{ext}}$ and $T_{c,\text{ext}}$), directly influence the achievable performance of real-world systems.

Importantly, the external temperatures are typically *more extreme* than the internal system temperatures because of practical constraints, such as thermal resistance in heat exchangers and imperfect thermal interfaces. This section highlights how incorporating external temperatures into performance calculations results in a more realistic maximum performance that is lower than the ideal values based solely on system properties.

6.10.1 Engine performance with external temperatures

(Figure 6.5) In an engine, the working substance exchanges heat with two distinct regions in the system:

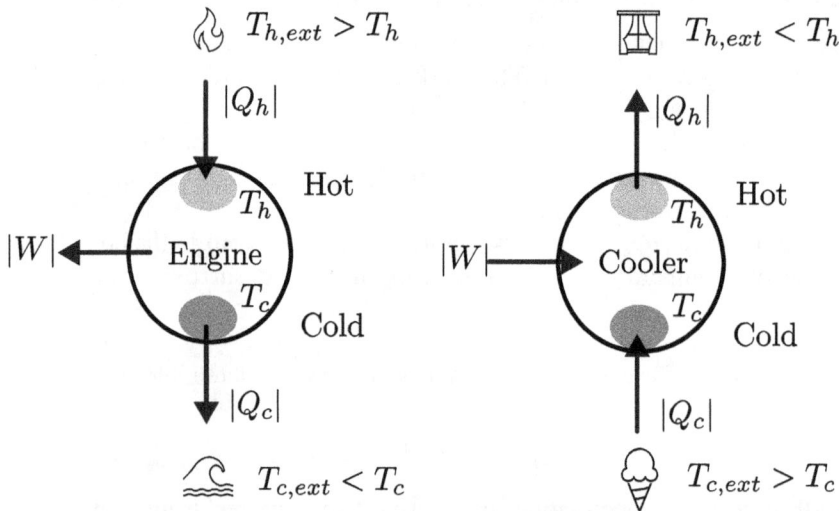

Figure 6.5. Schematic comparison of a heat engine (left) and a cooler or refrigerator (right), both interacting with external reservoirs. In the engine, heat is absorbed from a hot reservoir at $|T_h|$ and partially converted into work $|W|$, while the remaining heat $|Q_c|$ is rejected to a cold reservoir at $|T_c|$. In the cooler, work $|W|$ is supplied to extract heat $|Q_c|$ from a low-temperature reservoir and reject it as $|Q_h|$ to a higher-temperature sink. The external reservoirs ($T_{h,\text{ext}}$, $T_{c,\text{ext}}$) are shown to emphasize that they may differ from the system's internal temperatures $|T_h|$ and $|T_c|$.

- **Heat input** (Q_h): The working substance absorbs heat in the hot region at an internal temperature T_h. This heat originates from a high-temperature external source with a temperature $T_{h,\text{ext}}$.
- **Heat rejection** (Q_c): The working substance rejects heat in the cold region at an internal temperature T_c. This heat is transferred to a low-temperature external sink with a temperature $T_{c,\text{ext}}$.

Due to inefficiencies in heat exchangers and thermal resistances, the following relationships hold:

$$T_{h,\text{ext}} > T_h \text{ and } T_{c,\text{ext}} < T_c.$$

These more extreme external temperatures reduce the effective temperature difference driving the cycle. Consequently, when external temperatures are used to calculate the engine's efficiency, the maximum efficiency is lower than that derived from system properties alone. The realistic efficiency is expressed as

$$\eta_{\text{max}} = 1 - \frac{T_{c,\text{ext}}}{T_{h,\text{ext}}}.$$

This expression represents the upper limit of performance for real engines when accounting for external conditions.

6.10.2 Cooler performance with external temperatures

(Figure 6.5) In a cooler, work is used to transfer heat from a low-temperature region to a higher-temperature region. The working substance undergoes thermodynamic processes in the system's cold and hot regions:

- **Heat absorption** (Q_c): The working substance absorbs heat in the cold region at an internal temperature T_c. This heat originates from the surroundings at a higher temperature $T_{c,\text{ext}}$.
- **Heat rejection** (Q_h): The working substance rejects heat in the hot region at an internal temperature T_h, transferring it to the surroundings at a lower temperature $T_{h,\text{ext}}$.

Due to similar inefficiencies as in engines, the external temperatures satisfy the following relationships:

$$T_{h,\text{ext}} < T_h \text{ and } T_{c,\text{ext}} > T_c.$$

The COP of a cooler, accounting for external temperatures, is given by

$$\text{COP}_{\text{max}} = \frac{T_{c,\text{ext}}}{T_{h,\text{ext}} - T_{c,\text{ext}}}.$$

Note the difference between an engine and cooler.

6.11 Historical statements of the second law of thermodynamics

The second law of thermodynamics is often expressed through two classical formulations: the Clausius statement and the Kelvin–Planck statement. These statements describe fundamental limitations on heat and work interactions in cyclic processes. Although historically significant, these formulations are less emphasized in modern thermodynamics due to the deeper understanding provided by state variables such as entropy.

6.11.1 Clausius statement

The **Clausius statement** focuses on the behavior of heat transfer:

'It is impossible for a device operating in a cyclic process to transfer heat from a colder body to a hotter body without the input of external work.'

This principle explains the basic operation of refrigeration and heat pump systems. Heat naturally flows from high to low temperatures. To reverse this flow —moving heat from a cold region to a hot region—external work is required. For example:

- A refrigerator absorbs heat from its cold interior and rejects it to the warmer environment, powered by electrical work.
- A heat pump extracts heat from a colder outdoor space and transfers it to a warmer indoor area for heating, again requiring external energy.

The Clausius statement encapsulates the unidirectional nature of spontaneous heat transfer, emphasizing the necessity of work input to reverse this process.

6.11.2 Kelvin–Planck statement

The **Kelvin–Planck statement** focuses on the limits of energy conversion in heat engines:

'It is impossible to construct a device operating in a cyclic process that extracts heat from a single thermal reservoir and converts it entirely into work.'

This implies that no heat engine can have 100% efficiency. A portion of the heat absorbed from the hot region must always be rejected to a cold region. For instance:

- In a steam turbine, only part of the heat energy extracted from the high-temperature steam is converted to work; the rest is expelled as waste heat.

The Kelvin–Planck statement highlights the inherent inefficiency of heat engines, a direct consequence of the second law.

6.11.3 A modern perspective on these statements

While the Clausius and Kelvin–Planck statements were historically significant for understanding thermodynamic principles, modern thermodynamics, with its emphasis on *state variables* such as *entropy*, renders these formulations largely superfluous.

In this chapter, we derived the performance of thermodynamic cycles using the principle of entropy. For a complete cycle:

$$\Delta S_{\text{system}} = 0.$$

This ensures that the total entropy changes due to heat transfer and any entropy generated irreversibly must balance. From this single, fundamental principle:

- The unidirectional flow of heat described by the Clausius statement follows naturally, as entropy must increase when heat moves from hot to cold or external work must compensate otherwise.
- The Kelvin–Planck statement becomes self-evident because entropy changes impose a restriction on how much heat can be converted into work.

By focusing on entropy, modern thermodynamics provides a unified framework for understanding the second law without relying on separate, specialized statements.

6.12 Summary

This chapter provided an in-depth exploration of thermodynamic cycles, their mechanisms, and their practical implications. Key highlights include:

1. **Definition and purpose of cycles**: Thermodynamic cycles enable energy to be transformed repeatedly, making them indispensable in engines, power plants, refrigerators, and heat pumps.
2. **Ideal versus real cycles**: The Carnot cycle, representing the ideal benchmark for efficiency, was contrasted with real cycles, which include inefficiencies like friction and thermal losses.
3. **Analysis of key cycles**: Detailed analyses of the Brayton and Stirling cycles were presented. The Brayton cycle, used in gas turbines, operates with steady flow, while the Stirling cycle, common in cryogenic systems, involves oscillating flow and regenerator-based heat recycling.
4. **Performance metrics**: For engines, efficiency was discussed as the ratio of work output to heat input. For coolers and heat pumps, the coefficient of performance (COP) was introduced, emphasizing the desired cooling or heating effect relative to work input.
5. **Impact of external temperatures**: Practical performance limits were analysed by incorporating external temperatures into the efficiency and COP calculations, highlighting the importance of realistic assumptions in system design.

By connecting theoretical concepts with practical applications, this chapter lays a strong foundation for understanding and optimizing thermodynamic systems. It equips readers with the tools to evaluate the efficiency and performance of cycles in various engineering contexts.

IOP Publishing

A Classical Thermodynamics Toolkit

Srinivas Vanapalli

Chapter 7

Real gas properties and applications in thermodynamic cycles

7.1 Introduction

Thermodynamics extends far beyond idealized models, exploring the behavior of substances in realistic conditions where gases, liquids, and phase transitions present fascinating challenges. This chapter delves into the complexities of **real gas properties**, unveiling their relevance in critical industrial and scientific applications. From understanding deviations from ideal gas behavior to mastering the thermodynamic principles governing phase changes, this chapter lays the groundwork for tackling real-world engineering problems.

Key concepts such as **equations of state**, **phase diagrams**, and the generalization of thermodynamic laws to open systems are explored with a practical focus. Building on these foundations, we analyse essential thermodynamic cycles, such as the **vapor-compression refrigeration cycle** and the **Rankine cycle**, which form the backbone of modern cooling and power generation technologies. This chapter aims to equip readers with the tools to connect theory to practice, paving the way for solving challenges in fields such as refrigeration, energy systems, and cryogenics.

7.2 Limitations of the ideal gas model

The ideal gas law, $pV = nRT$, serves as a cornerstone in thermodynamics, providing a simplified relationship between pressure (p), volume (V), temperature (T), and the number of moles (n) of a gas. This model assumes that gas molecules have negligible volume and do not interact with one another, except during perfectly elastic collisions. While these assumptions are valid under many conditions, they break down in real-world applications where gases exhibit non-ideal behavior.

doi:10.1088/978-0-7503-6029-6ch7
7-1

7.2.1 When does the ideal gas model fail?

The ideal gas law works best at:
- **Low pressures**: Molecules are far apart, and their individual volumes are insignificant.
- **High temperatures**: The kinetic energy of molecules dominates any intermolecular forces.

However, at high pressures and low temperatures, real gases deviate significantly from ideal behavior. These deviations occur due to:
1. **Finite molecular volume**: As pressure increases, the volume of gas molecules becomes significant compared to the total volume of the gas.
2. **Intermolecular forces**: At low temperatures, attractive forces between molecules (e.g. van der Waals forces) cannot be ignored.

7.2.2 Why do we need real gas models?

The shortcomings of the ideal gas model necessitate more accurate representations of real gas behavior. These models account for:
- **Compressibility effects**, measured using the compressibility factor (Z).
- **Intermolecular attractions and repulsions**, captured through equations of state such as the van der Waals or Peng–Robinson equations.
- **Phase changes**, where gases transition to liquids or solids, requiring a completely different thermodynamic treatment.

By recognizing the limitations of the ideal gas law and adopting more comprehensive models, engineers and scientists can design systems that perform reliably under real-world conditions.

7.3 Equations of state for real gases

To address the limitations of the ideal gas law, more sophisticated models—referred to as equations of state—have been developed to describe the behavior of real gases. These equations incorporate corrections for molecular volume and intermolecular forces, providing a more accurate representation of gas behavior under non-ideal conditions.

7.3.1 The compressibility factor (Z)

One way to quantify deviations from ideal behavior is through the **compressibility factor**, defined as

$$Z = \frac{pV}{nRT}.$$

- For an ideal gas, $Z = 1$ at all conditions.
- For real gases, Z can deviate from unity, depending on pressure and temperature:

- $Z < 1$: Indicates that attractive forces dominate, causing the gas to compress more than predicted by the ideal gas law.
- $Z > 1$: Indicates that repulsive forces dominate, making the gas less compressible.

7.3.2 Common equations of state

Several equations of state have been developed to model real gas behavior. The most widely used include:

1. **van der Waals equation**: The van der Waals equation introduces two correction factors to the ideal gas law:

$$\left(P + \frac{a}{V^2}\right)(V - b) = RT.$$

- a: Accounts for attractive forces between molecules.
- b: Accounts for the finite volume of gas molecules.

This equation is a good starting point for understanding real gas behavior, although its accuracy is limited at extreme conditions.

2. **Redlich–Kwong equation**: An improvement over van der Waals, especially for gases at moderate temperatures:

$$P = \frac{RT}{V - b} - \frac{a}{T^{0.5}V(V + b)}.$$

3. **Peng–Robinson equation**: Widely used in engineering, particularly in the chemical and petroleum industries, this equation provides a more accurate representation of gas–liquid equilibria:

$$P = \frac{RT}{V - b} - \frac{a\alpha(T)}{T^{0.5}V(V + b)}.$$

- $\alpha(T)$: A temperature-dependent correction factor.

4. **Ideal versus real behavior at critical and saturation states**: Real gas equations are essential near the critical point, where small changes in temperature or pressure lead to significant density changes. They also help describe gas–liquid transitions during condensation or boiling.

7.3.3 Application of equations of state

Equations of state are vital for:

- **Designing compressors and turbines**: Ensuring accurate predictions of pressure, temperature, and volume changes.
- **Refrigeration systems**: Understanding the behavior of refrigerants under varying conditions.
- **Chemical processes**: Modeling reactions and separations where gases deviate significantly from ideal behavior.

7.4 Pressure–enthalpy (*p–h*) diagram

The *p–h* diagram is a powerful tool for visualizing the thermodynamic properties of substances across different phases. Figure 7.1 shows the *p–h* diagram for water illustrating the property landscape, providing insights into phase change behavior (figure 7.1).

7.4.1 Key features of the *p–h* diagram for water

1. **The two-phase dome**:
   - The dome-shaped region represents the liquid–vapor equilibrium. Within this dome, water coexists as a mixture of liquid and vapor.
   - The left boundary of the dome marks the saturated liquid line, where the first vapor bubbles form.
   - The right boundary marks the saturated vapor line, where the last liquid droplets evaporate.

2. **Critical point**:
   - At the top of the dome lies the critical point. Beyond this point, the liquid and vapor phases are indistinguishable, and the substance exhibits properties of both phases (supercritical fluid).

3. **Isotherms and phase transitions**:
   - Isotherms (constant temperature lines) show unique behaviors:
     - In the two-phase region, *p–h* isothermal lines are horizontal, representing constant pressure during phase change.
     - Outside the two-phase dome, the isotherms slope steeply, indicating single-phase (liquid or vapor) behavior.

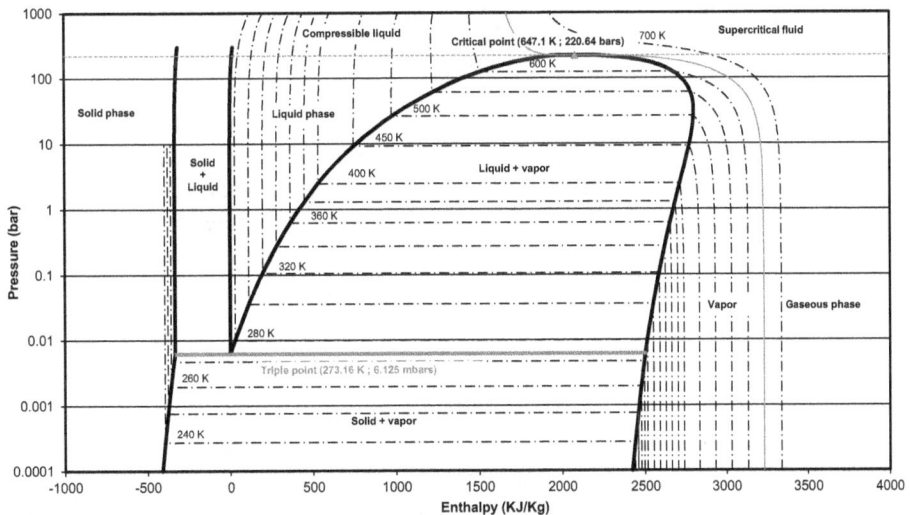

Figure 7.1. Pressure–enthalpy (*p–h*) diagram for water showing key thermodynamic regions, including subcooled liquid, two-phase mixture (liquid + vapor), superheated vapor, and supercritical fluid. The diagram also highlights the triple point (273.16 K; 6.125 mbar) and the critical point (647.1 K; 220.64 bar), along with selected isotherms and phase boundaries.

7.5 Pressure–temperature (*p–T*) diagram

The *p–T* diagram, often called the phase diagram, provides a comprehensive view of phase equilibrium. Figure 7.2 illustrates the *p–T* diagram for water, identifying key phase boundaries and points.

7.5.1 Key features of the *p–T* diagram for water

1. **Phase boundaries**:
   - The diagram delineates three main phase boundaries:
     - ○ **Sublimation line**: Separates the solid and vapor phases.
     - ○ **Melting line**: Separates the solid and liquid phases.
     - ○ **Boiling/condensation line**: Separates the liquid and vapor phases.
   - Each line represents equilibrium conditions where two phases coexist.
2. **Critical point**:
   - The critical point marks the upper limit of the boiling/condensation line, beyond which distinct liquid and vapor phases cease to exist.
3. **Triple point**:
   - The triple point is a unique state where solid, liquid, and vapor phases coexist in equilibrium. For water, this occurs at a pressure of 6.125 mbar and a temperature of 273.16 K.
4. **Solid region**:
   - The solid phase occupies a unique region in the *p–T* diagram. Depending on the temperature and pressure, water can form different solid structures, such as ice.

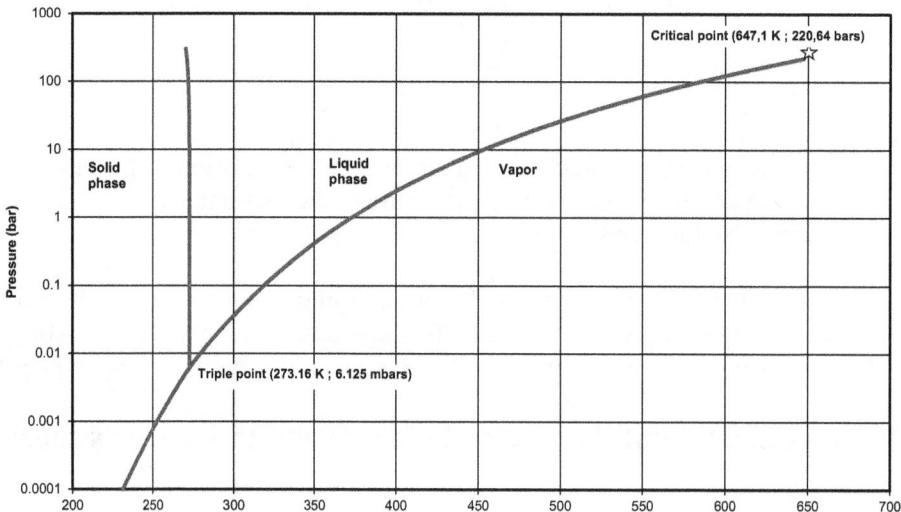

Figure 7.2. Pressure–temperature (*p–T*) diagram for water, illustrating the phase boundaries between solid, liquid, and vapor phases. The diagram highlights the triple point (273.16 K; 6.125 mbar) and the critical point (647.1 K; 220.64 bar), which represent key thermodynamic limits. The bold lines indicate equilibrium curves for phase transitions: sublimation, melting, and boiling/condensation.

7.6 Observations and real gas behavior

7.6.1 Key observations from p–T and p–h diagrams

1. **Real gas effects**:
   - In the supercritical region, water behaves neither as a typical liquid nor as a gas. The substance exhibits unique density and viscosity properties, useful in applications such as supercritical fluid extraction.
2. **Terminology**:
   - The term 'vapor' refers to a phase that can condense into a liquid under certain conditions. In contrast, the term 'gas' is often used to describe a phase that can be approximated by the ideal gas law.
3. **Application of phase diagrams**:
   - Engineers and scientists use these diagrams to design compressors, turbines, and heat exchangers that operate across single-phase and multi-phase regions.

The p–T and p–h diagrams are essential tools for understanding the thermodynamic properties of substances and analysing practical systems where real gas behavior and phase changes are significant.

7.6.2 A brief note on the Clapeyron equation

Phase change processes, such as boiling, condensation, melting, and sublimation, involve equilibrium between two phases of a substance. The conditions for this equilibrium are governed by the **Clapeyron equation**, which relates the slope of the phase boundary in a p–T diagram to the properties of the two coexisting phases:

$$\frac{dp}{dT} = \frac{\Delta h}{T \Delta v},$$

where:
- Δh is the enthalpy change (e.g. latent heat of vaporization or fusion),
- Δv is the specific volume difference between the two phases, and
- T is the absolute temperature.

   This equation is invaluable for understanding and predicting the behavior of substances during phase changes. While the equation is introduced here conceptually, its derivation and applications will be discussed in detail in the next chapter.

7.7 Generalization of the first and second laws to an open system

Thermodynamic systems often involve the flow of mass across their boundaries. Such systems, referred to as **open systems**, include components such as compressors, turbines, and heat exchangers. To analyse these systems, the first and second laws of thermodynamics must be adapted to account for the transport of energy and entropy with the entering and exiting mass.

7.7.1 First law of thermodynamics for an open system

The first law of thermodynamics states that energy cannot be created or destroyed. For an open system with mass flow, heat transfer ($\dot{Q}$), work transfer ($\dot{W}$), and energy transport by mass ($\dot{m}h$) must be considered. The general form of the first law is

$$\dot{Q} + \dot{W} + \sum \dot{m}_{in} h_{in} = \sum \dot{m}_{out} h_{out} + \frac{dE}{dt},$$

where:
- h is the specific enthalpy ($h = u + pv$),
- $\dot{m}_{in}$ and $\dot{m}_{out}$ are the mass flow rates into and out of the system, and
- dE/dt is the rate of change of the system's total energy.

In **steady-state conditions**, the system's energy does not change over time ($\frac{dE}{dt} = 0$), and the equation simplifies to

$$\dot{Q} + \dot{W} + \dot{m}_{in} h_{in} = \dot{m}_{out} h_{out}.$$

This form is commonly used for analysing flow devices such as turbines and compressors.

7.7.2 Second law of thermodynamics for an open system

The second law introduces entropy (S) as a measure of system disorder. For an open system, the general form of the second law is

$$\frac{\dot{Q}}{T} + \dot{m}_{in} s_{in} - \dot{m}_{out} s_{out} + \dot{s}_{gen} = \frac{dS}{dt},$$

where:
- s is the specific entropy,
- $\dot{s}_{gen}$ represents entropy generation due to irreversibilities.

In **steady-state conditions**, entropy does not accumulate in the system ($dS/dt = 0$), reducing the equation to

$$\frac{\dot{Q}}{T} + \dot{m}_{in} s_{in} + \dot{s}_{gen} = \dot{m}_{out} s_{out}.$$

7.8 Application of the laws

The generalized first and second laws are applied to analyse key components of thermodynamic systems. Each component plays a crucial role in energy conversion or transfer, and their analysis relies on the properties of the substance.

7.8.1 Heat exchangers

Heat exchangers transfer energy between two fluid streams without direct mixing. Consider the example in figure 7.3, where water enters the heat exchanger as liquid at 290 K and exits as saturated steam at 1 bar.

Using the first law of thermodynamics:

$$\dot{Q} = \dot{m}(h_2 - h_1).$$

From the steam tables:
- Enthalpy of liquid water (h_1) at 290 K = 70.8 J g^{-1}.
- Enthalpy of saturated steam (h_2) at 1 bar = 2674.9 J g^{-1}.

For a mass flow rate of 1 g s^{-1},

$$\dot{Q} = 1\ \text{g s}^{-1} \times (2674.9 - 70.8)\ \text{J g}^{-1} = 2604.1\ \text{W}.$$

This energy input transitions water from the subcooled liquid region into the two-phase dome, visible in the p–h diagram for water.

The entropy change during this process is

$$\dot{S} = \dot{m}(s_2 - s_1),$$

where:
- Entropy at the initial state (s_1) = 0.2513 J g^{-1} · K^{-1}.
- Entropy at the final state (s_2) = 7.3588 J g^{-1} · K^{-1}.

For the same mass flow rate:

$$\dot{S} = 1\ \text{g s}^{-1} \times (7.3588 - 0.2513)\ \text{J g}^{-1}\text{K}^{-1} = 7.1075\ \text{W K}^{-1}.$$

This example demonstrates how heat exchangers operate within the p–h property space and how energy and entropy are accounted for using the first and second laws.

7.8.2 Throttling devices

Throttling is an isenthalpic process, meaning that the enthalpy of the fluid remains constant across the throttling device. Since no heat (Q = 0) or work (W = 0) is exchanged with the surroundings, and changes in kinetic and potential energy are

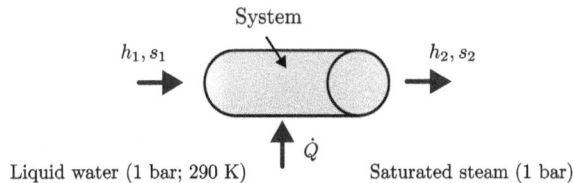

Figure 7.3. Schematic representation of a heat exchanger where liquid water at 1 bar and 290 K enters the system and is converted into saturated steam at the same pressure.

negligible, the first law of thermodynamics simplifies to $h_{\mathrm{in}} = h_{\mathrm{out}}$. Figure 7.4 shows examples of throttling devices.

Entropy increases in throttling due to irreversibilities. From the second law,

$$\dot{S}_e = \dot{S}_i + \dot{S}_{\mathrm{gen}}.$$

The p–h diagrams clearly illustrate the temperature and entropy changes, even as enthalpy remains constant.

7.8.3 Compressors and pumps

Compressors and pumps are designed to increase the pressure of gases and liquids (figure 7.5). In the p–h diagram for water:

- A pump moves a liquid from a low-pressure state (subcooled region) to a higher-pressure state.
- A compressor performs a similar function for gases, but the process involves significant temperature and entropy changes.

Using the first law of thermodynamics:

$$\dot{Q} + \dot{W} = \dot{m}(h_o - h_i).$$

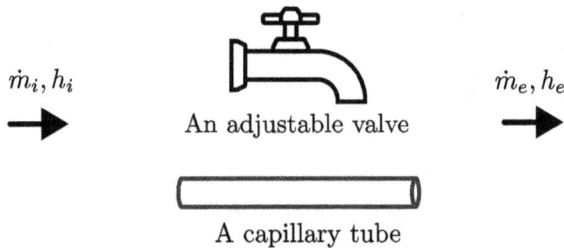

$\dot{m}_i, h_i$

An adjustable valve

$\dot{m}_e, h_e$

A capillary tube

Figure 7.4. Examples of throttling devices: an adjustable expansion valve and a capillary tube. In both cases, the mass flow rate and enthalpy remain constant across the device. These components operate without heat or work interaction with the surroundings, and the process is characterized by a significant drop in pressure and an increase in entropy due to irreversibilities.

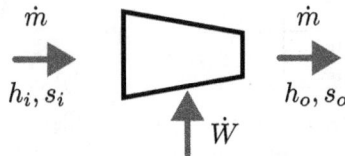

$\dot{m}$

h_i, s_i

$\dot{W}$

$\dot{m}$

h_o, s_o

Figure 7.5. Schematic of a compressor or pump, where a working fluid is pressurized by mechanical work input. The process typically results in increased enthalpy and, in real systems, also increased entropy due to irreversibilities.

If the process deviates from ideality, additional entropy generation occurs, reducing efficiency.

$$\dot{m}\dot{S}_o = \dot{m}\dot{S}_i + \dot{S}_{\text{gen}}.$$

7.9 Vapor-compression cycle

The vapor-compression cycle is the cornerstone of mechanical refrigeration systems, widely used in the food and pharmaceutical industries for freezing and cooling applications. It relies on a closed-loop cycle where the working fluid, called the refrigerant, undergoes thermodynamic processes to transfer heat from a low-temperature region (cooling) to a high-temperature region (heat rejection).

In this cycle, the refrigerant is the system, and its state evolves through four main processes: throttling, evaporation, compression, and condensation. These processes are best visualized on a pressure–enthalpy (p–h) diagram, as shown in figure 7.6. The thermodynamic properties of the refrigerant at the key states are summarized in table 7.1.

7.9.1 Throttling process (1 → 2)

- *Description*: The refrigerant at state 1 (high-pressure, saturated liquid) passes through an expansion valve. This is an irreversible process during which no heat ($Q = 0$) or work ($W = 0$) is exchanged with the surroundings.

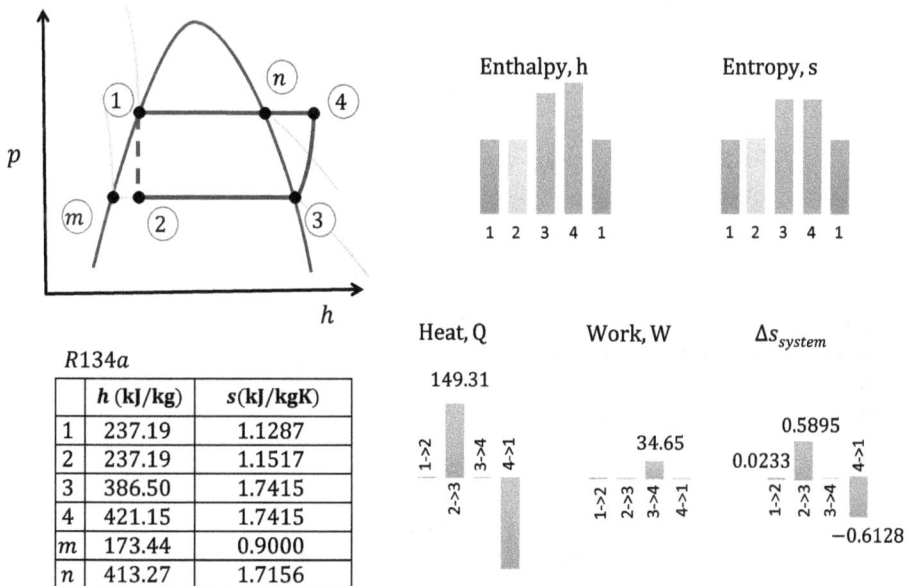

R134a

	h (kJ/kg)	s(kJ/kgK)
1	237.19	1.1287
2	237.19	1.1517
3	386.50	1.7415
4	421.15	1.7415
m	173.44	0.9000
n	413.27	1.7156

Figure 7.6. Thermodynamic analysis of a vapor-compression refrigeration cycle using R134a, shown on a pressure–enthalpy (p–h) diagram. The cycle consists of four main processes: throttling (1 → 2), evaporation (2 → 3), compression (3 → 4), and condensation (4 → 1). Bar charts illustrate the enthalpy and entropy values at each state. The accompanying table provides thermodynamic properties at key points, including intermediate states m and n. The plots for heat (Q), work (W), and system entropy change (Δ_{system}) quantify energy and entropy balances across each process.

Table 7.1. Thermodynamic properties of R134a at four key states in the vapor-compression refrigeration cycle.

State	p (bar)	T (K)	h (kJ kg^{-1})	s (kJ kg^{-1}K^{-1})
1	**7.0**	**300.00**	237.19	1.1287
2	**1.3**	**253.13**	237.19	1.1517
3	1.3	253.13	386.50	1.7415
4	7.0	307.68	421.15	1.7415

Table 7.2. Energy and entropy balances for each process in the vapor-compression refrigeration cycle using R134a. The table reports changes in enthalpy, entropy, heat transfer (Q), and work (W) per unit mass for each segment of the cycle. The total enthalpy and entropy changes over the full cycle are zero, confirming the cyclic nature of the system.

Process	Δh (kJ kg^{-1})	Δs (kJ kg^{-1} K^{-1})	Q (kJ kg^{-1})	W (kJ kg^{-1})
1–2	0	0.0233	0	0
2–3	149.31	0.5895	149.31	0
3–4	34.65	0	0	34.65
4–1	−183.96	−0.6128	−183.96	0
Cycle	**0**	**0**	−34.65	34.65

- *Energy analysis*: From the first law of thermodynamics,

$$Q + W = h_2 - h_1.$$

Since $Q = 0$ and $W = 0$,

$$h_2 = h_1.$$

The enthalpy remains constant ($h_1 = h_2 = 237.19$ kJ kg^{-1}). However, the refrigerant undergoes a significant drop in pressure, entering the two-phase region.

- *Entropy change*:

  During throttling, the entropy of the refrigerant increases ($s_2 > s_1$), even though no heat transfer occurs. This increase is due to irreversibilities inherent in the throttling process. As a result, the cycle cannot achieve thermodynamic maximum performance, as discussed in earlier chapters.

7.9.2 Evaporation process (2 → 3)

- *Description*: The refrigerant, now a low-pressure liquid–vapor mixture, enters the evaporator. It absorbs heat ($Q > 0$) from the surroundings, causing the liquid to vaporize completely into a saturated vapor at state 3. This is the cooling process.

- *Energy analysis*: From the first law,

$$Q + W = h_3 - h_2.$$

Since $W = 0$ (no work input/output),

$$Q = h_3 - h_2 = 386.50 - 237.19 = 149.31 \text{ kJ kg}^{-1}.$$

- *Entropy change*:

    The entropy of the refrigerant increases significantly during evaporation ($s_3 > s_2$) due to heat absorption. This can be observed in the bar chart for entropy in figure 7.6.

7.9.3 Compression process (3 → 4)

- *Description*: The refrigerant, now a low-pressure saturated vapor, is compressed adiabatically (isentropically) by a compressor to a high-pressure superheated vapor. This process requires work input ($W > 0$).
- *Energy analysis*: From the first law,

$$Q + W = h_4 - h_3.$$

Since $Q = 0$ (adiabatic process),

$$W = h_4 - h_3 = 421.15 - 386.50 = 34.65 \text{ kJ kg}^{-1}.$$

- *Entropy change*:

    For an ideal isentropic compression, the entropy remains constant ($s_4 = s_3$), as shown in the entropy bar chart. However, in real systems, some irreversibilities may slightly increase entropy.

7.9.4 Condensation process (4 → 1)

- *Description*: The high-pressure, superheated vapor enters the condenser, where it rejects heat ($Q < 0$) to the surroundings. The refrigerant transitions from a superheated vapor at state 4 to a high-pressure saturated liquid at state 1, completing the cycle.
- *Energy analysis*: From the first law,

$$Q + W = h_1 - h_4.$$

Since $W = 0$ (no work input/output),

$$Q = h_1 - h_4 = 237.19 - 421.15 = -183.96 \text{ kJ kg}^{-1}.$$

The negative sign indicates heat rejection.

- *Entropy change*:

    During condensation, the entropy of the refrigerant decreases ($s_1 < s_4$) as heat is removed. This is consistent with the entropy bar chart. The energy and entropy balances for each process are listed in table 7.2.

7.9.5 Entropy and enthalpy analysis of the cycle

The vapor-compression cycle highlights critical insights about the thermodynamic behavior of the refrigerant.

1. **System entropy**: The net entropy change of the refrigerant over a complete cycle is zero,

$$\Delta s_{\text{system}} = 0.$$

This is because the refrigerant returns to its initial thermodynamic state after completing one cycle. However, entropy generation due to irreversibilities ($s_{\text{irr}} > 0$) occurs, particularly during the throttling process ($1 \rightarrow 2$), where heat transfer is zero, but the entropy of the refrigerant increases.

2. **System enthalpy**: Similarly, the enthalpy of the refrigerant remains unchanged after completing the cycle,

$$\Delta h_{\text{system}} = 0.$$

Since the cycle is closed and the refrigerant returns to its initial state (state 1), there is no net change in enthalpy over the cycle.

7.9.6 The effect of irreversibilities on cycle performance

While the ideal vapor-compression cycle assumes isentropic compression and no irreversibilities, real systems experience losses that impact performance. Let us analyse the consequences of inefficiencies in the compressor and other components, focusing on entropy generation, enthalpy changes, and their thermodynamic implications.

7.9.6.1 Inefficient compression and its consequences

In an ideal compression process ($3 \rightarrow 4$), the refrigerant undergoes an **isentropic process**, meaning there is no entropy change ($s_4 = s_3$). However, in a real compressor with inefficiencies, irreversibilities lead to an increase in entropy ($s_4 > s_3$) at the outlet. This entropy generation has direct consequences on the thermodynamic state at point 4.

Impact on enthalpy at state 4

- With increased entropy ($s_4 > s_3$) and the same pressure p_4 (since the compressor still achieves the same high pressure as in the ideal case), the temperature of the refrigerant at state 4 is higher than in the ideal case.
- Higher temperature means higher enthalpy at state 4 ($h_4^{\text{real}} > h_4^{\text{ideal}}$).

Impact on compressor work (W)

- The work done by the compressor is the enthalpy difference between states 4 and 3,

$$W = h_4 - h_3.$$

- Since $h_4^{\text{real}} > h_4^{\text{ideal}}$, the work input required for compression increases in the presence of irreversibilities.
- Importantly, states 2 and 3 remain unchanged because the throttling and evaporation processes are unaffected by the compressor inefficiency. This means the required cooling ($Q_c = h_3 - h_2$) does not change.

Impact on heat rejection (Q_h)

- The condenser must reject the extra energy introduced into the system due to the increased compressor work.
- Heat rejection is calculated as

$$Q_h = h_4 - h_1.$$

- Since $h_4^{\text{real}} > h_4^{\text{ideal}}$, the enthalpy difference between states 4 and 1 increases, leading to higher heat rejection.

7.9.6.2 Key observations on work and heat transfer

1. **Work input increases**:

   Any irreversibilities in the system increase the required work input for a given cooling effect (Q_c). This is because the entropy generated in the system must be balanced by increased heat rejection to maintain the cyclic operation.

2. **Heat rejection increases**:

   With higher work input (W) due to irreversibilities, the condenser must reject more heat to the surroundings. This increases the enthalpy difference between states 4 and 1 ($h_4 - h_1$).

7.9.6.3 Example: Inefficient compression

Let us consider an example with an inefficient compressor where the outlet entropy increases:

- **Ideal case**: $s_3 > s_4^{\text{ideal}}$, $h_4^{\text{ideal}} = 421.15 \text{ kJ kg}^{-1}$.
- **Real case**: $s_4^{\text{real}} > s_4^{\text{ideal}}$, $h_4^{\text{real}} = 430.00 \text{ kJ kg}^{-1}$ (hypothetical).

In the real case:

- **Work input**:

$$W^{\text{real}} = h_4^{\text{real}} - h_3 = 430.00 - 386.50 = 43.50 \text{ kJ kg}^{-1}.$$

Compared to the ideal case,

$$W^{\text{ideal}} = h_4^{\text{ideal}} - h_3 = 421.15 - 386.50 = 34.65 \text{ kJ kg}^{-1}.$$

The compressor must now perform 8.85 kJ kg^{-1} more work for the same cooling effect.

- **Heat rejection**:

$$Q_h^{\text{real}} = h_4^{\text{real}} - h_1 = 430.00 - 273.19 = 192.81 \text{ kJ kg}^{-1}.$$

Compared to the ideal case,

$$Q_h^{\text{real}} = h_4^{\text{real}} - h_1 = 421.15 - 273.19 = 183.96 \text{ kJ kg}^{-1}.$$

Heat rejection increases by 8.85 kJ kg^{-1}, matching the increased work input.

Irreversibilities increase the total entropy of the system and surroundings, requiring additional work input to achieve the same cooling effect. Consequently, the **coefficient of performance (COP)** decreases because the same cooling effect now requires more work input:

$$\text{COP} = \frac{Q_c}{W} \left(\text{COP decreases as } W \text{ increases} \right).$$

7.9.7 Heat pumps and the vapor-compression refrigeration cycle

Although this chapter primarily describes the vapor-compression refrigeration cycle, the principles and thermodynamic analysis apply equally to heat pumps. The key distinction is in the intended benefit: while refrigeration focuses on extracting heat (Q_c) from a low-temperature region, heat pumps are designed to deliver heat (Q_h) to a high-temperature region as the useful output. The thermodynamic processes, including compression, condensation, throttling, and evaporation, remain identical, and the analysis using the first and second laws of thermodynamics is unchanged. The efficiency of a heat pump is measured by its COP for heating, given by

$$\text{COP}_{\text{heat}} = \frac{Q_h}{W}.$$

This highlights the flexibility of the vapor-compression cycle in serving both cooling and heating applications.

7.10 Rankine cycle

The Rankine cycle is a thermodynamic cycle commonly used in power plants to generate electricity from heat sources such as coal, natural gas, nuclear reactions, or solar energy. The working fluid, typically water, undergoes a series of processes, converting heat into mechanical work and completing a closed-loop cycle.

In a Rankine cycle, the working fluid transitions between liquid and vapor phases, making it a prime example of a two-phase thermodynamic cycle. The cycle consists of four main processes, which are shown on the pressure–volume ($p-V$) diagram provided in figure 7.7.

7.10.1 Main processes of the Rankine cycle

1. **Isentropic compression ($A \rightarrow B$):**
   - In this process, the liquid working fluid (water) is pumped from a low-pressure state (A) to a high-pressure state (B) using a pump.
   - The process is assumed to be adiabatic and isentropic, meaning no heat is transferred, and the entropy remains constant ($s_A = s_B$).
   - Work is done on the fluid by the pump: $W_{\text{pump}} = h_B - h_A$.

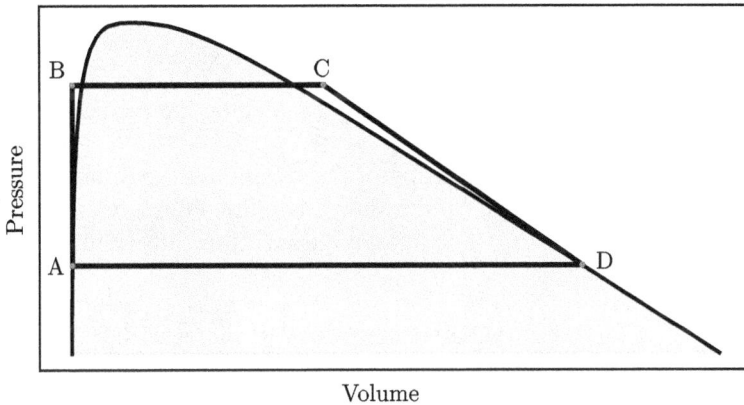

Figure 7.7. Pressure–volume (*p–V*) diagram of the Rankine cycle, illustrating the four main processes: isentropic compression (*A → B*), isobaric heat addition (*B → C*), isentropic expansion (*C → D*), and isobaric heat rejection (*D → A*). While the Rankine cycle is often represented in pressure–enthalpy (*p–h*) diagrams for engineering calculations, this figure uses the *p–V* format to emphasize that thermodynamic cycles can be visualized on different property planes to aid conceptual understanding.

2. **Isobaric heat addition** (*B → C*):
   - The high-pressure liquid enters the boiler, where it absorbs heat (Q_{in}) from an external heat source.
   - The liquid is converted into high-pressure, high-temperature vapor (state *C*).
   - The enthalpy increase is: $Q_{in} = h_C - h_B$.
3. **Isentropic expansion** (*C → D*):
   - The vapor expands through a turbine, producing work ($W_{turbine}$).
   - This process is also assumed to be adiabatic and isentropic ($s_C = s_D$).
   - The work *output* is $W_{turbine} = h_C - h_D$.
4. **Isobaric heat rejection** (*D → A*):
   - The low-pressure vapor enters the condenser, where it rejects heat (Q_{out}) to the surroundings, condensing into a saturated liquid.
   - The enthalpy *decrease* is $Q_{out} = h_D - h_A$.

7.10.2 Worked example

- *Assumptions*:

  - Working fluid: water.
  - Pressure at state *A*: 0.1 bar (condenser pressure).
  - Pressure at state *B*: 30 bar (boiler pressure).
  - Turbine inlet temperature: 500 °C.
  - All processes are ideal (isentropic compression and expansion).

- *Using steam tables*:

  ○ At state A ($p_A = 0.1$ bar, saturated liquid),
    $h_A = 191.81$ kJ kg^{-1}, $s_A = 0.6492$ kJ kg^{-1} K^{-1}.
  ○ At state B ($p_B = 30$ bar, subcooled liquid),
    approximate using pump work,

$$W_{\text{pump}} = v_A(p_B - p_A),$$

  with $v_A = 0.001043$ m^3 kg^{-1},

$$W_{\text{pump}} = 0.001043 \times (30 \times 10^5 - 0.1 \times 10^5) = 3.13 \text{ kJ kg}^{-1}$$

$$h_B = h_A + W_{\text{pump}} = 191.81 + 3.13 = 194.94 \text{ kJ kg}^{-1}.$$

  ○ At state C ($p_C = 30$ bar, $T_C = 500$ °C),
    from steam tables,

$$h_C = 3408.1 \text{ kJ kg}^{-1}, \quad s_C = 6.597 \text{ kJ kg}^{-1} \text{K}^{-1}.$$

  ○ At state D ($p_D = 0.1$ bar, isentropic expansion $s_D = s_C = 6.597$ kJ kg^{-1} K^{-1}),
    from steam tables,

$$h_D = 2206.5 \text{ kJ kg}^{-1}.$$

7.10.3 Energy analysis
Pump work:

$$W_{\text{pump}} = 3.13 \text{ kJ kg}^{-1}.$$

Turbine work:

$$W_{\text{turbine}} = h_C - h_D = 3408.1 - 2206.5 = 1201.6 \text{ kJ kg}^{-1}.$$

Heat input:

$$Q_{\text{in}} = h_C - h_B = 3408.1 - 194.94 = 3213.2 \text{ kJ kg}^{-1}.$$

Heat rejected:

$$Q_{\text{out}} = h_D - h_A = 2206.5 - 191.81 = 2014.7 \text{ kJ kg}^{-1}.$$

7.10.4 Thermal efficiency
The thermal efficiency of the Rankine cycle is

$$\eta = \frac{W_{\text{net}}}{Q_{\text{in}}},$$

where

$$W_{net} = W_{turbine} - W_{pump} = 1201.6 - 3.13 = 1198.5 \, \text{kJ kg}^{-1}$$

$$\eta = \frac{1198.5}{3213.2} = 0.373 = 37.3\%.$$

7.11 Summary

In this chapter, we examined the critical role of **real gas properties** in thermodynamics and their application to engineering systems. The highlights include:

1. **Real gas behavior**:
   - We began by exploring when and why real gases deviate from the ideal gas model due to finite molecular volume and intermolecular forces.
   - Common equations of state, such as van der Waals and Peng–Robinson, were introduced to accurately predict real gas behavior.

2. **Phase changes and thermodynamic analysis**:
   - Phase diagrams (p–T and p–h) were presented as powerful tools for understanding phase transitions.
   - The thermodynamic properties during boiling, condensation, and supercritical transitions were described, highlighting their practical implications.

3. **First and second laws for open systems**:
   - The adaptation of thermodynamic laws to open systems such as compressors, turbines, and heat exchangers was discussed.
   - Entropy generation and its impact on system performance were analysed in detail, showcasing their relevance in engineering design.

4. **Thermodynamic cycles**:
   - The **vapor-compression refrigeration cycle** was systematically explained, emphasizing the refrigerant as the system. Key processes, heat and work transfer, and entropy changes were analysed to understand its limitations and performance.
   - The **Rankine cycle**, a cornerstone of power generation, was introduced. A worked example illustrated how energy flows and thermal efficiency are calculated for a realistic system.

5. **Impact of irreversibilities**:
   - Irreversibilities, particularly in components such as compressors and throttling devices, were shown to degrade cycle performance.
   - Strategies for minimizing entropy generation to optimize efficiency were discussed.

Chapter 8

Free energy and the spontaneity of a process

8.1 Introduction

This chapter provides an in-depth exploration of the fundamental thermodynamic concepts necessary for understanding the spontaneity of physical and chemical processes. Free energy, particularly Gibbs free energy, plays a pivotal role in predicting whether a process can occur spontaneously. This chapter integrates key thermodynamic principles, such as entropy, enthalpy, and work, to create a framework that allows for the systematic analysis of processes, from phase transitions to chemical reactions.

We begin by discussing the concept of spontaneity through the example of the melting of ice. This sets the stage for the introduction of Gibbs free energy and how it can be used to determine the spontaneity of various processes. Further sections expand on the application of Gibbs free energy in both physical and chemical processes, providing examples such as the combustion of hydrogen, fuel cell reactions, and electrolysis.

Additionally, the chapter introduces the Clausius–Clapeyron relation, a critical equation for understanding phase transitions, particularly in the context of vaporization and sublimation. The relation allows for the prediction of changes in vapor pressure with temperature and pressure, aiding in the understanding of phase equilibria.

The chapter concludes by discussing the Helmholtz free energy, which is used in scenarios where systems are at constant volume and temperature, in contrast to the Gibbs free energy's application in constant pressure and temperature systems. Through examples and derivations, this chapter equips readers with the tools to solve practical and theoretical thermodynamics problems in various domains such as material science, chemical engineering, and environmental science.

8.2 Spontaneity example: ice melting to water

To understand the concept of free energy and the spontaneity of a process, let us begin with a familiar example: the melting of ice into water. On Earth, at a

doi:10.1088/978-0-7503-6029-6ch8

temperature of 20 °C, this process occurs naturally and spontaneously. However, if we consider the same process on Mars, where the ambient temperature is well below 0 °C, the outcome is very different. By analysing these cases, we can develop a systematic understanding of the conditions for spontaneity and introduce the concept of Gibbs free energy.

8.2.1 The role of entropy in spontaneity

The total entropy change (ΔS_{total}) for a process can be expressed as the sum of the entropy changes of the system (ΔS_{sys}) and the surroundings (ΔS_{surr}):

$$\Delta S_{total} = \Delta S_{sys} + \Delta S_{surr}.$$

For processes occurring at constant pressure and temperature, the entropy change of the surroundings is directly related to the heat exchanged with the system (Q),

$$\Delta S_{surr} = -\frac{Q}{T_{surr}},$$

where T_{surr} is the temperature of the surroundings, and Q is the heat absorbed or released by the system at constant pressure. Using the enthalpy change of the system ($\Delta H_{sys} = Q$),

$$\Delta S_{surr} = -\frac{\Delta H_{sys}}{T_{surr}}.$$

Substituting this relationship into the expression for ΔS_{total}, we obtain

$$\Delta S_{total} = \Delta S_{sys} - \frac{\Delta H_{sys}}{T_{surr}}.$$

A process is considered **spontaneous** if the total entropy change is positive, i.e.

$$\Delta S_{total} > 0.$$

8.2.2 Example: Ice melting on Earth and Mars

Let us evaluate the spontaneity of ice melting on Earth (20 °C) and Mars (−50 °C). Assume the following thermodynamic values for water:

$$\Delta H_{fus} = 334\ \text{kJ kg}^{-1},\ \ \Delta S_{fus} = 1.22\ \text{kJ (kg} \cdot \text{K)}^{-1}.$$

8.2.2.1 Case 1: On Earth (T = 293 K)
The total entropy change (ΔS_{total}) for the melting process is calculated as

$$\Delta S_{total} = \Delta S_{sys} - \frac{\Delta H_{fus}}{T_{surr}}.$$

Substituting the values,

$$\Delta S_{sys} = \Delta S_{fus} = 1.22 \text{ kJ (kg} \cdot \text{K)}^{-1}, T_{surr} = 293 \text{ K}, \Delta H_{fus} = 334 \text{ kJ kg}^{-1}$$

$$\Delta S_{total} = 1.22 - \frac{334}{293}.$$

Calculate

$$\Delta S_{total} = 1.22 - 1.14 = 0.08 \text{ kJ (kg} \cdot \text{K)}^{-1}.$$

Since $\Delta S_{total} > 0$, the process is *spontaneous* at this temperature.

8.2.2.2 Case 2: On Mars (T = 223 K)

The total entropy change is calculated similarly,

$$\Delta S_{total} = \Delta S_{sys} - \frac{\Delta H_{fus}}{T_{surr}}.$$

Substituting the values,

$$\Delta S_{sys} = 1.22 \text{ kJ (kg} \cdot \text{K)}^{-1}, T_{surr} = 223 \text{ K}, \Delta H_{fus} = 334 \text{ kJ kg}^{-1}.$$

$$\Delta S_{total} = 1.22 - \frac{334}{223}.$$

Calculate

$$\Delta S_{total} = 1.22 - 1.50 = -0.28 \text{ kJ (kg} \cdot \text{K)}^{-1}.$$

Since $\Delta S_{total} < 0$, the process is *not spontaneous* at this temperature.

8.3 Reformulation in terms of free energy

To simplify the analysis of spontaneity, we introduce the concept of **Gibbs free energy** (G) as a thermodynamic property defined by

$$G = H - TS,$$

where H is the enthalpy, T is the temperature, and S is the entropy of the system.

For processes at constant temperature (T = constant), the change in Gibbs free energy (ΔG) is given by

$$\Delta G = \Delta H - T\Delta S.$$

We now consider the total entropy change of the universe during the process, which consists of the entropy change of the system and that of the surroundings:

$$\Delta S_{total} = \Delta S_{sys} + \Delta S_{surr}$$

Assuming that the surroundings act as an ideal thermal reservoir at a constant temperature T_{surr}, and that the only interaction with the surroundings is through heat transfer Q, the entropy change of the surroundings is given by:

$$\Delta S_{surr} = - \frac{Q}{T_{surr}}.$$

Substituting this into the total entropy expression, we obtain:

$$\Delta S_{\text{total}} = \Delta S_{\text{surr}} = -\frac{Q}{T_{\text{surr}}}.$$

Applying the First Law of Thermodynamics for a closed system, the heat exchanged can be written as:

$$Q = \Delta H_{\text{sys}} - W$$

which leads to:

$$\Delta S_{\text{total}} = \Delta S_{\text{sys}} - \frac{\Delta H_{\text{sys}} - W}{T_{\text{surr}}}.$$

Multiplying both sides of the equation by T_{surr}, we have:

$$T_{\text{surr}}\Delta S_{\text{total}} = T_{\text{surr}}\Delta S_{\text{sys}} - \left(\Delta H_{\text{sys}} - W\right).$$

Now, under the assumption that the system and surroundings are at the same temperature, i.e., $T_{\text{surr}} = T$, this expression simplifies to:

$$T\Delta S_{\text{total}} = T\Delta S_{\text{sys}} - \left(\Delta H_{\text{sys}} - W\right).$$

Substituting the definition of $\Delta G = \Delta H_{\text{sys}} - T\Delta S_{\text{sys}}$, we obtain

$$T\Delta S_{\text{total}} = -\Delta G + W.$$

From the **second law of thermodynamics**, the total entropy change ΔS_{total} for a process must satisfy

$$\Delta S_{\text{total}} \geqslant 0.$$

This condition implies that

$$T\Delta S_{\text{total}} \geqslant 0.$$

Rearranging the earlier equation

$$-\Delta G + W \geqslant 0 \text{ or equivalently } \Delta G \leqslant W.$$

Thus, the Gibbs free energy change (ΔG) serves as the upper limit of the useful work (W) that can be extracted from a process under constant temperature and pressure conditions. When the process is **reversible**, $\Delta S_{\text{total}} = 0$, and the equality holds:

$$\Delta G = W.$$

For **irreversible processes**, where $\Delta S_{\text{total}} > 0$, the inequality holds strictly:

$$\Delta G < W.$$

Finally, when no useful work is done ($W = 0$) the condition for spontaneity ($\Delta S_{\text{total}} > 0$) becomes

$$\Delta G < 0.$$

This reformulation shows that for a process at constant temperature, the spontaneity of the process is determined by the Gibbs free energy. Specifically:

- If $\Delta G < 0$, the process is **spontaneous**.
- If $\Delta G > 0$, the process is **non-spontaneous**.

8.4 Heat transfer and the role of work in spontaneity

The concepts of spontaneity and entropy can be extended to processes involving heat transfer and work, such as those in heat engines or refrigerators. Let us analyse three distinct cases to understand how spontaneity, entropy changes, and work interplay.

8.4.1 Case 1: Spontaneous heat transfer (without work)

When energy transfers as heat from a warm reservoir to a cold reservoir (figure 8.1), the process occurs spontaneously. This process is governed by the second law of thermodynamics, as it increases the total entropy of the system and surroundings ($\Delta S_{total} > 0$):

- The warm reservoir loses heat (Q_{warm}), reducing its entropy.
- The cold reservoir gains heat (Q_{cold}), increasing its entropy by a greater amount, due to its lower temperature.
- As a result, the net entropy change is positive:
$$\Delta S_{total} = \Delta S_{warm} + \Delta S_{cold} > 0.$$

This **irreversible process** is a natural flow of energy, requiring no work input or output.

8.4.2 Case 2: Spontaneous heat transfer with work generation (heat engine)

If a heat engine operates between the same warm and cold reservoirs (figure 8.1), it harnesses part of the energy transferred as heat to perform work. In this idealized case (no irreversibilities), the total entropy change is zero ($\Delta S_{total} = 0$):

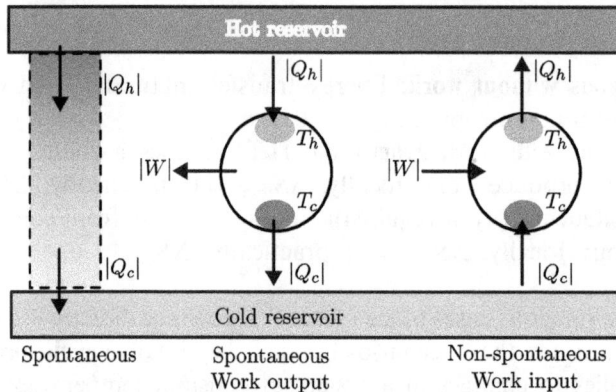

Figure 8.1. Illustration of heat transfer between hot and cold reservoirs. Three scenarios are depicted: (1) spontaneous heat flow from hot to cold (left), (2) spontaneous heat engine operation generating work (center), and (3) non-spontaneous refrigeration requiring work input (right).

- Heat Q_{in} flows from the warm reservoir to the engine.
- Part of this energy is converted into work (W_{out}).
- The remainder is rejected as heat (Q_{out}) to the cold reservoir.
- The entropy gain in the cold reservoir offsets the entropy loss in the warm reservoir, resulting in

$$\Delta S_{total} = \Delta S_{warm} + \Delta S_{cold} = 0.$$

In practice, real engines have irreversibilities (e.g. friction, heat transfer), leading to a slightly positive total entropy change ($\Delta S_{total} > 0$). Even so, the entropy increase is less than in the case of spontaneous heat transfer without work, as the engine utilizes some of the heat flow to produce work.

8.4.3 Case 3: Non-spontaneous heat transfer (work input required)

When energy is transferred from a cold reservoir to a warm reservoir (figure 8.1), the process is **non-spontaneous** because it decreases the total entropy of the system and surroundings ($\Delta S_{total} < 0$) if left to occur naturally. However, this process can be achieved by performing work, as in the case of refrigerators or heat pumps:
- Work (W_{in}) is applied to drive heat (Q_{cold}) from the cold reservoir to the warm reservoir.
- The system's total entropy change remains non-negative because the work done compensates for the entropy reduction in the heat transfer.

For a refrigerator operating ideally:

$$\Delta S_{total} = \Delta S_{cold} + \Delta S_{warm} = 0.$$

In reality, irreversibilities make $\Delta S_{total} > 0$, requiring slightly more work to overcome these inefficiencies.

8.4.4 Summary of the three cases

1. **Spontaneous without work**: Energy transfers naturally from warm to cold, increasing total entropy ($\Delta S_{total} > 0$).
2. **Spontaneous with work generation**: Heat engines harness part of the heat transfer to produce work. Ideally, $\Delta S_{total} = 0$; practically, $\Delta S_{total} > 0$.
3. **Non-spontaneous (work required)**: Energy transfers from cold to warm with work input. Ideally, $\Delta S_{total} = 0$; practically, $\Delta S_{total} > 0$.

Note: Unlike the previous cases of ice melting, we cannot directly use the Gibbs free energy concept (ΔG) in these scenarios because they involve **cyclic processes**. Gibbs free energy applies to changes in a system at constant temperature and pressure, whereas cycles involve multiple stages with varying conditions. Instead, the analysis relies on entropy changes and energy balances over the entire cycle.

8.5 Gas expansion, compression: entropy, and Gibbs free energy

This example combines the concepts of entropy and Gibbs free energy to analyse gas expansion and compression.

8.5.1 Case 1: Spontaneous expansion (no work)

In this scenario, gas expands freely from a high-pressure reservoir to the ambient. This is a **spontaneous process** because it occurs naturally without requiring work input.

- Entropy discussion:
  - As the gas expands, its internal energy decreases, and its entropy increases ($\Delta S_{system} > 0$) due to the increase in volume and dispersal of energy.
  - There is no heat exchange ($Q = 0$), so the surroundings do not gain or lose entropy ($\Delta S_{surroundings} = 0$).
  - The total entropy change is positive:

$$\Delta S_{total} = \Delta S_{system} + \Delta S_{surroundings} > 0.$$

- Gibbs free energy:
  - The Gibbs free energy change for the system is **negative**, reflecting the spontaneity of the process:

$$\Delta G < 0.$$

  - Since there is no mechanism to harness the released energy, no useful work is produced ($W = 0$).

8.5.2 Case 2: Controlled expansion with work generation

Here, the gas expands through a turbine, extracting useful work. The process consists of two stages: adiabatic expansion and heat exchange with the surroundings.

- **Stage 1**. Adiabatic expansion in the turbine:
  - The gas expands through the turbine, cooling as it does work on the turbine.
  - Since the process is adiabatic, no heat is exchanged ($Q = 0$), and the entropy of the gas remains constant:

$$\Delta S_{system} = 0 \text{ (adiabatic and ideal)}.$$

  - The gas's internal energy decreases, and its volume increases.
- **Stage 2**. Heat exchange with the surroundings:
  - After expansion, the gas passes through a heat exchanger where it absorbs heat from the surroundings to restore its temperature to the initial state.
  - During this stage, the entropy of the gas increases ($\Delta S_{system} > 0$), while the surroundings lose entropy ($\Delta S_{surroundings} < 0$).

- Total entropy change:
  - For an *ideal process*, the entropy gained by the gas exactly offsets the entropy lost by the surroundings, so the total entropy change is zero:

    $$\Delta S_{\text{total}} = 0.$$

  - *In practice*, irreversibilities such as friction and turbulence result in entropy generation, causing a slight increase in total entropy:

    $$\Delta S_{\text{total}} > 0.$$

- Gibbs free energy:
  - The maximum work obtainable from the system corresponds to the Gibbs free energy change:

    $$W_{\text{max}} = -\Delta G.$$

  - *In practice*, irreversibilities reduce the actual work output:

    $$W_{\text{actual}} < -\Delta G.$$

8.5.3 Case 3: Non-spontaneous compression

In this scenario, work is required to compress the gas from ambient pressure back into the high-pressure reservoir. As case 2, the process consists of two stages: adiabatic compression and heat rejection to the surroundings.

- **Stage 1**. Adiabatic compression in the compressor:
  - The gas is compressed, increasing its internal energy and temperature.
  - Since the process is adiabatic, no heat is exchanged ($Q = 0$), and the entropy of the gas remains constant:

    $$\Delta S_{\text{system}} = 0 \text{ (adiabatic and ideal)}.$$

  - The gas's volume decreases, and its pressure increases.
- **Stage 2**. Heat rejection to the surroundings:
  - After compression, the gas passes through a heat exchanger where it releases heat to the surroundings, returning to its initial temperature.
  - During this stage, the gas loses entropy ($\Delta S_{\text{system}} < 0$), while the surroundings gain entropy ($\Delta S_{\text{surroundings}} > 0$).

- Total entropy change:
  - For an *ideal process*, the entropy lost by the gas is exactly offset by the entropy gained by the surroundings, so the total entropy change is zero:

    $$\Delta S_{\text{total}} = 0.$$

  - *In practice*, irreversibilities such as heat generation and friction lead to a slight increase in total entropy:

    $$\Delta S_{\text{total}} > 0.$$

- Gibbs free energy:
  - The minimum work required to compress the gas corresponds to the Gibbs free energy change:

$$W_{\min} = \Delta G.$$

  - *In practice*, irreversibilities increase the actual work required:

$$W_{\text{actual}} > \Delta G.$$

8.5.4 Why Gibbs free energy is 'free' energy

Gibbs free energy represents the energy available to do *useful work* at constant temperature:
- **Case 2** (expansion): $-\Delta G$ represents the maximum useful work the system can deliver.
- **Case 3** (compression): ΔG represents the minimum work required to drive the process.

It is called 'free' energy because it excludes the heat transfer, focusing solely on the energy that can be utilized or must be supplied.

8.5.5 Summary of the three cases

- **Case 1** (spontaneous expansion):
  $\Delta S_{\text{total}} > 0$, $\Delta G < 0$, and no work is produced ($W = 0$).
- **Case 2** (controlled expansion):
  $\Delta S_{\text{total}} = 0$ ideally (>0 in practice), $\Delta G < 0$, and work is generated ($W_{\text{actual}} < -\Delta G$).
- **Case 3** (non-spontaneous compression):
  $\Delta S_{\text{total}} = 0$ ideally (>0 in practice), $\Delta G > 0$, and work is required ($W_{\text{actual}} > \Delta G$).

8.6 Treating physical and chemical processes systematically

Thermodynamic principles allow us to analyse both **physical** and **chemical** processes systematically, determining their spontaneity, direction, and associated energy exchanges. By expressing such processes as reactions, we can use the tools of thermodynamics to calculate the heat transfer, work involved, and the spontaneity of the process.

8.6.1 Physical processes as reactions

A familiar example of a physical process is the **melting of ice**, which can be expressed as

$$\text{Ice} \rightarrow \text{Water}.$$

This representation allows us to treat melting as a 'reaction', just as we would with a chemical process. By applying thermodynamic principles, we can determine whether the process is spontaneous. For instance:

- At 293 K (20 °C), melting is spontaneous because the Gibbs free energy change is negative ($\Delta G < 0$).
- At -10 °C, freezing is spontaneous as the reverse reaction (Water $\rightarrow$ Ice) has $\Delta G < 0$.

Through this approach, we can quantify the heat transfer, entropy changes, and even the maximum or minimum work involved.

8.6.2 Chemical processes as reactions

In addition to physical processes, chemical reactions can also be analysed systematically. For example:

1. **Formation of rust** (engineering example):

$$4Fe + 3O_2 \rightarrow 2Fe_2O_3.$$

   This reaction occurs spontaneously in the presence of oxygen and moisture, leading to the formation of rust—a process of interest in material science and corrosion engineering.

2. **Combustion of methane** (energy example):

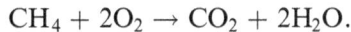

$$CH_4 + 2O_2 \rightarrow CO_2 + 2H_2O.$$

   This reaction is of paramount importance in energy engineering, as it represents the combustion of natural gas to produce energy.

3. **Formation of water** (fuel cell example):

$$2H_2 + O2 \rightarrow 2H_2O.$$

   This process is the basis of hydrogen fuel cells, which generate electricity through a controlled reaction between hydrogen and oxygen.

These examples, drawn from physics and engineering contexts, allow us to systematically analyse chemical processes to determine whether they are spontaneous and calculate their energy changes.

8.6.3 Spontaneity and systematic analysis

The central thermodynamic question is: *is the process spontaneous?*

To answer this question, we express the process as a reaction and calculate:

1. **Reaction enthalpy** (ΔH_r):

$$\Delta H_r = \sum_i^{\text{products}} n_i H_i - \sum_i^{\text{reactants}} n_i H_i.$$

   This represents the total energy change of the system ($Q + W$) due to the reaction, including contributions from bond breaking and formation.

2. **Reaction entropy** (ΔS_r):

$$\Delta S_r = \sum_i^{\text{products}} n_i S_i - \sum_i^{\text{reactants}} n_i S_i.$$

The entropy change tells us about the energy dispersal during the reaction.

3. **Reaction Gibbs free energy** (ΔG_r):

$$\Delta G_r = \sum_i^{\text{products}} n_i G_i - \sum_i^{\text{reactants}} n_i G_i.$$

The sign of ΔG_r determines whether the process is spontaneous ($\Delta G_r < 0$), non-spontaneous ($\Delta G_r > 0$), or at equilibrium ($\Delta G_r = 0$).

For example:
- A **spontaneous reaction** will proceed in the forward direction (e.g. ice melting at 293 K).
- A **non-spontaneous reaction** requires external energy (e.g. water freezing at 20 °C).
- At **equilibrium**, both forward and reverse reactions occur at the same rate, as seen in reversible processes.

8.6.4 Heat transfer and work in reactions

By systematically representing a process as a reaction, we can determine the associated energy interactions, including heat transfer (Q) and work (W):

1. **Heat transfer** (Q): At constant pressure, the heat transfer for a reaction depends on the enthalpy and work interactions. If no other forms of work (W_{other}) are involved, the heat transfer equals the reaction enthalpy:

$$Q = \Delta H_r \text{ (if } W_{\text{other}} = 0).$$

However, in cases like a **hydrogen fuel cell**, electrical work is extracted as part of the process. In such cases, the relationship changes:

$$Q = T\Delta S_r.$$

Here:
- ΔS_r is the entropy change of the reaction.
- T is the temperature of the system, assumed constant.

This equation highlights that the heat transfer depends on the entropy change of the reaction, rather than the enthalpy directly. For instance, in a fuel cell, the electrical work extracted reduces the energy available as heat, making $Q \neq \Delta H_r$.

2. **Work** (W): The Gibbs free energy change determines the work interactions:
   - **Maximum work (expansion processes)**: The maximum useful work delivered is

$$W_{\text{max}} = -\Delta G_r.$$

This occurs in reversible processes, where all energy from the reaction is efficiently converted to work.

○ **Minimum work (compression processes)**: The minimum work required to drive a non-spontaneous reaction is

$$W_{min} = \Delta G_r.$$

In practical processes, irreversibilities lead to deviations from these ideal values.

8.6.5 Summary

- For reactions without work interactions ($W_{other} = 0$), $Q = \Delta H_r$.
- For reactions with significant work interactions, $Q = T\Delta S_r$, with the remaining energy extracted as work ($-\Delta G_r$).
- The systematic treatment of ΔH_r, ΔS_r, and ΔG_r provides a complete picture of energy and entropy changes, enabling precise predictions of heat transfer, work, and spontaneity.

8.7 Distinguishing between the Gibbs free energy of a substance and a reaction

In thermodynamics, understanding the behavior of substances and reactions is fundamental when estimating the properties of a system under varying conditions. One crucial concept is Gibbs free energy, which helps determine whether a process will proceed spontaneously. However, it is important to distinguish between the Gibbs free energy of a substance and the Gibbs free energy of a reaction, as their interpretations and implications differ significantly. In this section, we will explore these differences using two key examples: the Gibbs free energy of a single substance (water) in different phases and the Gibbs free energy of a reaction, specifically the phase change of water from liquid to vapor.

8.7.1 Gibbs free energy of a substance

The Gibbs free energy of a substance describes how its energy varies with temperature and pressure under equilibrium conditions. It provides insights into the phase transitions and the stability of the substance at different conditions. The behavior of Gibbs free energy for water across its various phases—ice, liquid water, and steam—can be visualized in figure 8.2.

The graph shows the Gibbs free energy (G) of water as a function of temperature (T) at a constant pressure of 1 bar. In this plot, we can observe three distinct lines representing the different phases of water: ice, liquid water, and steam. The graph clearly shows the following:

1. **The negative slope of each curve**: This reflects the fact that the entropy (S) of the system is positive. As temperature increases, the entropy increases, and since the Gibbs free energy is defined as $G = H - TS$, the negative slope indicates that the system's Gibbs free energy decreases with increasing temperature at a constant pressure.

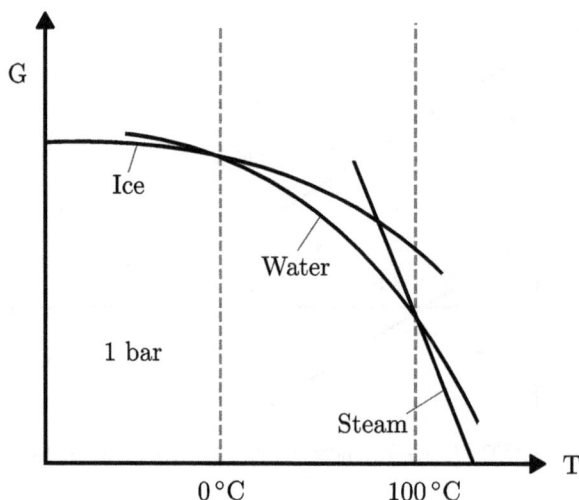

Figure 8.2. The behaviour of Gibbs free energy for water across its various phases.

2. **The curve is not a straight line**: The curves for each phase (ice, water, steam) are not linear because the enthalpy (H) is also a function of temperature. In other words, as the temperature changes, the heat capacity (Cp) of the substance changes, which affects the enthalpy and leads to the non-linear behavior of the Gibbs free energy curve.

The intersection of these curves represents the phase transitions. For instance, at 0 °C, the curve for ice intersects the curve for water, indicating the melting point, and at 100 °C, the curve for liquid water intersects with the steam curve, indicating the boiling point.

8.7.2 Gibbs free energy of a reaction

In contrast, the Gibbs free energy of a reaction represents the overall change in free energy during a chemical or physical transformation, such as a phase transition or a chemical reaction. This is not simply the Gibbs free energy of a substance but rather the difference in Gibbs free energy between the reactants and products, as the reaction progresses.

Let us now consider an example: the phase transition of water from liquid to vapor. The second graph (figure 8.3) shows the Gibbs free energy of the reaction $H_2 O$ (l) → $H_2 O$ (g), which is the vaporization of water at different temperatures and pressures. This process is explained in detail in the next chapter.

This graph is characteristic of a phase change reaction, where the Gibbs free energy is calculated for the system transitioning between two states—liquid and vapor. The important details are:

1. **The decrease in Gibbs free energy as temperature increases**: As we increase the temperature, the Gibbs free energy of the vapor phase (G_{vapor}) becomes lower

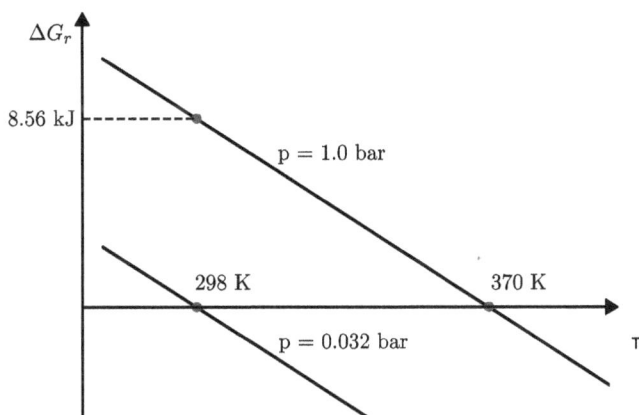

Figure 8.3. The Gibbs free energy of the reaction H_2O (l) $\rightarrow$ H_2O (g).

than that of the liquid phase (G_{liquid}), indicating that the vapor phase becomes more stable at higher temperatures.

2. **The transition point**: At the boiling point, the two phases—liquid and vapor—have the same Gibbs free energy, meaning the system is at equilibrium. At this point, any additional heat added to the system will cause the phase transition to occur without a change in temperature, leading to the vaporization of the liquid water.

3. **The pressure dependence**: The phase change from liquid to vapor is also influenced by pressure. In the graph, the transition occurs at different temperatures for different pressures, demonstrating how the Gibbs free energy of the reaction depends on both temperature and pressure. For example, at higher pressures, the boiling point of water increases, which is consistent with the Clausius–Clapeyron relation.

8.7.3 Key differences between Gibbs free energy of a substance and a reaction

- **Substance**: The Gibbs free energy of a substance, represents its potential energy and its tendency to change phase based on temperature. The curve (figure 8.2) reflects how the substance behaves at a given pressure and temperature, and the intersection points of the curves signify phase transitions. The negative slope indicates that, as temperature increases, the system's entropy increases, and the Gibbs free energy decreases.

- **Reaction**: On the other hand, the Gibbs free energy of a reaction, reflects the change in free energy as the system progresses from one phase (liquid) to another (vapor). This curve (figure 8.3) shows the relationship between the Gibbs free energy of the reactants and products during the transformation, and it demonstrates how external factors such as pressure and temperature affect the spontaneity of the phase change or reaction.

In the case of the phase transition of water from liquid to vapor, the reaction is spontaneous at higher temperatures because the Gibbs free energy of the vapor phase becomes lower than that of the liquid phase, and the transition occurs when their Gibbs free energies become equal.

The key distinction between the Gibbs free energy of a substance and that of a reaction lies in what they represent. The Gibbs free energy of a substance reflects how the energy of a single phase evolves with temperature and pressure, and it allows us to predict phase transitions. In contrast, the Gibbs free energy of a reaction reflects the change in free energy during a transformation between different phases or chemical states, indicating whether the transformation is spontaneous or not under given conditions.

8.8 The Clausius–Clapeyron relation and phase transitions

In thermodynamics, the Clausius–Clapeyron relation is crucial for understanding phase transitions, particularly in the context of vaporization, sublimation, and melting. It links the change in Gibbs free energy (ΔG) with changes in pressure and temperature during a phase transition. In this section, we will focus on the derivation and application of the Clausius–Clapeyron relation for phase transitions.

8.8.1 Derivation of the Clausius–Clapeyron relation

Consider a phase transition between two phases of a substance, for example, liquid to vapor, at constant temperature and pressure. At equilibrium, the change in Gibbs free energy (ΔG) for this transition must be zero:

$$\Delta G_r = \Delta H_r - T\Delta S_r = 0,$$

where:
- ΔH_r is the enthalpy change of the phase transition (latent heat),
- ΔS_r is the entropy change of the transition, and
- T is the temperature at which the phase transition occurs.

This simplifies to

$$\Delta H_r = T\Delta S_r.$$

For a phase transition occurring at constant temperature, the change in Gibbs free energy between the two phases can be expressed in terms of the volume change (ΔV_r) and pressure (p):

$$d\Delta G_r = -\Delta S_r dT + \Delta V_r dp = 0.$$

Since we are considering the equilibrium condition for the phase transition, the total differential change in Gibbs free energy is zero. This implies that the change in Gibbs free energy is related to both the entropy and volume changes during the phase transition.

Now, rearranging the equation, we obtain

$$\frac{\Delta S_r}{\Delta V_r} = \frac{dp}{dT}.$$

To isolate the relationship between pressure and temperature, we rewrite this equation in a form that relates the vapor pressure (p) to the temperature (T). Using the fact that $\Delta H_r = T\Delta S_r$, we can express the latent heat of the phase transition as

$$\Delta H_r = T\Delta V_r \frac{dp}{dT}.$$

Now, to proceed further, let us treat the vapor as an ideal gas. The ideal gas law is given by

$$pV = nRT.$$

For the vapor phase, we can approximate the volume change ΔV_r as the volume of the vapor phase, since the volume of the liquid phase is negligible compared to that of the vapor phase. Thus, $\Delta V_r \approx \frac{RT}{p}$, which is derived from the ideal gas law.

Substituting this into the earlier equation,

$$\Delta H_r = T\left(\frac{RT}{p}\right)\frac{dp}{dT}.$$

Simplifying,

$$\Delta H_r = \frac{RT^2}{p}\frac{dp}{dT}.$$

Now, let us isolate $\frac{dp}{dT}$:

$$\frac{dp}{dT} = \frac{p\Delta H_r}{RT^2}.$$

Taking the natural logarithm of both sides to express the pressure–temperature relationship, we obtain

$$\frac{d \ln p}{dT} = \frac{\Delta H_r}{RT^2}.$$

This is the Clausius–Clapeyron relation, which describes how the vapor pressure p of a substance changes with temperature T.

8.8.2 Integration of the Clausius–Clapeyron relation

To integrate this equation, we assume that the latent heat of the phase transition (ΔH_r) is constant over the temperature range, which is often a valid approximation for small temperature intervals. The integrated form of the Clausius–Clapeyron relation becomes

$$\ln\left(\frac{p_2}{p_1}\right) = \frac{\Delta H_r}{R}\left(\frac{1}{T_1} - \frac{1}{T_2}\right),$$

where:
- p_1 and p_2 are the vapor pressures at temperatures T_1 and T_2, respectively.

This form of the Clausius–Clapeyron relation allows you to calculate the change in vapor pressure (p) for a substance over a given temperature range, given the latent heat of the phase transition. This equation is particularly useful for determining the boiling point of liquids under different pressures, such as predicting the boiling point of water at varying altitudes.

8.9 Thermodynamic analysis of hydrogen reactions

To further illustrate the interplay between enthalpy, entropy, and Gibbs free energy in thermodynamics, we consider three reactions involving hydrogen: combustion of hydrogen, operation of a hydrogen fuel cell, and electrolysis of water. These cases demonstrate spontaneous and non-spontaneous processes and how energy transfers can be quantified.

We focus on **one mole of water** in all reactions to ensure consistency and simplify the analysis (see table 8.1 for a summary of the thermodynamic values).

8.9.1 Case 1: Combustion of hydrogen

- *Reaction*:

$$H_2\ (g) + 1/2O_2\ (g) \rightarrow H_2O\ (l).$$

- *Thermodynamic parameters*:
   - Enthalpy change (ΔH_r),

$$\Delta H_r = -285.8\ \text{kJ mol}^{-1}\ \text{of water}.$$

Table 8.1. Summary of the thermodynamic values for hydrogen reactions (per mole of water).

Process	ΔH_r (kJ mol^{-1})	ΔS_r (kJ mol^{-1} · K^{-1})	ΔG_r (kJ mol)$^{-1}$	Spontaneity	Work interaction
Combustion of hydrogen	−285.8	−0.163	−237.2	Spontaneous	No work produced
Hydrogen fuel cell	−285.8	−0.163	−237.2	Spontaneous	$W_{\text{electrical}} = 237.2\ \text{kJ mol}^{-1}$
Electrolysis of water	+ 285.8	+ 0.163	+ 237.2	Non-spontaneous	$W_{\text{electrical}} = 237.2\ \text{kJ mol}^{-1}$

○ Entropy change (ΔS_r),

$$\Delta S_r = -163 \text{ J mol}^{-1}\text{ K}^{-1} = -0.163 \text{ kJ mol}^{-1}\text{ K}^{-1}.$$

○ Gibbs free energy change (ΔG_r), at 298 K:

$$\Delta G_r = \Delta H_r - T\Delta S_r.$$

$$\Delta G_r = -285.8 - (298 \times -0.163).$$

$$\Delta G_r = -285.8 + 48.6 = -237.2 \text{ kJ mol}^{-1}.$$

- *Interpretation*:
  ○ The negative ΔG_r indicates that combustion is *spontaneous*.
  ○ The reaction is highly exothermic, releasing significant heat.

8.9.2 Case 2: Hydrogen fuel cell

- *Reaction*:

$$H_2 \text{ (g)} + \frac{1}{2}O_2 \text{ (g)} \rightarrow H_2O \text{ (l)}.$$

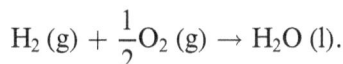

This reaction occurs electrochemically in a fuel cell to generate electricity.
- *Thermodynamic parameters*:
  ○ Enthalpy change (ΔH_r),

$$\Delta H_r = -285.8 \text{ kJ mol}^{-1} \text{ of water}.$$

  ○ Entropy change (ΔS_r),

$$\Delta S_r = -0.163 \text{ kJ mol}^{-1}\text{ K}^{-1}$$

  ○ Gibbs free energy change (ΔG_r),

$$\Delta G_r = -237.2 \text{ kJ mol}^{-1}.$$

- *Energy distribution*:
  ○ Electrical work: The maximum electrical work equals the Gibbs free energy change,

$$W_{\text{electrical}} = -\Delta G_r = 237.2 \text{ kJ mol}^{-1}.$$

  ○ Heat released: The remaining energy is dissipated as heat,

$$Q = \Delta H_r - W_{\text{electrical}} = -285.8 - (-237.2) = -48.6 \text{ kJ mol}^{-1}.$$

- *Interpretation*:
  ○ The reaction is *spontaneous*, as $\Delta G_r < 0$.

○ The fuel cell efficiently converts part of the energy into electrical work, with the rest released as heat.

8.9.3 Case 3: Electrolysis of water

- *Reaction*:

$$H_2O \text{ (l)} \rightarrow H_2 \text{ (g)} + \frac{1}{2}O_2 \text{ (g)}.$$

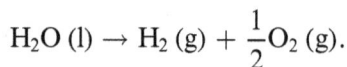

This is the reverse of the combustion reaction and requires energy input.
- *Thermodynamic parameters*:
  ○ Enthalpy change (ΔH_r),

$$\Delta H_r = +285.8 \text{ kJ mol}^{-1} \text{ of water.}$$

  ○ Entropy change (ΔS_r),

$$\Delta S_r = +0.163 \text{ kJ mol} \cdot \text{K}^{-1}.$$

  ○ Gibbs free energy change (ΔG_r),

$$\Delta G_r = \Delta H_r - T\Delta S_r.$$

$$\Delta G_r = 285.8 - (298 \times 0.163).$$

$$\Delta G_r = 285.8 - 48.6 = 237.2 \text{ kJ mol}^{-1}.$$

- *Energy requirements*:
  ○ Electrical work: The minimum electrical work required equals the Gibbs free energy change,

$$W_{\text{electrical}} = \Delta G_r = 237.2 \text{ kJ mol}^{-1}.$$

  ○ Heat absorbed: Additional heat is absorbed to supply the entropy-related energy,

$$Q = \Delta H_r - W_{\text{electrical}} = 285.8 - 237.2 = 48.6 \text{ kJ mol}^{-1}.$$

- *Interpretation*:
  ○ The positive ΔG_r confirms that electrolysis is *non-spontaneous* and requires external energy.
  ○ Efficient electrolysis minimizes electrical energy input by utilizing heat from the surroundings.

8.10 Phase change process: graphite to diamond and vice versa

The transformation between graphite and diamond is a fascinating phase change process that exemplifies the interplay of thermodynamic properties such as enthalpy (ΔH), entropy (ΔS), and Gibbs free energy (ΔG). Both graphite and diamond are

Table 8.2. Summary of the thermodynamic values for the graphite to diamond and diamond to graphite reactions.

Reaction	ΔH (kJ mol^{-1})	ΔS (kJ mol^{-1} ° K^{-1})	ΔG (kJ mol^{-1})	Spontaneity
Graphite to diamond	+1.9	−0.0034	+ 2.91	Non-spontaneous
Diamond to graphite	−1.9	+0.0034	−2.91	Spontaneous

solid phases of carbon, but their thermodynamic stability depends on the pressure and temperature conditions (see table 8.2 for a summary of the thermodynamic values).

8.10.1 Graphite to diamond ($C_{graphite} \rightarrow C_{diamond}$)

- *Reaction*:

$$C_{graphite} \rightarrow C_{diamond}.$$

- *Thermodynamic parameters, at standard conditions (298 K, 1 atm)*:
  ○ Enthalpy change (ΔH),

$$\Delta H = +1.9 \text{ kJ mol}^{-1}.$$

  Diamond has slightly higher enthalpy than graphite, indicating it is less stable energetically under standard conditions.
  ○ Entropy change (ΔS),

$$\Delta S = -3.4 \text{ J mol K}^{-1} = -0.0034 \text{ kJ mol}^{-1} \text{K}^{-1}.$$

  The negative entropy change reflects the increased structural order of diamond compared to graphite.
  ○ Gibbs free energy change (ΔG),

$$\Delta G = \Delta H - T\Delta S.$$

  At 298 K

$$\Delta G = 1.9 - (298 \times -0.0034) = 1.9 + 1.01 = +2.91 \text{ kJ mol}^{-1}.$$

  The positive ΔG indicates that the transformation of graphite to diamond is *non-spontaneous* under standard conditions.
- *Interpretation*:
  ○ Graphite is the more stable phase at standard conditions (1 atm, 298 K).

8.10.2 Diamond to graphite ($C_{diamond} \rightarrow C_{graphite}$)

- *Reaction*:

$$C_{diamond} \rightarrow C_{graphite}.$$

- *Thermodynamic parameters*:
  - Enthalpy change (ΔH),

  $$\Delta H = -1.9 \, \text{kJ mol}^{-1}.$$

  The reverse transformation releases energy as graphite is more stable.
  - Entropy change (ΔS),

  $$\Delta S = +0.0034 \, \text{kJ mol}^{-1} \, \text{K}^{-1}.$$

  The positive entropy change reflects the increase in disorder when diamond converts to graphite.
  - Gibbs free energy change (ΔG),

  $$\Delta G = -2.91 \, \text{kJ mol}^{-1}.$$

- *Interpretation*:
  - The negative ΔG indicates that diamond spontaneously converts to graphite at standard conditions.
  - However, the process is extremely slow because of the high activation energy required to break the strong covalent bonds in diamond's crystalline structure. This is why diamonds remain stable over geological timescales.

8.11 Helmholtz free energy

The Helmholtz free energy (F) is another thermodynamic potential, defined as

$$F = U - TS,$$

where U is the internal energy, T is the temperature, and S is the entropy of the system. The Helmholtz free energy is particularly useful for analysing systems under varying volume and temperature conditions.

8.11.1 When to use Helmholtz free energy

Unlike the Gibbs free energy, which is used for processes at varying pressure and temperature, the Helmholtz free energy is applicable when:
1. The system's *volume is varying*.
2. The system's *temperature is varying*.

Under these constraints:

- The change in Helmholtz free energy (ΔF) represents the **maximum useful work** that can be extracted from the system.
- Spontaneity is determined by the sign of ΔF:
  - ○ $\Delta F < 0$: The process is spontaneous.
  - ○ $\Delta F = 0$: The system is at equilibrium.
  - ○ $\Delta F > 0$: The process is non-spontaneous.

8.11.2 Relation to Gibbs free energy

The analysis of processes using Helmholtz free energy follows a similar framework to Gibbs free energy, with the key difference being the constraints:

- Gibbs free energy is suitable for systems at *varying pressure and temperature*.
- Helmholtz free energy is suitable for systems at *varying volume and temperature*.

For example:

- In studying chemical reactions or phase changes at constant pressure (common in open systems such as Earth's atmosphere), Gibbs free energy is the appropriate tool.
- For systems such as confined gases in a sealed, rigid container (constant volume), phase transitions in a closed space, or molecular simulations of small systems, Helmholtz free energy becomes more relevant because it quantifies the maximum work obtainable at constant volume and temperature. Examples include studying gas behavior in sealed chambers, analysing protein folding in solvents, modeling phase transitions in crystalline solids, or exploring the thermodynamics of plasmas and polymers under fixed volume constraints.

8.12 Summary

In this chapter, we developed a systematic understanding of the spontaneity of processes using thermodynamic potentials, particularly Gibbs free energy. Through examples such as ice melting on Earth and Mars, we demonstrated how **Gibbs free energy** changes with temperature and pressure and how it governs the spontaneity of processes. The chapter emphasizes the importance of phase transitions and the relationship between enthalpy, entropy, and free energy in determining whether a process will occur spontaneously.

We also introduced the **Clausius–Clapeyron relation**, which is used to model phase transitions such as liquid–vapor or sublimation transitions. The Clausius–Clapeyron relation provides a direct link between the change in Gibbs free energy and pressure/temperature changes during phase transitions. This relation was derived and applied to practical examples, such as water boiling at varying altitudes and pressures.

Furthermore, the chapter discussed the **Helmholtz free energy**, which applies to systems at constant volume and temperature, contrasting it with the Gibbs free energy, which is used for constant pressure systems. Helmholtz free energy helps in analysing systems such as confined gases or small systems in simulations.

Finally, through comprehensive examples—such as hydrogen combustion, hydrogen fuel cells, and the electrolysis of water—we illustrated the relationship between Gibbs free energy, enthalpy, entropy, and the maximum work achievable or required in various chemical processes. These examples serve to deepen the understanding of how thermodynamics applies to real-world applications.

By synthesizing the concepts of entropy, enthalpy, Gibbs free energy, and Helmholtz free energy, this chapter provides the theoretical foundation and practical tools necessary for analysing and predicting the behavior of systems in diverse thermodynamic processes.

IOP Publishing

A Classical Thermodynamics Toolkit

Srinivas Vanapalli

Chapter 9

Thermodynamic potentials—applications under non-standard temperature and pressure conditions

9.1 Introduction

In this chapter, we delve deeper into the concept of thermodynamic potentials, focusing on their behavior under varying conditions. The chapter builds upon the foundational thermodynamic principles established earlier, particularly concerning free energy and the spontaneity of processes. It introduces key thermodynamic potentials—internal energy (U), enthalpy (H), and Gibbs free energy (G)—and explores their intricate relationships with temperature, pressure, and volume.

These potentials provide a comprehensive framework for understanding the energy exchanges that occur in both reversible and irreversible processes. Through their partial derivatives with respect to state variables, such as entropy, volume, pressure, and temperature, we can derive important thermodynamic quantities. This enables a systematic approach to analysing various physical processes and reactions, including those that involve phase transitions, chemical reactions, and work performed during expansion or compression.

A central theme in this chapter is the application of these potentials in practical scenarios, especially in processes involving work. We explore the various types of work expressions such as $-pdV$ work and Vdp work, highlighting their differences in thermodynamic contexts. Furthermore, we discuss how to calculate and interpret the Gibbs free energy change for both substances and reactions, extending these concepts to conditions where temperature and pressure deviate from standard values.

doi:10.1088/978-0-7503-6029-6ch9

9.2 Thermodynamic potentials and their derivatives

Thermodynamic potentials, such as internal energy (U), enthalpy (H), and Gibbs free energy (G), are fundamental properties that describe the energy state of a system. These potentials depend on specific state variables and provide valuable insights into the behavior of systems under different thermodynamic conditions. By exploring their mathematical relationships, we can derive useful expressions that connect physical properties like temperature, pressure, and entropy.

9.2.1 The internal energy $U(S, V)$

The internal energy U is a function of entropy S and volume V. From the thermodynamic identity,

$$dU = TdS - pdV.$$

We see that the partial derivatives of U are

$$\left(\frac{\partial U}{\partial S}\right)_V = T \text{ and } \left(\frac{\partial U}{\partial V}\right)_S = -p.$$

These relationships indicate that:
- T: The temperature of the system is the partial derivative of internal energy with respect to entropy at constant volume.
- $-p$: The negative pressure is the partial derivative of internal energy with respect to volume at constant entropy.

9.2.2 The enthalpy $H(S, p)$

The enthalpy H is defined as

$$H = U + pV.$$

Differentiating, we obtain

$$dH = dU + d(pV) = TdS - pdV + pdV + Vdp = TdS + Vdp.$$

Here, H is a function of entropy S and pressure p. From the expression

$$dH = TdS + Vdp.$$

The partial derivatives are

$$\left(\frac{\partial H}{\partial S}\right)_p = T \text{ and } \left(\frac{\partial H}{\partial p}\right)_S = V.$$

These relationships reveal that:
- T: The temperature is the partial derivative of enthalpy with respect to entropy at constant pressure.
- V: The volume of the system is the partial derivative of enthalpy with respect to pressure at constant entropy.

9.2.3 The Gibbs free energy $G(p, T)$

The Gibbs free energy G is defined as

$$G = H - TS.$$

Differentiating,

$$dG = dH - d(TS) = TdS + Vdp - TdS - SdT = Vdp - SdT.$$

Here, G is a function of pressure p and temperature T. From the expression,

$$dG = Vdp - SdT.$$

The partial derivatives are

$$\left(\frac{\partial G}{\partial p}\right)_T = V \text{ and } \left(\frac{\partial G}{\partial T}\right)_p = -S.$$

These relationships indicate that:
- V: The volume is the partial derivative of Gibbs free energy with respect to pressure at constant temperature.
- $-S$: The negative entropy is the partial derivative of Gibbs free energy with respect to temperature at constant pressure.

9.2.4 Summary table

To summarize these relationships, we present table 9.1.

These relationships demonstrate how thermodynamic potentials and their derivatives provide a complete description of the thermodynamic behavior of systems under various constraints. The clarity of this framework simplifies the analysis of practical thermodynamic processes.

Table 9.1. Summary of thermodynamic potentials, their natural independent variables, and the resulting partial derivatives. These identities are fundamental to evaluating temperature, pressure, volume, and entropy from the internal energy U, enthalpy H, and Gibbs free energy G, depending on the chosen representation of the system.

Potential X	Independent variables	$\left(\frac{\partial X}{\partial a}\right)_b$	Result
$U(S, V)$	S, V	$\left(\frac{\partial U}{\partial S}\right)_V$	T
		$\left(\frac{\partial U}{\partial V}\right)_S$	$-p$
$H(S, p)$	S, p	$\left(\frac{\partial H}{\partial S}\right)_p$	T
		$\left(\frac{\partial H}{\partial p}\right)_S$	V
$G(p, T)$	p, T	$\left(\frac{\partial G}{\partial p}\right)_T$	V
		$\left(\frac{\partial G}{\partial T}\right)_p$	$-S$

9.3 Examples to illustrate the use of thermodynamic relationships

This section presents detailed examples based on the thermodynamic relationships $dU = TdS - pdV$, $dH = TdS + Vdp$, and $dG = Vdp - SdT$. These examples demonstrate how thermodynamic potentials are used under specific conditions to calculate energy changes and work.

9.3.1 Example 1: $S = $ constant, $dU = -pdV$

This example highlights an isentropic process ($S = $ constant) where the internal energy change (dU) equals the pV-work done by or on the system.

Scenario: A gas compressor compresses a gas in a piston–cylinder system.

- *Initial and final conditions*: The gas starts at a low pressure and high volume. As the piston compresses the gas, its volume decreases while its pressure increases.
- *Key process assumption*: The process is isentropic ($S = $ constant), meaning no entropy change occurs.
- *Analysis*: $dU = TdS - pdV$. With $S = $ constant, $dS = 0$, simplifying to $dU = -p \, dV$. The internal energy change equals the work done on the gas during compression.

9.3.2 Example 2: $S = $ constant, $dH = Vdp$

This example shows how the change in enthalpy (dH) corresponds to the work associated with pressure changes during an isentropic flow process.

Scenario: A gas compressor increases the pressure of a gas as it flows through the machine.

- *Initial and final conditions*: The gas enters the compressor at a low pressure and exits at a higher pressure. The flow process is steady and involves zero heat exchange.
- *Key process assumption*: The process is isentropic ($S = $ constant).
- *Analysis*: $dH = TdS + Vdp$. With $S = $ constant, $dS = 0$, simplifying to: $dH = Vdp$. The increase in enthalpy corresponds to the work done by the compressor to increase the gas pressure.

9.3.3 Example 3: $T = $ constant, $dG = Vdp$

This example illustrates that under constant temperature conditions, the change in Gibbs free energy (dG) equals the flow work resulting from pressure changes.

Scenario: A gas compressor increases the pressure of a gas as it flows through the machine.

- *Initial and final conditions*: The gas enters the compressor at a low pressure and exits at a higher pressure. The flow process is steady and involves zero temperature change.
- *Key process assumption*: The temperature at the entry and exit is the same ($T_{\text{entry}} = T_{\text{exit}}$).

- *Analysis*: $dG = Vdp - SdT$. Since the temperature is the same at the entry and exit ($dT = 0$), the equation simplifies to: $dG = Vdp$. The change in Gibbs free energy reflects the work associated with the pressure increase under constant temperature conditions.

9.3.4 Example 4: $p = $ constant, $dH = TdS$

This example demonstrates that for a constant pressure process, the change in enthalpy (dH) corresponds to the heat transfer if no other work is involved.

Scenario: Heating a gas at constant pressure.
- *Initial and final conditions*: The gas is heated, increasing its temperature and entropy while maintaining constant pressure.
- *Key process assumption*: The pressure is constant ($p = $ constant).
- *Analysis*: $dH = TdS + Vdp$. With $p = $ constant, $dp = 0$, simplifying to $dH = TdS$. The enthalpy change corresponds to the heat transfer that increases the system's entropy.

9.4 Understanding the difference between $-pdV$ and Vdp work

In thermodynamics, the work done by or on a system is typically a result of pressure and volume changes. The two main expressions for work, $-pdV$ and Vdp, describe different scenarios. To illustrate these differences clearly, let us consider the sequence of events in a piston–cylinder system where both the volume and pressure of the gas change during the process.

In figure 9.1, the gas is expanding in a piston–cylinder system, pushing the piston outward. This results in an increase in the volume of the gas, and the work done by the system is captured by the expression $-pdV$, where p is the pressure and dV is the change in volume. This represents **positive work** done by the gas on the surroundings. The pressure remains relatively constant during the expansion process, and the system's volume increases as the gas pushes the piston outward. The work is directly proportional to the pressure and the change in volume.

Figure 9.1. A schematic of a piston–cylinder system under the influence of an external weight m. The weight exerts a constant force on the piston, maintaining a fixed pressure inside the system. This setup is often used to conceptualize reversible processes and study thermodynamic work interactions at constant pressure.

9.4.1 Understanding the Vdp work using figure 9.2

9.4.1.1 Figure 9.2, panel 1: Initial set-up (valve closed)
In this initial scenario, the system consists of a piston with the valve closed. The pressure in the system is at P_{in}, and gas is not allowed to leave because the valve is closed. The system starts at an initial state with a certain volume.
- *Description*: The pressure is constant at P_{in}, and no work is done at this stage, as the volume is not changing.
- *Work expression*: No work is done in this configuration because there is no volume change, and the system is static.

9.4.1.2 Figure 9.2, panel 2: Gas expansion (piston moving right, valve still closed)
In this panel, the piston moves outward, expanding the gas inside. The pressure remains constant at P_{in} as the gas pushes against the piston. The valve is still closed, so no gas can exit the system.
- *Description*: As the piston moves outward, the system's volume increases, and the pressure remains constant at P_{in}.
- *Work expression*: The work done by the gas is given by the expression $W = -\int_p dV$. The work is *negative* since the volume increases ($dV > 0$) during expansion.

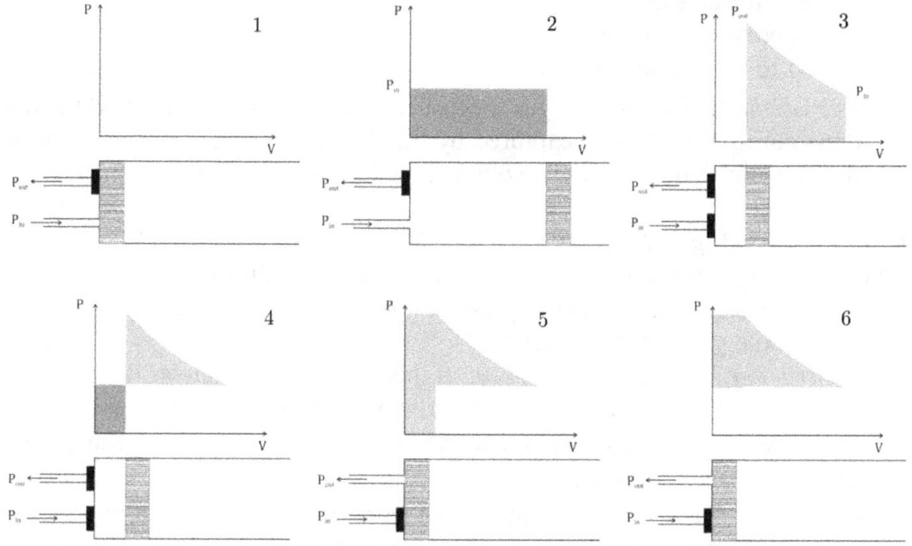

Figure 9.2. Six different quasistatic processes involving an ideal gas in a piston–cylinder system with open valves. Each configuration results in boundary work. The difference lies in how the piston moves and how the inlet and outlet valves are operated. In all cases, the pressure inside the cylinder remains nearly uniform, and the expansion or compression is slow enough to ensure a quasistatic process.

9.4.1.3 Figure 9.2, panel 3: Compression (piston moving left, valve closed)
In this step, the piston starts moving inward, compressing the gas. The volume decreases, and the pressure increases within the system since the valve is still closed, and the gas cannot escape.

- *Description*: The gas is compressed, and the volume decreases while the pressure increases. The valve is closed, so the gas remains trapped inside the system.
- *Work expression*: The work done is still represented by $W = -\int_p dV$, but since the volume decreases ($dV < 0$), the work done on the system is *positive*.

9.4.1.4 Figure 9.2, panel 4: Valve opens, no change in work
In this panel, we are showing the work done so far, but nothing new happens at this point. The system is still in the same state, with the gas having been compressed up to this point. The work done thus far is represented by the area under the curve.

- *Description*: There is no further change in pressure or volume in this step; it is just showing the work that has accumulated up to this point.
- *Key insight*: We are simply displaying the accumulated work until the valve opens. The graph shows the work so far, but no new energy transfer occurs.

9.4.1.5 Figure 9.2, panel 5: Valve opens, gas moves out at constant pressure
In this step, the valve opens, and the gas begins to move out of the system. The pressure remains constant during this process, and the gas leaves the system as the piston continues to move inward.

- *Description*: The valve opens, and the gas starts exiting the system. The pressure remains constant as the gas leaves, and the system now undergoes a flow process. The volume decreases as the gas moves out, and the work done is related to the pressure–volume relationship, represented by the integral $\int_V dp$.
- *Key insight*: In this case, the gas is expelled at a constant pressure, and the work done during this phase is captured by the flow work, where Vdp characterizes the pressure change while the volume remains approximately constant.

9.4.1.6 Figure 9.2, panel 6: Net work as Vdp area under the graph
In this final step, panel 6 shows the net work done during the process. The figure represents the area under the pressure–volume curve, capturing the net flow work done as the gas exits the system.

- *Description*: No new changes occur in this step; it simply shows the total work accumulated up to this point. The net work is represented by the area under the pressure–volume curve, reflecting the integral $\int_V dp$ from the pressure drop while the gas exits the system.
- *Key insight*: This figure highlights the net work done during the entire process, as reflected in the area under the curve, but no new events or changes occur in this step.

9.4.1.7 Summary
This sequence demonstrates the difference between $-pdV$ and Vdp work:
1. $-pdV$ describes work done during compression or expansion in a closed system where both pressure and volume change.

2. Vdp describes work done during a flow process where the pressure changes, but the volume remains nearly constant as gas exits or enters the system.

The panels in figure 9.2 provide a clear illustration of how the work is done in different thermodynamic scenarios.

9.5 Distinguishing between substance and reaction: thermodynamic properties across conditions

In thermodynamics, understanding the behavior of substances and reactions is fundamental, especially when estimating the properties of a system at different temperatures and pressures. A **single substance** refers to the study of the properties of a material, such as water, at various conditions. These properties include Gibbs free energy (G), enthalpy (H), and entropy (S), and they provide insights into how the material behaves as temperature and pressure change. For example, we can observe how the enthalpy of water changes when its temperature is increased or when it is subjected to different pressures. In this context, we are not concerned with any transformation or phase change of the material, just how its intrinsic properties evolve with the environment.

On the other hand, a **reaction** refers to a process where the system undergoes a transformation from one phase or state to another. A common example is the phase change of water from liquid to gas during boiling. During such a reaction, the system's enthalpy, entropy, and Gibbs free energy change as the substance undergoes a phase transition due to changes in temperature and pressure. This process involves energy exchange with the surroundings and may also be influenced by external conditions, such as heating or cooling rates.

Understanding the difference between a substance and a reaction is crucial because while substances are studied individually for how their properties respond to changing conditions, reactions focus on how entire systems evolve during transitions between different states. This subtle difference is important and often not well clarified in standard texts. In our book, we aim to provide a clear explanation of both cases and introduce tools to estimate the thermodynamic properties—enthalpy, entropy, and Gibbs free energy—for both individual substances and chemical or physical reactions.

To achieve this, we will present methods for calculating these properties in different environments. For substances, we will show how to determine enthalpy, entropy, and Gibbs free energy at various temperatures and pressures based on known data. For reactions, we will introduce the concept of reaction enthalpy, entropy, and Gibbs free energy, and how they vary with temperature and pressure during phase transitions or chemical processes.

9.6 Calculating enthalpy, entropy, and Gibbs free energy at non-standard pressure and temperature

Up to this point, we have primarily considered reactions under standard conditions of pressure and temperature. However, the thermodynamic principles we have discussed are equally valid for systems under non-standard conditions. This is particularly important when designing systems for environments such as Mars, deep

under the Earth's surface, or in a cruising airplane, where the temperature and pressure conditions differ from standard atmospheric values. Other examples include high-pressure reactors, deep-sea environments, or industrial applications operating under varying temperatures and pressures.

Thermodynamic properties such as enthalpy (H), entropy (S), and Gibbs free energy (G) at specific pressures and temperatures can often be extracted from databases. However, for quick estimates, it is helpful to have a method for calculating approximate values of these properties at different pressures and temperatures, based on data from standard conditions. In this section, we will present a procedure for calculating enthalpy, entropy, and Gibbs free energy at any given temperature, both for pure substances and reactions.

9.6.1 Determining approximate enthalpy at any temperature

The heat capacity at constant pressure (C_p) is defined as the rate of change of enthalpy with respect to temperature:

$$C_p = \left(\frac{\partial H}{\partial T}\right)_p.$$

To calculate the enthalpy at a temperature T, we can integrate this expression from a reference temperature T_0:

$$H(T) = H(T_0) + \int_{T_0}^{T} C_p(T)\, dT.$$

If we assume that the heat capacity is constant and independent of temperature, the equation simplifies to

$$H(T) = H(T_0) + C_p(T - T_0).$$

Here, $H(T_0)$ represents the standard enthalpy value at the reference temperature.

9.6.2 Determining approximate entropy at any temperature

Entropy can be defined as

$$dS = \frac{Q}{T} = \frac{C_p(T)\, dT}{T}.$$

Integrating this expression gives the entropy at temperature T:

$$S(T) = S(T_0) + \int_{T_0}^{T} \frac{C_p(T)}{T}\, dT.$$

If we assume that the heat capacity is constant and independent of temperature, this simplifies to

$$S(T) = S(T_0) + C_p \ln\left(\frac{T}{T_0}\right).$$

Here, $S(T_0)$ represents the standard entropy value at the reference temperature.

9.6.3 Determining approximate Gibbs free energy at any temperature

The Gibbs free energy at any temperature is related to the enthalpy and entropy by

$$G(T) = H(T) - T \cdot S(T).$$

Thus, the Gibbs free energy can be calculated by substituting the values of $H(T)$ and $S(T)$ derived above. These equations provide a straightforward way to estimate the Gibbs free energy at different temperatures based on known values at standard conditions.

9.6.4 Calculating the Gibbs free energy of a reaction at any temperature

When considering reactions, we are particularly interested in how the enthalpy, entropy, and Gibbs free energy change with temperature and pressure. For example, consider a general reaction

$$2A + B \longrightarrow 3C + D.$$

The enthalpy of the reaction at standard conditions is given by

$$\Delta H_r(p_0, T_0) = 3H_C(p_0, T_0) + H_D(p_0, T_0) - 2H_A(p_0, T_0) - H_B(p_0, T_0).$$

At any temperature T, the enthalpy change is

$$\Delta H_r(p_0, T) = 3H_C(p_0, T) + H_D(p_0, T) - 2H_A(p_0, T) - H_B(p_0, T).$$

By expanding the enthalpy of each substance at a different temperature, we obtain

$$\Delta H_r(p_0, T) = 3[H_C(p_0, T_0) + C_{p,C}(T - T_0)] + [H_D(p_0, T_0) + C_{p,D}(T - T_0)]$$
$$- 2[H_A(p_0, T_0) + C_{p,A}(T - T_0)] - [H_B(p_0, T_0) + C_{p,B}(T - T_0)].$$

Similarly, for the entropy change of the reaction, we start with the standard condition

$$\Delta S_r(p_0, T_0) = 3S_C(p_0, T_0) + S_D(p_0, T_0) - 2S_A(p_0, T_0) - S_B(p_0, T_0).$$

At any temperature T, the entropy change is

$$\Delta S_r(p_0, T) = 3\left[S_C(p_0, T_0) + C_{p,C}\ln\left(\frac{T}{T_0}\right)\right] + \left[S_D(p_0, T_0) + C_{p,D}\ln\left(\frac{T}{T_0}\right)\right]$$
$$- 2\left[S_A(p_0, T_0) + C_{p,A}\ln\left(\frac{T}{T_0}\right)\right] - \left[S_B(p_0, T_0) + C_{p,B}\ln\left(\frac{T}{T_0}\right)\right].$$

Finally, the Gibbs free energy of the reaction at standard conditions is

$$\Delta G_r(p_0, T_0) = 3G_C(p_0, T_0) + G_D(p_0, T_0) - 2G_A(p_0, T_0) - G_B(p_0, T_0).$$

At any temperature T, the Gibbs free energy change is

$$\Delta G_r(p_0, T) = 3[H_C(p_0, T) - TS_C(p_0, T)] + [H_D(p_0, T) - TS_D(p_0, T)] - 2[H_A(p_0, T) - TS_A(p_0, T)]$$
$$- [H_B(p_0, T) - TS_B(p_0, T)].$$

Alternatively, this expression simplifies to

$$\Delta G_r(p_0, T) = \Delta H_r(p_0, T) - T\Delta S_r(p_0, T).$$

This simplification provides a more straightforward way to compute the Gibbs free energy change at non-standard temperatures.

9.6.5 Heat capacity of a reaction

The heat capacity of the reaction can be expressed as

$$C_{p,r} = 3C_{p,C} + C_{p,D} - 2C_{p,A} - C_{p,B}.$$

Thus, the enthalpy and entropy of a reaction can be rewritten in terms of the reaction's heat capacity:

$$\Delta H_r(p_0, T) = \Delta H_r(p_0, T_0) + C_{p,r}(T - T_0)$$

$$\Delta S_r(p_0, T) = \Delta S_r(p_0, T_0) + C_{p,r} \ln\left(\frac{T}{T_0}\right).$$

These equations simplify the calculations for reactions under varying temperatures, enabling easier estimation of the thermodynamic properties during the reaction process.

9.7 Determining Gibbs free energy of a reaction at any pressure

Let us consider the same reaction as before but focus on how pressure affects the Gibbs free energy.

In order to account for the pressure dependence, we need to identify the phase of each substance in the reaction—solid (s), liquid (l), or gas (g):

$$2A\ (s) + B\ (g) \longrightarrow 3C\ (g) + D\ (l).$$

For each substance, the change in Gibbs free energy $dG(p, T)$ is given by the equation

$$dG(p, T) = -SdT + Vdp.$$

Since we are interested in the pressure dependence, the temperature term (dT) becomes zero. Therefore, we have

$$dG(p, T) = Vdp.$$

For a reaction, the change in Gibbs free energy $\Delta G_r(p, T)$ is related to the change in volume of the system,

$$d\Delta G_r(p, T) = \Delta V_r dp.$$

The volume change in the reaction, ΔV_r, is the difference in volumes between the products and reactants:

$$\Delta V_r = 3V_C + V_D - 2V_A - V_B.$$

Since the volume of a gas is much larger than that of a solid or liquid, we can generally neglect the volume contributions of solids and liquids when gases are present in a reaction. However, it is important to note that if the reaction involves only solids and liquids, we should not neglect their volumes. For example, in the reaction where graphite converts to diamond (both solids), the volume of both substances must be considered.

Returning to our reaction, we can neglect the volumes of substances A and D, which are solids or liquids. For gases, we use the ideal gas expression for one mole of a substance. Therefore, the volume change in the reaction becomes

$$\Delta V_r = 3RT/p + 0 - 2 \cdot 0 - RT/p = 2RT/p.$$

Now, the change in Gibbs free energy for the reaction at any pressure is

$$d\Delta G_r = 2RT/p \, dp.$$

Integrating this expression gives the total change in Gibbs free energy between the pressures p_0 and p:

$$\Delta G_r(p) - \Delta G_r(p_0) = \int_{p_0}^{p} 2RT/p \, dp = 2RT \ln\left(\frac{p}{p_0}\right).$$

This approach shows how to determine the pressure dependence of Gibbs free energy for a reaction. By understanding the volume change in the reaction and applying the ideal gas law, we can derive an expression for the Gibbs free energy that is applicable at any pressure. This method is essential for understanding how reactions behave under varying pressure conditions.

9.8 Gibbs free energy and reaction behavior under non-standard temperature and pressure

In thermodynamics, understanding how a reaction behaves under various pressure and temperature conditions is critical for predicting its spontaneity and feasibility. To start, it is important to recall the fundamental equation for Gibbs free energy:

$$\Delta G_r = \Delta H_r - T\Delta S_r.$$

For the sake of simplicity in the following analysis, we will assume that the temperature dependence of enthalpy (H) and entropy (S) is negligible. This will allow us to focus on the key relationship between Gibbs free energy and temperature, but it is worth noting that this is a rough approximation. If we plot this equation, with Gibbs free energy (ΔG_r) on the y-axis and temperature on the x-axis, we obtain a straight line. The Y-intercept represents the change in enthalpy (ΔH_r), and the slope is given by $-\Delta S_r$ (see figure 9.3).

This graphical approach is extremely useful because it offers a quick way to predict how reactions behave at different temperatures. Although this method does not provide exact values, it guides us in the right direction and forms a useful first tool for thermodynamic investigations.

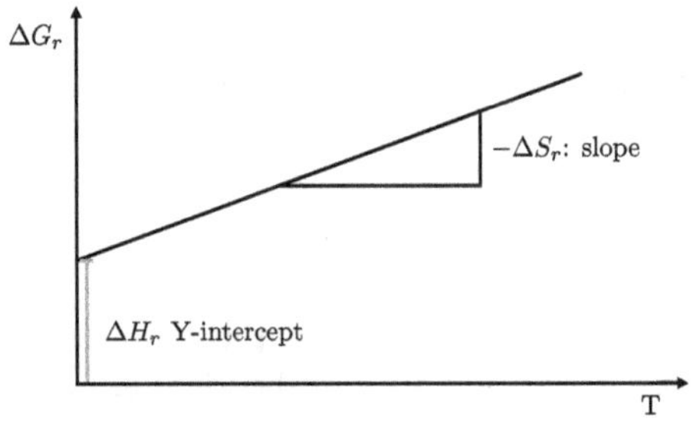

Figure 9.3. Temperature dependence of the Gibbs free energy change of a chemical reaction, ΔG_r, at constant pressure. The plot illustrates the linear relation $\Delta G_r = \Delta H_r - \Delta S_r$, where the y-intercept corresponds to the enthalpy change of reaction ΔH_r, and the slope equals the negative of the entropy change of reaction ΔS_r.

9.8.1 Case studies: reaction behavior under various conditions

Now, let us apply this graphical analysis to a few example reactions to observe how their behavior changes under different pressure and temperature conditions. Each case is an example of a chemical reaction where the entropy and enthalpy changes are different, impacting the spontaneity of the process.

9.8.1.1 Case 1: $\Delta S_r > 0$; $\Delta H_r < 0$
Consider the reaction

$$C\,(s) + O_2\,(g) \longrightarrow CO_2\,(g).$$

The relevant data for the substances involved in this reaction are shown in the table on the left-hand side of figure 9.4.

For this reaction, the entropy change (ΔS_r) is positive, and the enthalpy change (ΔH_r) is negative. This means the reaction is always spontaneous at all temperatures. The slope of the Gibbs free energy versus temperature plot is negative because the entropy is increasing, and the Y-intercept is negative because the reaction is exothermic (releases heat) (see the graph on the right-hand side of figure 9.4).

9.8.1.2 Case 2: $\Delta S_r > 0$; $\Delta H_r > 0$
Now, consider the reaction

$$PCl_5\,(g) \longrightarrow PCl_3\,(g) + Cl_2\,(g).$$

The relevant data for this reaction are shown in the table on the left-hand side of figure 9.5.

In this case, the slope of the ΔG_r versus temperature plot is negative (due to positive entropy), but the Y-intercept is positive because the reaction starts with a positive enthalpy change. The plot crosses the x-axis at 518 K, which is the transition temperature where the process becomes spontaneous. Below this temperature, the reaction is non-spontaneous, while above it, it becomes spontaneous (see the graph on the right-hand side of figure 9.5).

	H(kJ)	G(kJ)	S(J/K)
C	0	0	5.74
O_2	0	0	205.03
CO_2	−393.5	−394.36	213.63
	−393.5	−394.36	2.86

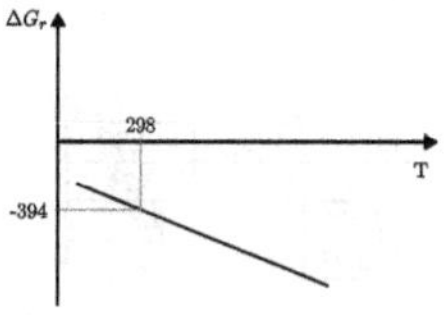

Figure 9.4. Temperature dependence of the Gibbs free energy change for the formation of carbon dioxide. The plot is based on standard thermodynamic data shown in the table. The negative slope indicates that the reaction becomes more favorable at higher temperatures.

	H(kJ)	G(kJ)	S(J/K)
Cl_2	0	0	223
PCl_3	−287	−268	312
PCl_5	−375	−305	365
	88	37	170

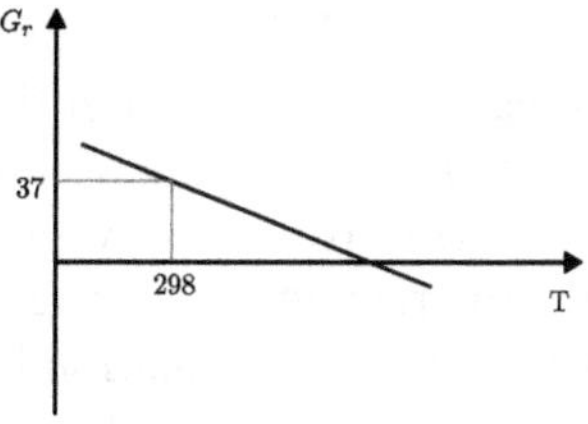

Figure 9.5. Gibbs free energy change versus temperature for the formation of phosphorus pentachloride (PCl_5) from phosphorus trichloride (PCl_3) and chlorine (Cl_2), based on standard thermodynamic data. The reaction becomes spontaneous above a certain temperature.

	H(kJ)	G(kJ)	S(J/K)
NH_3	−46.11	−16.48	192.34
HCl	−92.31	−95.30	186.80
NH_4Cl	−314.43	−202.97	94.70
	−176.01	−91.19	−284.44

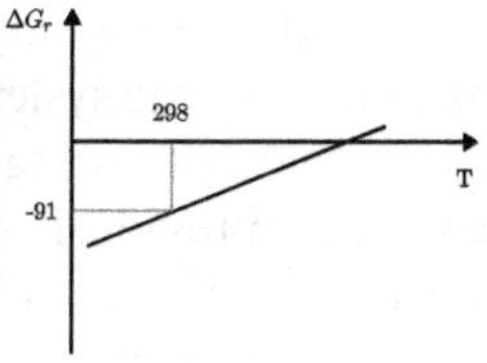

Figure 9.6. Temperature dependence of the Gibbs free energy change for the formation of ammonium chloride (NH_4Cl) from ammonia (NH_3) and hydrogen chloride (HCl), based on standard thermodynamic data.

9.8.1.3 Case 3: $\Delta S_r < 0$; $\Delta H_r < 0$
Consider the reaction

$$NH_3 \text{ (g)} + HCl \text{ (g)} \longrightarrow NH_4Cl \text{ (s)}.$$

The data for this reaction are shown in the table on the left-hand side of figure 9.6.

In this case, the reaction has a negative entropy change ($\Delta S_r < 0$) and a negative enthalpy change ($\Delta H_r < 0$). The curve has a positive slope because the entropy is negative. The reaction is initially spontaneous at lower temperatures, but as the temperature increases, the system reaches a transition point around 619 K, beyond which the reaction becomes non-spontaneous (see the graph on the right-hand of figure 9.6).

9.8.1.4 Case 4: $\Delta S_r < 0$; $\Delta H_r > 0$
Consider the reaction

$$C \text{ (graphite)} \longrightarrow C \text{ (diamond)}.$$

	H(kJ)	G(kJ)	S(J/K)
C(g)	0	0	5.74
C(d)	1.895	2.900	2.377
	1.895	2.900	−3.363

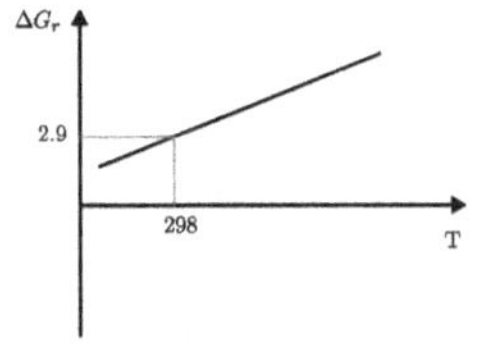

Figure 9.7. Temperature dependence of the Gibbs free energy change for the conversion of graphite (C(g)) to diamond (C(d)), illustrating the thermodynamic favorability of graphite at room temperature.

The data for this reaction are shown in the table on the left-hand side of figure 9.7.

This reaction involves the transformation of graphite into diamond. As the reaction involves a negative entropy change ($\Delta S_r < 0$) and a positive enthalpy change ($\Delta H_r > 0$), the reaction will never occur spontaneously at any temperature. The transition from graphite to diamond is not driven by temperature but may be influenced by extreme pressure conditions, which we will discuss in the next section.

9.8.2 Temperature dependence of enthalpy and entropy

While the graphical analysis above provides a useful first step in understanding the behavior of reactions at different temperatures, we have neglected the temperature dependence of enthalpy and entropy. By incorporating this dependency, we can obtain a better approximation of the system's behavior, though the result remains an estimate due to the assumption of constant heat capacities.

For example, in case 3, we previously calculated the enthalpy and entropy at the transition temperature of 619 K. Including the temperature dependence of heat capacities yields more accurate values for enthalpy and entropy at this temperature, refining our estimate of the Gibbs free energy,

$$\Delta G_r = \Delta H_r - T\Delta S_r.$$

This approach is still an approximation because we assume heat capacities are constant, but it provides more accurate predictions of the reaction's behavior at different temperatures.

9.8.3 Example with hydrogen

Hydrogen is a key player in the ongoing energy transition and is of particular interest for various energy applications. To illustrate the concepts of Gibbs free energy and its implications, we will examine hydrogen in several reactions, considering three main cases: combustion, fuel cells, and electrolysis.

9.8.4 Combustion of hydrogen

The combustion of hydrogen is a straightforward and important reaction:

$$H_2 \text{ (g)} + 0.5O_2 \text{ (g)} \longrightarrow H_2O \text{ (l)}.$$

The thermodynamic data for this reaction are shown in table 9.2.

Table 9.2. Standard enthalpy, Gibbs free energy, and entropy values for hydrogen, oxygen, and water at 298 K.

Substance	H (kJ)	G (kJ)	S (J K^{-1})
H_2	0.00	0.00	130.68
O_2	0.00	0.00	102.57
H_2O	−285.83	−237.13	69.91

The Gibbs free energy for the reaction can be calculated using the equation

$$\Delta G_r = \Delta H_r - T\Delta S_r.$$

Substituting the values,

$$\Delta G_r = -285.83 - (298 \times (-163.34 \text{ J K}^{-1})).$$

This results in

$$\Delta G_r = -237.16 \text{ kJ mol}^{-1} \text{ of } H_2.$$

It is important to note that this Gibbs free energy is calculated per mole of hydrogen or half mole of oxygen or one mole of water. If you are asked to calculate the Gibbs free energy per mole of oxygen, you need to multiply the value by two, which gives −474.32 kJ mol^{-1} of O_2.

This reaction falls under case 3 from earlier discussions, where the process is spontaneous at room temperature. However, this reaction will have a **transition temperature** where the process becomes non-spontaneous. In this case, that temperature would be around 1750 K. The question arises: why does hydrogen still burn on the Sun despite the high temperature? The answer lies in the extremely high pressure on the Sun's surface (although the Sun technically does not have a solid surface, its outer layers exhibit high pressure), which significantly increases the transition temperature.

In this case, the work done by the system (W_{other}) is zero, meaning that all the energy released is in the form of heat. Therefore, the heat transfer is

$$Q = \Delta H_r - W_{other} = -285.83 \text{ kJ mol}^{-1} \text{ of } H_2.$$

This is an exothermic reaction, which releases energy. A famous (albeit dangerous) application of this reaction is hydrogen-filled balloons, as seen in the Zeppelin disaster—hydrogen's flammability can be highly destructive.

9.8.5 Hydrogen fuel cell

The hydrogen fuel cell operates on a similar principle to hydrogen combustion but with an important difference: it generates electrical energy by transporting electrons through a circuit. The reaction in a fuel cell is

$$H_2 \text{ (g)} + 0.5O_2 \text{ (g)} + 2e^- \longrightarrow H_2O \text{ (l)} + 2e^-.$$

Here, we have the same chemical reaction, but it also involves the movement of electrons, producing work. The energy output is captured as electrical work, and the maximum work $|W_{other}|$ is equal to the Gibbs free energy:

$$|W_{other}| = \Delta G_r = 237.16 \text{ kJ mol}^{-1} \text{ of } H_2.$$

This is a **negative work** value, meaning the system is delivering energy to the surroundings in the form of electrical work.

The heat transfer in this case is

$$Q = T\Delta S_r = -48.67 \text{ kJ mol}^{-1} \text{ of } H_2.$$

This indicates that the reaction is exothermic.

What happens if the fuel cell operates at a higher pressure, say 100 bar? The change in Gibbs free energy with pressure can be determined by the volume change of the reaction

$$d\Delta G_r = \Delta V_r dp.$$

The volume change for the reaction is given by

$$\Delta V_r = V_{H_2O} - V_{H_2} - 0.5 V_{O_2}.$$

Since V_{H_2O}, the volume of liquid water, is negligible, we use the ideal gas law to approximate the volume change of hydrogen and oxygen. The volume change is then

$$\Delta V_r = -1.5 \cdot RTp.$$

Integrating this expression gives

$$\Delta G_r(p) - \Delta G_r(p_0) = \int_{p_0}^{p} 1.5RT \ln\left(\frac{p}{p_0}\right) dp.$$

At 100 bar, the Gibbs free energy is

$$\Delta G_r(p = 100 \text{ bar}) = -254 \text{ kJ mol}^{-1}.$$

Since the absolute value of ΔG_r is larger at 100 bar compared to 1 bar, the work output of the fuel cell is larger, resulting in higher voltage.

9.8.6 Hydrogen electrolysis

Hydrogen electrolysis is the reverse of the fuel cell reaction, where water is split into hydrogen and oxygen. The reaction is

$$H_2O + 2e^- \longrightarrow H_2 \text{ (g)} + 0.5O_2 \text{ (g)} + 2e^-.$$

This is a non-spontaneous reaction at room temperature, meaning it requires external work to proceed. The minimum work required is

$$\Delta G_r = 237.16 \text{ kJ mol}^{-1} \text{ of } H_2.$$

This shows that electrolysis requires energy input, which is why we use electrical power to split water into hydrogen and oxygen.

9.8.7 Hydrogen combustion at higher pressure

When hydrogen combusts at higher pressures, the Gibbs free energy changes with pressure. As pressure increases, the transition temperature at which the reaction changes from spontaneous to non-spontaneous also increases. At extremely high pressures, such as those on the Sun's surface (estimated at 265 billion bar), the transition temperature becomes very large, making hydrogen combustion feasible even at temperatures far beyond Earth's typical conditions.

9.9 Understanding water's phase diagram

The principles we have applied to chemical reactions can also be used for physical phase changes. Let us take the example of the phase change of water from liquid to vapor (boiling). The reaction is

$$H_2O \text{ (l)} \longrightarrow H_2O \text{ (g)}.$$

The Gibbs free energy change for this process determines the boiling point of water at a given pressure. Using data for enthalpy and entropy at standard conditions, we can calculate the transition temperature or boiling point. For water, this transition temperature is typically around 373.13 K at 1 bar, but it can be influenced by pressure.

By applying the Gibbs free energy equation,

$$\Delta G_r = \Delta H_r - T\Delta S_r.$$

We can analyse the boiling point of water at various pressures. For example, if we calculate the Gibbs free energy at 370.08 K, we find $\Delta G_r = -1.55$ kJ, which still slightly deviates from the expected value due to the abrupt change in heat capacity near the boiling point. However, using the data at 298 K and solving for the pressure at which water boils, we find

$$p = 0.032 \text{ bar}.$$

This shows that water will boil at 298 K at a much lower pressure than the standard boiling point.

This analysis of hydrogen and water's behavior under varying pressures and temperatures shows how the principles of Gibbs free energy can be applied not only to chemical reactions but also to physical phase changes. Understanding these concepts is essential for predicting reaction spontaneity and studying phase transitions, both of which are crucial in a wide range of industrial and scientific applications.

9.10 Calculating electromotive force (EMF) in electrochemical cells

The electromotive force (EMF) of an electrochemical cell represents the maximum potential difference between its two electrodes when no current is flowing. This potential difference drives the movement of electrons through an external circuit, facilitating the electrochemical reaction. Understanding how to calculate EMF is essential for analysing and designing electrochemical systems such as batteries and fuel cells.

9.10.1 Fundamental relationship between Gibbs free energy and EMF

The relationship between the Gibbs free energy change (ΔG) of a reaction and the EMF of an electrochemical cell is given by

$$\Delta G = -nFE_{cell},$$

where:
- ΔG is the Gibbs free energy change (in joules).
- n is the number of moles of electrons transferred in the reaction.
- F is the Faraday constant (96485 C mol^{-1}).
- E_{cell} is the EMF of the cell (in volts).

This equation indicates that the EMF of a cell is directly related to the spontaneity of the electrochemical reaction: a positive EMF corresponds to a negative ΔG, signifying a spontaneous reaction.

9.10.1.1 Step 1: Calculate the Gibbs free energy change for the reaction
First, determine the standard Gibbs free energy change (ΔG°) for the reaction. This is typically done using standard thermodynamic data (enthalpy and entropy) for the substances involved.

The standard Gibbs free energy change is given by

$$\Delta G^\circ = \Delta H^\circ - T\Delta S^\circ,$$

where:
- ΔH° is the standard enthalpy change (in joules, J).
- T is the temperature (in kelvins, K).
- ΔS° is the standard entropy change (in joules per kelvin, J K^{-1}).

9.10.1.2 Step 2: Relate Gibbs free energy to EMF
Once you have the standard Gibbs free energy change ΔG° for the reaction, use the equation relating Gibbs free energy to EMF:

$$\Delta G^\circ = -nFE_{cell}^\circ.$$

Rearrange this equation to solve for the EMF:

$$E_{cell}^\circ = -\frac{\Delta G^\circ}{nF}.$$

9.11 Pressure dependence of the graphite–diamond transition

In chapter 8, we discussed the transformation from graphite to diamond under standard conditions, where the Gibbs free energy change (ΔG) was positive, indicating that the transformation is non-spontaneous at 1 atm pressure and room temperature. However, pressure plays a significant role in this phase transition. At higher pressures, the reaction becomes more favorable, and the transition from graphite to diamond can occur spontaneously.

The formula we will use to describe the pressure dependence of this reaction is

$$d\Delta G_\gamma = \Delta V_\gamma \, dp - \Delta S_\gamma \, dT.$$

9.11.1 Step 1: Understanding the volume and entropy change

For the graphite to diamond transformation, the volume change (ΔV_γ) is negative, as diamond is denser than graphite. Therefore, the volume change ΔV_γ is a key factor in driving the reaction forward under high pressure.

The entropy change (ΔS_γ) for the reaction is negative because diamond has a more ordered structure than graphite. However, at high pressures, the contribution from the $\Delta_\gamma dp$ term becomes much more significant and helps drive the reaction forward, making the transformation spontaneous.

9.11.2 Step 2: Calculating Gibbs free energy change at standard pressure

At standard conditions (298 K and 1 atm), we can calculate the Gibbs free energy change for the transformation from graphite to diamond. From earlier discussions, we know that

$$\Delta H_\gamma = 1.895 \text{ kJ mol}^{-1}, \quad \Delta S_\gamma = -3.363 \text{ J K}^{-1}\text{·mol}^{-1}.$$

The Gibbs free energy change at standard pressure and temperature can be calculated as

$$\Delta G_\gamma = \Delta H_\gamma - T\Delta S_\gamma.$$

Substituting the values,

$$\Delta G_\gamma = 1.895 \text{ kJ mol}^{-1} - \left(298 \text{ K} \times \left(-3.363 \text{ J} \quad \text{K}^{-1}\text{·mol}^{-1} \times \frac{1}{1000 \text{ J kJ}^{-1}}\right)\right)$$

$$\Delta G_\gamma = 1.895 \text{ kJ mol}^{-1} + 1.006 \text{ kJ mol}^{-1}$$

$$\Delta G_\gamma = 2.901 \text{ kJ mol}^{-1}.$$

This positive value shows that the transformation is non-spontaneous at standard conditions.

9.11.3 Step 3: Calculating Gibbs free energy change at high pressure

Now, we consider a high-pressure environment, such as 5 GPa, which is typical for synthetic diamond production. We will calculate how the pressure dependence affects the Gibbs free energy change.

9.11.3.1 Volume change and pressure contribution
The volume change for the graphite to diamond transformation is approximately

$$\Delta V_\gamma \approx -5.7 \times 10^{-6} \text{ m}^3 \text{ mol} - 1.$$

At a pressure of 5 GPa, the pressure contribution to the Gibbs free energy change is

$$\Delta V_\gamma \Delta p = (-5.7 \times 10^{-6}\,\mathrm{m^3\,mol^{-1}}) \times (5 \times 10^9\,\mathrm{Pa})$$

$$\Delta V_\gamma \Delta p = -28.5\,\mathrm{kJ\,mol^{-1}}.$$

9.11.3.2 Temperature contribution

Since we are assuming room temperature ($T = 298$ K), and there is no significant change in temperature ($\Delta T = 0$), the temperature term in the equation is zero:

$$\Delta T = 0.$$

9.11.3.3 Total Gibbs free energy change

Now, using the pressure contribution and the earlier values for enthalpy and entropy, we calculate the total Gibbs free energy change at high pressure:

$$\Delta G_\gamma = 1.895\,\mathrm{kJ\,mol^{-1}} - 1.006\,\mathrm{kJ\,mol^{-1}} + (-28.5\,\mathrm{kJ\,mol^{-1}})$$

$$\Delta G_\gamma = -27.611\,\mathrm{kJ\,mol^{-1}}.$$

At 5 GPa, the Gibbs free energy change is negative, indicating that the graphite to diamond transformation is *spontaneous* at this pressure.

This analysis shows that while the graphite to diamond transformation is non-spontaneous at standard pressure, it becomes spontaneous at high pressures due to the significant contribution of the volume change term $\Delta V_\gamma \Delta p$ in the Gibbs free energy equation. This effect is crucial in the synthetic production of diamonds, where high pressures are used to drive the reaction. By using the correct thermodynamic equation, we can predict that the transformation from graphite to diamond becomes favorable under high pressure, and this provides insight into the conditions needed for diamond synthesis in both natural and industrial settings.

9.12 Summary

In this chapter, we explored the central role of thermodynamic potentials in understanding the behavior of systems under various conditions. The relationships between internal energy, enthalpy, and Gibbs free energy were discussed in detail, with emphasis on their mathematical formulations and physical significance.

We demonstrated how the first and second laws of thermodynamics can be applied to calculate energy changes and work done in various systems. The chapter addressed work expressions, showing how they differ in systems undergoing compression and expansion (using $-pdV$ work) versus those involving flow processes (using Vdp work). By clearly distinguishing these work terms, we emphasized their relevance to different practical processes, including engines, compressors, and refrigerators.

A significant portion of the chapter was dedicated to examining the pressure and temperature dependence of reactions. We considered the transformation of graphite to diamond as a case study, demonstrating how pressure influences the spontaneity of phase transitions. The use of Gibbs free energy and its relationship with pressure allowed us to predict when such transformations become favorable.

The latter sections of the chapter delved into the methods for calculating enthalpy, entropy, and Gibbs free energy at non-standard pressures and temperatures. We extended these concepts to reactions and phase changes, such as the combustion of hydrogen and the boiling of water, emphasizing how these properties evolve with changes in environmental conditions.

Finally, we explored the application of these principles to electrochemical cells, particularly focusing on the calculation of electromotive force (EMF) and its dependence on Gibbs free energy. We highlighted how understanding these relationships can aid in the design of efficient energy systems, such as fuel cells and batteries, which are essential for modern technological advancements.

In summary, this chapter provided a thorough understanding of thermodynamic potentials, their derivatives, and how to apply them to real-world scenarios. Through examples and case studies, we illustrated how these principles guide us in predicting the behavior of substances and reactions under varying temperature and pressure conditions. The knowledge gained here is vital for anyone involved in fields that deal with energy transformations, chemical processes, or material properties under different environmental constraints.

IOP Publishing

A Classical Thermodynamics Toolkit

Srinivas Vanapalli

Chapter 10

Conclusion: where to go from here

Thermodynamics begins simply—with energy, temperature, and pressure—but quickly reveals itself to be a universal language. In this book, we have walked through that language together, chapter by chapter, building a foundation rooted in intuition and clarity.

We started with energy—the mighty internal energy U—the strength within a system. Then we brought in entropy S, which asks: *How is that energy arranged?* From there, we met enthalpy H, which accounts for the energy needed not just to grow, but to push back the world—a concept essential in systems open to the atmosphere. Finally, we encountered the Gibbs free energy G, the portion of energy available to do useful work when all other requirements are met.

To conclude, consider this playful analogy with elephants (figure 10.1)—each one representing a thermodynamic potential:
- The **first elephant** stands tall, full of internal energy U.
- The **second elephant**, now grown and pushing against the air, needs more—the total $H = U + pV$.
- The **third elephant**, destined for transformation, shows that the maximum useful work is not all of H, but only the portion we call G.

With this journey, we've not only met these characters but followed them through processes, cycles, and phase transitions. We have seen that thermodynamics is not confined to heat engines or classical systems. It applies just as much to modern technologies—in microelectronics, chemistry, renewable energy, and even quantum systems.

Yet, this is not the end. Some essential topics lie beyond the scope of this book, particularly those that are crucial in the context of sustainability and energy systems. Concepts like **availability**, **exergy**, and **irreversibility accounting** are necessary tools to evaluate the *quality* of energy and to determine how much can truly be harnessed

doi:10.1088/978-0-7503-6029-6ch10
10-1

I am strong and have energy U
S: How is U arranged in me?

Mama told me I was very tiny
when I was born, and as I grew,
I had to do work
against the air around me.

If you destroy me the
total energy you will
get back is equal to H.
Out of this the maximum
useful work is G

Figure 10.1. A playful analogy using elephants to illustrate thermodynamic potentials: the first elephant represents internal energy U, the second shows enthalpy H, accounting for work done against atmospheric pressure, and the third illustrates Gibbs free energy G, the portion of energy available to do useful work.

before it is lost to entropy. These ideas are foundational to next-generation thinking in energy system design and deserve a dedicated study.

For deeper explorations, classic textbooks such as Van Wylen and Sonntag or Moran and Shapiro offer rigorous mathematical treatments. They are valuable companions as you move from this conceptual introduction into more advanced or domain-specific applications—whether in mechanical engineering, chemical thermodynamics, or materials science.

The next step is up to you. Perhaps you will apply these ideas in designing cryogenic systems, optimizing fuel cells, modeling climate interactions, or even interpreting biological processes. No matter the direction, the core principles remain the same—powerful, universal, and deeply insightful.

Let this be your springboard. The tools are in your hands. The elephants have spoken.

Further reading

The following books provide complementary perspectives on thermodynamics and its applications. Each emphasizes different aspects of the subject, from conceptual clarity to mathematical rigor, and from classical foundations to modern cryogenic practice. Readers are encouraged to consult them for deeper study and alternative explanations of key ideas introduced in this book.

An Introduction to Thermal Physics, Daniel V Schroeder, 2021
A beautifully written text for physics students, offering an accessible and intuitive introduction to energy, entropy, and temperature. Its greatest strength lies in the statistical interpretation of entropy and the seamless connection between microscopic and macroscopic descriptions. However, the treatment of open systems, other forms of work, and thermodynamic cycles is limited. Topics such as irreversible and

quasistatic processes or spontaneity at non-standard conditions are not explored in depth, but the book remains one of the clearest conceptual text to thermal physics.

Fundamentals of Thermodynamics (SI Version, 7th Edition), Claus Borgnakke and Richard E Sonntag

A cornerstone of engineering thermodynamics education, this text provides a comprehensive introduction supported by numerous worked examples and problems. It follows the engineering sign convention (heat in and work out are positive). The book is particularly strong in its systematic development of properties, cycles, and energy balances. While its focus is primarily on closed systems and engineering processes, discussions of non-quasistatic processes and spontaneity at non-standard conditions are limited. Nevertheless, it remains an indispensable companion for anyone seeking an engineering-oriented approach to classical thermodynamics.

Concepts in Thermal Physics, Stephen J Blundell and Katherine M Blundell

A modern and engaging text that bridges thermodynamics and statistical mechanics, written with clarity for physics students. It covers advanced topics such as Fermi–Dirac and Bose–Einstein statistics, phase transitions, and quantum effects in matter. However, it does not devote much attention to the thermodynamic treatment of irreversibility, other work interactions, or spontaneous processes, aspects emphasized in this book. It serves as an excellent second text for readers wishing to extend their understanding beyond classical thermodynamics.

Fundamentals of Thermal-Fluid Sciences, Yunus A Çengel, John M Cimbala, and Afshin J Ghajar

This book combines thermodynamics, fluid mechanics, and heat transfer, offering a unified perspective for engineering students. It provides an extensive range of examples and exercises that connect theory with practice. The coverage of energy conservation and cycle analysis is broad, though the conceptual discussions on irreversibility and spontaneity remain brief. As an introductory resource, it is particularly valuable for those wishing to see how thermodynamic principles extend to transport and flow processes.

Cryogenic Helium Refrigeration for Middle and Large Powers, Guy Gistau, 2019

For readers interested in low-temperature engineering, this book provides a detailed overview of helium-based cryogenic refrigeration systems used in large-scale applications such as accelerators, fusion reactors, and space missions. It introduces advanced thermodynamic cycles, component-level descriptions, and system integration methods. The text blends theoretical analysis with practical engineering experience, making it an essential reference for those venturing into applied cryogenics.

Helium Cryogenics, Steven W Van Sciver, 2nd Edition, Springer, 2012

A classic reference that captures both the classical and quantum aspects of helium, the only element that remains liquid down to absolute zero at normal pressures. The book

combines detailed thermophysical property data with rigorous analyses of refrigeration and liquefaction systems. It illustrates how helium behaves as a bridge between macroscopic thermodynamics and quantum phenomena, covering subjects such as superfluidity, heat transport, and cryogenic component design. Together with *The Handbook of Cryogenic Engineering* (edited by J Weisend II), it forms a foundational resource for advanced students and professionals in low-temperature science and engineering.

www.ingramcontent.com/pod-product-compliance
Lightning Source LLC
Chambersburg PA
CBHW080546220326
41599CB00032B/6378